Botanical Progress,
Horticultural Innovations
and Cultural Changes

Botanical Progress,
Horticultural Innovations
and Cultural Changes

Botanical Progress,
Horticultural Innovations
and Cultural Changes

Botanical Progress,
Horticultural Innovations
and Cultural Changes

Dumbarton Oaks Colloquium on the History of Landscape Architecture XXVIII

Held at Dumbarton Oaks May 6–8, 2004

Botanical Progress, Horticultural Innovation and Cultural Change

edited by Michel Conan and W. John Kress

Published by Dumbarton Oaks Research Library and Collection
Washington, D.C.

Distributed by Harvard University Press, 2007.

© 2007 Dumbarton Oaks
Trustees for Harvard University
Washington, D.C.

Published by Dumbarton Oaks Research Library and Collection and
Spacemaker Press. Distributed by Harvard University Press, 2007

Cataloging-in-Publication Data for this volume
is on file with the Library of Congress.

ISBN 0-88402-327-3
978-0-88402-327-2

Printed in China

Contents

*Botanical Progress,
Horticultural Innovations
and Cultural Changes*

*Botanical Progress,
Horticultural Innovations
and Cultural Changes*

*Botanical Progress,
Horticultural Innovations
and Cultural Changes*

*Botanical Progress,
Horticultural Innovations
and Cultural Changes*

A Historical View of Relationships Between Humans and Plants

Michel Conan, *Dumbarton Oaks*
&
W. John Kress,[1] *Smithsonian Institution Department of Botany*

These studies of "Botanical Progress, Horticultural Innovations and Cultural Changes" constitute a renewed attempt at understanding the history of relationships between plants and humans in response to urgent concerns about contemporary landscape design. Present-day environmentalism in landscape architecture recommends that humans not interfere with plants at all, and we certainly wish, at the same time, to preserve much of wild nature, and use all sorts of cultivated plants in daily life. This begs a question that is not answered by environmental discourse: What kind of relationships should humans entertain with cultivated plants?

Botanical Progress

However, before we go any further, we should take into account that the knowledge of plants and the view humans take of their existence and number have changed over the last centuries. Humanity's view of nature and the accumulation of knowledge about the natural world can be traced back through many ancient cultures, and certainly the use of plants has varied over time and across continents. In Western cultures, museums, botanical gardens, and herbaria have traditionally been the academic centers for the study of plant life. Early botanical gardens established in Italy in the Renaissance, and Spain before the tenth century A.D.[2] set the stage for the development of later natural history institutions in Europe. In the eighteenth and nineteenth centuries, botanical institutions began to flourish in Europe and North America as exploratory expeditions returned from far away lands bringing with them many plant species new to science. The age of discovery of nature in turn gave way to the development of new schemes for classifying plants and eventually an understanding of the evolutionary principles responsible for the process of speciation and their origin.

At the start of the great age of exploration by European naturalists in the eighteenth and nineteenth centuries, nature seemed mysterious, immense, and infinite. Expeditions to explore uncharted regions of the world were sent out by governments, monarchies, and wealthy patrons to survey and acquire new lands, to bring back new plant products, such as spices and medicines, and to collect natural history specimens for newly established national museums as well as private collections. Most of the preserved and living specimens of plants that were brought back from Africa, South America, and Asia to museums and botanic gardens by such explorers as Alexander von Humboldt, James Cook, Charles Darwin, Ernest Henry

Wilson, Frank Kingdon-Ward, and Charles Wilkes, to name only a few, were new to science and to horticulture. Discovery and description of biodiversity proceeded at a pace as if the natural world were limitless, enduring, and permanent. Botanists at botanic gardens at Kew, Edinburgh, Madrid, Berlin, St. Petersburg, Singapore, and Bogor, at natural history museums in Paris, London, Leiden, and later Washington, as well as universities at Uppsala, Oxford, and Harvard proceeded to describe new species of plants at a frenzied pace, especially from the tropical regions of the world. The great age of exploration starting in the eighteenth century resulted in an explosion of discovery and documentation of biodiversity in the nineteenth and twentieth centuries.

The initial period of global exploration was soon followed by the colonization of these regions in the Americas, Africa, and Asia at an unprecedented rate by Europeans. In Europe the expanding populations, disease epidemics, economic hardships, and religious persecution sent hundreds of thousands of people to these newly opened regions that promised limitless opportunities and riches. The massive, unspoiled landscapes encountered by European settlers to North America, for example, were viewed as an endless, natural garden to be cultivated and exploited, regardless of the native peoples that inhabited these lands.[3] Yet at the same time this wilderness was forbidding and frightening to these early colonists. In response to both of these perceptions, limitless bounty but frightening wilderness, the woodlands were felled, crops were planted, towns and cities grew at the great expense of the natural landscapes. The slow but steady threat to the survival of native species in the New World had begun.

As the nineteenth and twentieth centuries progressed through unbridled expansion of human populations throughout the world, biologists began to realize that natural habitats, landscapes, and even species were indeed limited, transitory, and ephemeral. The abundance of discoveries of species that started in the eighteenth century led to an intense period of descriptions and taxonomic analysis in the nineteenth and twentieth centuries. The tremendous influx of new species being described required an overhaul of the earlier classification system of Linnaeus (1753).[4] Major new classification systems of plants to incorporate the new discoveries were proposed first by the French botanist de Jussieu (1789),[5] followed by the British taxonomists Bentham and Hooker (1862–1883)[6] and later the Germans Engler and Prantl (1887–1915).[7]

After the turn of the nineteenth century, Darwin's theory of evolution through natural selection and the developing field of genetics preoccupied a different set of biologists in their investigations of the natural world. In the 1940s and 1950s a significant decrease took place in the description of new species of plants, perhaps as a result of the international effects of World War II. This decrease of new discoveries was coupled with an increase in the reanalysis of the taxonomic hierarchy and relationship of taxa to reflect new ideas on the nature of species resulting from a new synthesis of evolutionary ideas led by biologists Theodosius Dobzhansky, Ernst Mayr, George Gaylord Simpson, and George Ledyard Stebbins Jr.[8] The intense interest by evolutionary biologists in understanding how species are related to each other, initiated in the 1940s and termed phylogenetics, has persisted to the present day. New technological advances using DNA sequence data have revolutionized our concepts on the phylogenetic relationships of plants and a new classification of flowering plants is gaining wide acceptance.[9]

At the same time that taxonomists and evolutionists were trying to understand the evolution and classification of plants, ecologists and environmentalists were beginning to assess the relationship of people to natural habitats. Aldo Leopold and others in the 1940s were early advocates who clearly saw the threat of unbridled human expansion to natural environments and the species that inhabited them. It was not until the 1970s that a significant realization was made by most biologists, ecologists, and taxonomists that the natural world was under threat and in trouble. In the last three decades of the twentieth century the urgent need to understand and protect the Earth's habitats and organisms has resulted in an explosion of new academic programs aimed at studying the environment, of new professional societies (e.g., the Society for Conservation Biology) and local activist groups to unite scientists and citizens in taking action, and even new legislation (e.g., the Endangered Species Act) to turn concern for the environment into law.

At the turn of the twenty-first century, it has become clear to biologists, conservationists, and a significant segment of the general public that a major extinction of plants, animals, and microorganisms caused by human activities is not only possible but probable unless immediate action is taken. This threat has resulted in the creation of many local, regional, and national government and nongovernment organizations devoted to halting and reversing these activities. One international response to this imminent extinction as a result of increasing degradation of the environment was the Convention on Biological Diversity (CBD) authorized at the Earth Summit in Rio de Janeiro, Brazil, in 1992. This treaty initiated a revolution in the value placed on biodiversity and the intellectual property rights attached to nature. According to the Convention, biodiversity should be conserved, sustainably used, and its benefits shared among all parties. Since the Rio Summit, 188 countries have signed the treaty and a host of additional resolutions, national strategies, and work plans have been developed and implemented. Although not the initial intent, one result of the CBD was that nature, like other commercial commodities, has now become internationalized. The ownership of nature, whether it be for natural product development through "bioprospecting," establishing logging concessions on indigenous people's land, or collecting plant specimens for scientific study, is now a matter of worldwide concern and law.

The globalization of nature and biodiversity coupled with increased species extinction has significantly changed the way that modern-day plant explorers and taxonomists pursue their activities. As habitat destruction accelerates, the pace of discovery, identification, and description of new species of plants has not sped up. The number of newly described species of plants only slightly increased in the 1980s and 1990s over the previous four decades. Unlike the predominant perceptions of the seventeenth and eighteenth centuries that nature and species were infinite and limitless, we now know this assessment is not true. Estimates of the number of plant species currently present on Earth range from 220,000 to over 420,000[10]. By extrapolation from what we have already described and what we estimate to be present it is possible that at least ten percent of all vascular plants are still to be discovered and described.[11] This number suggests that a considerable amount of work still needs to be done by botanists to find them.

The marriage of biology and advanced technology is leading to the development of novel tools with the potential to transform current methods of plant collecting, such as image recognition software to be employed in electronic field guides and on-the-spot, rapid DNA sequencing, termed DNA bar coding[12] for species identification. These technological tools will define and drive the new plant explorers of the future. The documentation of the remaining species of plants with the aid of these new tools will provide a solid basis for the precise identification of the species-rich areas of the world for immediate assessment, conservation, and protection. The social, economic, political, and technological changes of the last few decades have ushered in the final age of plant exploration and conservation in the twenty-first century. We can also see that cultural and scientific changes such as new developments of genetics allow the creation of new cultivated plants, and new forms of the exploitation of plants. This will have a great impact on landscape architecture, as much as economy or medicine. Conversely, we know that large agricultural changes in eighteenth-century England, for instance, had a profound impact on English society, and some of its colonies. Clearly, mutual relationships between cultivated plants and human cultures have developed in many ways during the past millennia, and might develop in different ways in our own future.

Turning to the Study of Agriculture

Yet we lack a broad historical perspective on the relationships between humans and plants that would allow us to think about possible futures. This is not, however, revelation. The question of these relationships goes back to the end of the nineteenth century with the groundbreaking work on the origins of cultivated plants by Alphonse de Candolle, a study of archival documents and ancient languages published in Geneva in 1882.[13] This impressive work is at the origin of the diffusion theory of agriculture that rests on the idea that hunter collectors had been pushed by hunger towards agriculture after they had depleted their hunting and collecting grounds of most of their natural resources.[14] This thesis contributed to focusing scholarly research upon the quest for the origin of agriculture, diverting it from broader ranging studies of relationships between plants and humans.

This approach, however, ran into difficulties that invited approaching the question from a completely different point of view, putting horticulture rather than agriculture, and historical rather than prehistorical periods at the center of our discussions. Studies of the origin of agriculture went through different phases. Several scientific disciplines contributed one after the other, bringing new facts to light and overthrowing old ones. The necessity to struggle with scarce evidence seems to have made the mystery of origin more obscure as research progressed, and led to the development of imaginary models of the social use of plants in archaic times. Let us see how.

The use of botanical tours and comparative linguistics, in the 1880s, limited the time range explored by Candolle to 4–3000 B.P. Between 1921 and 1943, Nicolay Ivanovich Vavilov, his team and their followers[15] turned to botany.[16] They called on modern botany of their time to understand the processes of transformation of wild into cultivated plants. And they replaced scientifically obsolete categories of pure, mongrel, or hybrid races of degeneration or atavism that were and still are used in lay discussions of domesticated plants, with concepts borrowed from genetics, such as dominant or recessive genes, homozygous or heterozygous plants. It enabled them to study the role of human intervention in the transformation of the biological life of plants. It also steered discussions of plant biology away from the dreadful discourse of race in human affairs. They wanted to discover on site the origins of plants that were used to develop agriculture. They also thought—like Candolle—that the sites of the origins of cultivated plants would coincide in the majority of cases with the places where agriculture was initiated. After a very large number of botanical investigations all around the world, they concluded that agriculture originated in a small number of geographical regions of the world. These were mountainous areas on the fringes of tropical and subtropical regions in East Asia, Southeast Asia, the Caucasus and Asia Minor, Central American highlands, and the Andes. In 1943 André Haudricourt and Louis Hédin, following in his footsteps, published a study of "the tight and reciprocal relationship between Man and cultivated plants."[17] They defined agriculture as "the intentional planting of seeds or tubers towards a larger crop," and located its origin at the beginning of the Neolithic, about eight to ten thousand years ago. Their definition of agriculture covers horticulture as well, and many scholars accepted without further inquiry that horticulture developed as a branch of agriculture after this date.

The botanical approach of plant variety to determine the center from which agriculture originated and to determine the origin of specific crops stimulated a great amount of research, which ultimately produced contradictory reports. These contradictions led to a search for hard archaeological evidence that would establish the true origin of agriculture, and researchers lost sight of the broader question about the history of mutual relationships between humans and plants. In 1994, only ten years ago, Daniel Zohary and Maria Hopf published a survey of more than twenty different archaeological types of evidence. They concluded that: " . . . it is evident that the crops domesticated in the Near East nuclear area were also the initiators of food production in Europe, central Asia, the Indus Basin and the Nile Valley . . . All over those vast areas the start of food production depended on the same Near Eastern crops."[18] Thus archaeology seemed to confirm the classical diffusion theory. Yet, even on archaeological grounds, this conclusion was debatable since there is evidence of grain grinding implements from 20,000–15,000 B.P. in the Nile valley in Africa. And, in 1993 David Phillipson, using botanical evidence, concluded that "Most of the crops which are or have been cultivated in Africa are species indigenous to that continent (. . .) which must presumably have been first cultivated there."[19] Moreover, the theory did not apply easily to the development of agriculture in America.

Did Cultural Changes Matter?

Competing directions of research had been pursued during the same period. Carl Sauer already argued against the classical diffusion theory in 1952, saying " . . . People living in the shadow of famine do not have the time to undertake the slow and leisurely experimental steps out of which a better food supply is to develop in a somewhat distant future . . . Needy and miserable societies are not inventive."[20] Instead, he pictured a shift from hunting and nomadism to fishing in sedentary places that fostered the development of a relationship with trees providing fibers for making nets and fishing implements as well as baskets, and with poisonous plants for stunning fishes. He proposed that women had begun a special relationship with plants mostly for fibers and possibly medicinal and magical purposes long before any plant would have been an important staple in a new food economy. His model no longer centered on the origins of European culture, but started with garden cultivation in south Asia and south China along rivers, lakes and estuaries. This suggested a new line of investigation that called on the reconstruction of social and cultural dynamics of relationships between humans and plants. Unfortunately, it rested more on imagination than observation of actual human behaviors. Instead, rather than try to imagine the relationship to nature of small societies ten thousand years ago, it seemed wise to carry field observations of present-day preagricultural societies.

Eric Higgs and Margaret Jarman in 1969 and 1972 also rejected the economic line of reasoning that interpreted change in the relationships between man and nature as a response to nutritional needs.[21] They proposed instead that changes in the relationships between man and nature had taken place incrementally in response to local or regional changes in the environment, climate, demography, warfare, religion, or skills. Moreover, they called attention to the reversibility of change, noting that agriculturists under political pressure could revert to hunting and gathering as it happened in South America under the duress created by the European colonization. Their criticism also exposed the anachronism that was built into the question of the origins of agriculture. In order to answer that question it had been necessary to outline a number of criteria that should be met for agriculture to exist. Three conditions were commonly accepted: (1) the existence of domesticated plants, (2) systematic planting practices, and (3) the use of specialized planting areas. Of course, these conditions make sense as a way of defining contemporary agriculture. If we think that the course of history was guided by economic necessity toward the present form of exploitation of natural resources it is perfectly reasonable to search for the moment in the past when this model was first introduced. And, conversely, when forcing past evidence to fit this model we inadvertently introduce the assumption that our form of exploitation of nature results from a necessary development of the relationships between men and nature. In fact, we do even more. When describing the steps of human evolution as a series of necessary passages from one form of relationships between men and nature, from scavenger to farmer in the early Neolithic, we claim that the present industrial agriculture is the necessary result of a ten-thousand-year-old course of development, and thus cannot be replaced. We create a myth that supports the belief in the necessity of the present model of use of nature.

Several anthropologists, many of them ethnobotanists, have done so. Their works raise fascinating issues. Philippe Descola in his study of the Achuar in 1986 showed them to be hunters, fishermen, and horticulturists.[22] He highlights the large botanical knowledge transmitted across generations, and the specialization of women's botanical knowledge of cultivated plants. Men alone practiced hunting, men and women fishing, and women gardening, while men cultivated a few poisonous plants for hunting and fishing outside of the gardens. In fact men could not enter the gardens for magical reasons, much less manage them. Moreover, each plant was cared for, as a child of hers, by a woman who knew the relevant magic to apply in any circumstance. We can conclude with other ethnobotanists that "Knowledge of planting in and of itself does not mandate a social transformation of so-called hunter-gatherers to farmers."[23] The Sawiyamö are also hunters and horticulturists, cultivating taro, yams, sweet potato, sago palm, plantains, coconut, pandanus, betel, and vine in several swidden lots in the forest. It is interesting to note that the garden lots are not fenced and are used as bait to hunt wild pigs coming from the forest to forage into them. But I should insist on the commonly observed fact that gardening—like hunting and fishing—seems to demand magic practices and cultivation of plants for magic, to a degree that varies greatly from one culture to another and that eludes any known pattern. The understanding of relationships between plants and humans in non-Western civilizations demands a detailed understanding of local cultures and social organization as demonstrated by many studies since Malinowski's "Coral Gardens and their Magic,"[24] in 1935. Yet these studies suggest so many different models of man-to-nature relationships that it would be vain to choose one rather than the others to account for the origins of early agriculture. The question of the origins of agriculture should be laid to rest for a while!

Returning to Garden Studies with New Questions

By contrast, these studies provide fascinating insights into the mutual relationships between the knowledge of plants, their cultivation, and local culture in the anthropological sense. There is, however, no reason to believe that we could learn about the mutual relationships of humans to plants only by turning to examples outside of Western society. Instead, we have much to learn from a systematic study of all aspects of the history of relationships between plants and humans that we can retrieve from the past, paying much attention to cultural changes. We should roam through the whole of the historic periods of mankind, including the present, without any preconception of a question on which all thinking should be focused. Rather we should explore very different approaches to the relationships between humans and plants in very different cultural, environmental, and historical contexts, hopefully to discover some promising questions for further explorations.

To help discover new questions, this book puts horticulture rather than agriculture at the center of inquiries. How should we then define horticulture? In a rather intriguing paper on "Domestication of animals, cultivation of plants and management of people," André Haudricourt suggested a radical difference between agriculture and horticulture.[25] Horticulture expresses respectful care for each plant, acts mostly indirectly upon them, and always handles them individually, as opposed to agriculture, which proceeds to mass-management of plants, acts directly upon them, and engages in violent treatment such as reaping and flaying. Then, he went on to propose that agriculture and pastoralism were correlated to the Indo-European political style of management, in the same way that horticulture was correlated to the Chinese political style of management. His view of horticulture was exemplified in the cultivation of taro in Melanesia, and by the art of Chinese gardens. It replaces Euro-centered with Asian-centered perspectives. As a result, he saw the French garden as typical of the agricultural management of nature, putting into question the relationship between gardens and horticulture. Such a definition of horticulture, however, is not helpful because it unduly generalizes from an Asian practice of the relationships between humans and plants. It should serve as a warning that we should be open to the varieties of mutual relationships between plants and humans that have developed, and use whenever possible local or regional categories for the description of places, plants and planting. Moreover, we should not restrict the domain of our discussions to edible plants, or to any category of plants but, rather, remain open to studying any plant that has been associated with humans.

We wish to seek how cultural modes of understanding of plants and societies may interact, and how they relate to the history of different practices in gardening, religion, literature, economy, or politics, as well as in region or environment. Rather than follow Haudricourt's structuralist interpretations, we should examine how cultural modes of understanding of plants and societies may interact, and how they relate to the history of different practices in gardening, religion, literature, economy or politics, as well as region or environment. This is why this volume is far-ranging in the periods and the issues of human history it addresses, and attempts to unravel different kinds of relationships between the construction of knowledge, the cultivation of plants, and the processes of cultural and social change. It is divided into three directions of exploration of linkages between botany, horticulture, and cultural changes. The chapters are presented in historical order for reasons of clarity and certainly not to suggest any pattern of cultural diffusion across the world.

The first part revisits ancient links between culture, botany and horticulture. Addressing the period of the late Republic and Early Roman Empire, Alain Touwaide shows the importance of plant classification for the organization of Roman gardens. The Islamic world further developed the knowledge of plants and gardening, and Elliot Wolfson for Medieval Andalusia, and Maria Subtelny for the Persian world, show how these early forms of botany and horticulture opened the way to extraordinary developments of mystical culture and poetry. Nurhan Atasoy, clearly setting apart the Ottoman from the Andalusian, Arabic, and Persian garden traditions, shows a completely different realm of mutual influence between flower cultivation, gardening, and developments in the visual arts. Then she takes her topic a step further into a study of the flower's role in cultural encounters and exchanges between Western Europe and the Ottoman Empire. In the fifth chapter, Susan Toby Evans, guiding our attention from the Mediterranean to the Meso-American world just before its dramatic encounter with European powers, further illustrates the variety of links between culture and horticulture.

The second part explores links between horticultural and political changes. The topic seems narrower and yet the variety of links that can be observed still appears considerable. Here again, the chapters each concentrate on some very different aspects of gardening. Yizhar Hirschfeld shows the importance of perfumes for Palestine in the Roman times, while Mohammed El Faïz, turning to the long history of Muslim Spain, shows the dynamic relationships between politics, experimental development of horticulture, and progress of plant knowledge and collecting. In a completely different vein, Wybe Kuitert shows how selection and hybridization of plum and cherry blossoms in medieval Japan was directly linked to political changes. Georges Métailié provides another completely unexpected example of such links, with the development of a science of grafting in Yuan China; and Saúl Alcántara Onofre provides a similar interpretation for the development of floating gardens in Tenochtitlan, and later Mexico D.F. Thus, we are invited to acknowledge the economic or symbolic importance of technical gestures in horticulture—hybridization, grafting, making floating gardens—and to recognize the role of political choices for their development in each historical context.

The last part offers an attempt at exploring an even more limited set of linkages, turning to horticulture's contribution to economic and cultural changes. Mauro Ambrosoli sets the tone for this part, showing the role of vernacular gardens and gardening in Italian agricultural estates since the times of the Renaissance; Michel Conan examines the complex relationships between horticultural and agricultural modernization before the French revolution and the ensuing development of utopian views of social reform predicated on new discoveries in horticulture and the introduction of new plants coming from America. Therese O'Malley shows under which conditions, in the period just following the American Revolution, horticulture and garden design became established on this side of the Atlantic. Returning to the Islamic world, Daniel Martin Varisco shows how a specific plant provided a drug that has fashioned the economy and the culture of Yemen. It introduces one of the wonderful mysteries of this history of relationships between humans and plants that invites us to reflect upon the huge variety of possible relationships: this plant has become an integral part of Yemeni culture but did not originate in Yemen and failed to spread beyond the borders of the country. Finally, Peter del Tredici turns to the contemporary views of an ideal relationship between humans and plants that would be based on a rehabilitation of the ecologies that prevailed before humans began interfering. He shows that this is just another haphazard form of horticulture. Even left to themselves, natural species tend to compete in unpredictable ways. And in any case, humans cannot be active in preserving nature and pretend that other actions by the rest of society can be bracketed out. Relationships between humans and nature demand much more research before we can claim a horticultural knowledge that could establish sustainable relationships between nature and societies all around the world.

Plants all over the world display the formidable impact that human life has had upon the course of other natural beings. All beliefs in a simple course of change that could be explained by reducing human life to a functional—mostly economic—necessity have to be discarded. Plants, like *Rosa centifolia* in Persia, *Prunus mume* and *Prunus serrulata* in Japan, or the maize and the *cempazúchi* in the chinampas of Tenochtitlan, have migrated into all sorts of aspects of local cultures in symbolic forms that derived from their cultivation for economic reasons, and their history—still to be told—will not be disentangled from human history. Domesticated plants have provided some of the building blocks from which human cultures have been constructed, and for that reason played a deep role in the changes of their arts, economies, and polities. Many chapters of this book, however, testify to long-lasting relationships between different plants and their host society of humans that depended on the stability of a highly significant cultural form. Some changes in relationships between humans and plants were willfully planned as the multiplication of citrus fruit trees in the large gardens of Andalusia, but many plans did fail. In that respect, the present utopia of ecological rehabilitation of the earth seems to simply invert the enlightenment utopia of a ubiquitous presence of all plant species in everybody's garden! Yet, we can see, as in the case of spreading plants in the Italian countryside, that many changes were involuntary, and resulted from the unpredictable interlacing of human passions. There is no tentative conclusion to this volume. On the contrary, we think of it as inviting further discussions and research, before some fruitful questions can be raised for investigation. We hope at least you will find it tantalizing to start thinking about ways of better understanding or creating better relationships between humans and plants.

NOTES

[1] Part of this joint essay has been adapted from the final chapter of Gary A. Krupnick and W. John Kress (ed.), *Plant Conservation: A Natural History Approach* (Chicago: University of Chicago Press, 2005).

[2] See Mohammed El Faïz (Cedimes University, Morocco), *Horticultural Changes and Political Upheavals in Middle-Age Andalusia,* in this volume.

[3] Philipp Shabecoff, *A Fierce Green Fire: The American Environmental Movement* (New York: Hill and Wang, 1993).

[4] Carl Linnaeus, (1753) *Species Plantarum* (Stockholm: 2001).

[5] Antonii Laurentii De Jussieu, *Genera plantarum secundum ordines naturales disposita: juxta methodum in horto Regio Parisiensi* (Parisii: Apud Viduam Herissant et Theophilum Barrois, 1789).

[6] George Bentham and Joseph Dalton Hooker, *Genera plantarum ad exemplaria imprimis in herbariis kewensibus servata definita;* (London Reeve and Co. v.1, pt. 2–v. 3. 1862–1883).

[7] Adolf Engler and Karl Anton. Prantl, *Die Natürlichen Pflanzenfamilien nebst ihren Gattungen und wichtigeren Arten, insbesondere den Nutzpflanzen, unter Mitwirkung zahlreicher hervorragender Fachgelehrten.* Nachträge II zum II.-IV. Teil, über die Jahre 1897 und 1898 (Leipzig: Wilhelm Engelmann, 1900).

[8] Ernst Mayr and William B. Provine, *The Evolutionary Synthesis: Perspectives on the Unification of Biology* (Cambridge, Mass.: Harvard University Press, 1980).

[9] Angiosperm Phylogeny Group [APG]. "An update of the Angiosperm Phylogeny Group classification for the orders and families of flowering plants: APG II." *Botanical Journal of Linnea Society* 141:399–436. 2003.

[10] Rafael Govaerts, "How many species of seed plants are there?" *Taxon* 50(4):1085–1090. 2001. Robert W Scotland, and Alexandra H. Wortley, "How many species of seed plants are there?" *Taxon* 52:101–104. 2003.

[11] W. John Kress and Ellen Farr, unpublished data.

[12] W. John Kress, "Paper floras: How long will they last? A review of *Flowering Plants of the Neotropics.*" *Amer. J. Bot.* 91: 2124–2127. 2004.

[13] Alphonse de Candolle, *Origin of cultivated plants* (New York: Hafner Pub. Co., 1959). Alphonse de Candolle: 1806–1893.

[14] Candolle, *Origine des plantes cultivées,* Genève, Slatkine, 1984 (2–3). "Let me come back to the species that savages may be inclined to cultivate. Sometimes they find them in their own country, but they often get them from neighboring people better served by natural conditions or already engaged in some civilization process. . . . History shows that wheat, corn and sweet potato (. . .) as well as several other plants quickly spread before historical times. (2) (. . .) The different causes that favored or acted against the beginnings of agriculture well explain why there have been agriculturists in some regions since thousands of years, while other regions are still inhabited by nomadic tribes. . . . This eventually led to the formation of centers from which the most useful species have spread out. In the north of Asia, Europe and America, temperatures are unfavorable and indigenous plants are not very productive, but since hunting and fishing provided resources, agriculture could only penetrate at a later time, and people could do without good southern species without suffering too much." (3)

[15] André G. Haudricourt and Louis Hédin, *L'Homme et les plantes cultivées,* Paris: A.M. métailié, 1943, ré-édition avec une préface de Michel Chauvet, (Paris, A. M. Métailié, 1987).

[16] Nicolay Ivanovich Vavilov, *The origin, variation, immunity and breeding of cultivated plants: selected writings* (Waltham, Mass.: Chronica Botanica Co., 1951). N. I. Vavilov (ed.), *Bulletin of Applied Botany and Plant Breeding,* in Russian with English summaries, N. I. Vavilov (ed.), *Theoretical Bases of Plant Breeding,* (Moscow-Leningrad: State agricultural publishing house, 1935–1943).

[17] See note 15.

[18] Daniel Zohary and Maria Hopf, *Domestication of Plants in the Old World* (Oxford: Clarendon, 1994) quoted in D. J. McConnell, *The Forest Farms of Kandy, and other gardens of complete design* (Burlington (USA): Ashgate Studies in Environmetal Policy and Practice, 2003), (428).

[19] David W. Phillipson, *African Archaeology* (Cambridge: Cambridge University Press, 1993) quoted in McConnell, *The Forest Farms of Kandy . . .* (443).

[20] Carl Sauer, *Agricultural Origins and Dispersals* (New York: American Geographical Society, 1952) quoted by McConnell, (469).

[21] Eric Sydney Higgs, Margaret R. Jarman, and others, *Early European Agriculture: Its Foundations and Development,* Written in Honour of Eric Higgs (British Academy, Major Research Project in the Early History of Agriculture. Cambridge: Cambridge University Press, 1982).

[22] Philippe Descola, *Les lances du crépuscule: relations jivaros* (Haute Amazonie, Paris: Plon, ca. 1993). Philippe Descola and Gísli Pálsson (ed.), *Nature and society: anthropological perspectives* (New York: Routledge, 1996).

[23] Philip Guddemi, "When horticulturists are like hunter-gatherers: The Sawiyanö of Papua New Guinea," *Ethnology* 31: 303–314 Quoted in McClellan, *The Forest Farms of Kandy . . .* (466).

[24] Bronislaw Malinowski, *Coral Gardens and their Magic: A Study of the Methods of Tilling the Soil and of Agricultural Rites in the Trobriand Islands* (London, G. Allen & Unwin, 1935).

[25] André G. Haudricourt, "Domestication des animaux, culture des plantes, et traitement d'autrui," in *L'Homme,* 1962, #1, (40–50).

*Botanical Progress,
Horticultural Innovations
and Cultural Changes*

*Botanical Progress,
Horticultural Innovations
and Cultural Changes*

*Botanical Progress,
Horticultural Innovations
and Cultural Changes*

*Botanical Progress,
Horticultural Innovations
and Cultural Changes*

Visionary Rose:
Metaphorical Interpretation of
Horticultural Practice in Medieval Persian
Mysticism

Maria E. Subtelny

In a highly evocative tale he relates in the *Makhzan al-asrār* ("Treasury of Secrets"), the twelfth-century Persian poet, Niẓāmī, whose oeuvre is an acknowledged repository of Iranian myth and legend, illustrates the way in which the rose was perceived in the medieval Persian imagination. According to the tale, two court physicians challenged each other to a duel by poisons. The first physician made his rival swallow a potion so powerful that, in the words of the poet, it would have melted the Black Stone, the sacred meteorite housed at the Meccan shrine of the Ka'ba. The second physician was able, however, to produce an appropriate antidote to save himself from the poison's lethal effects. He, in his turn, picked a rose from a nearby garden, cast a spell on it, and handed it to his opponent to smell.[1] The latter became so terrified at the prospect of inhaling a poison for which no antidote could possibly be found, that he immediately collapsed and died.[2] Although the tale points primarily to the power of psychological suggestion, what is of significance for our topic is that it is the mysterious nature of scent, and of the scent of the rose in particular, that serves as the metaphorical vehicle in this classic text of medieval Persian poetic narrative.[3]

A Culture of Flowers

According to ancient Iranian legendary accounts, the rose was introduced into Iran by the mythical king, Farīdūn, who figures prominently in the Persian national epic, *Shāh-nāmeh* ("The Book of Kings"), and to whom were attributed a number of innovations connected with the rise of early Iranian civilization, such as the founding of cities, the construction of irrigation works, animal husbandry, and the laying out of gardens.[4] Firmly embedded in Persian popular culture, such accounts not only point to the antiquity of rose cultivation in Iran but also confirm the link that has been made by social anthropologists such as Jack Goody between the cultural use of flowers and complex societies based on advanced irrigation agriculture and characterized by social stratification and sophisticated record-keeping.[5]

The culture of flowers in Iran was always closely linked with the cultivation of the Persian garden, which in the economy of the intensive irrigation agriculture practiced in most regions of Iran belonged as much to the domains of architecture and hydraulic construction as it did to those of horticulture and arboriculture. The medieval Persian garden, particularly in the form of the quadripartite architectural garden (*chahārbāgh*), was the direct descendent of the ancient Persian "paradise" (*paridaiza*) of the Achaemenid kings, which had formed an intrinsic part of the imperial palace institution.[6] Even the advent of Islam to Iran did not exert a negative impact on Persian garden culture. On the contrary, Islamic civilization, which in the high caliphal period was based largely on Persian ideas of kingship and imperial administration, and which sought to emulate the refined customs of the Sasanian courts, served to disseminate Persian horticultural practices and concepts of garden design even to far-flung regions such as Andalusia.[7]

Horticulture and arboriculture were thus associated in Persian culture not with luxury or even aesthetics, but rather with ancient Iranian conceptions of kingship, according to which the king's role was perceived as a civilizing one.[8] By expanding the irrigation network, introducing new botanical species, founding urban settlements, and dispensing justice to his subjects, the king assumed a cosmic role as guarantor of the fertility of the land and of the progress of sedentary culture.[9] Viewed from this perspective, the legendary account of the introduction of the rose into Iran by one of its most charismatic mythical kings assumes greater cultural significance than would appear at first blush.

It was no doubt along with the model of the Persian *paridaiza* in the Hellenized Near East, which even adopted the term in its Greek form, *paradeisos*, as attested by a variety of sources from Ptolemaic Egypt to Palestine of the Hasmonean dynasty, that the cultivation of roses spread throughout the region.[10] The antiquity of rose cultivation in Iran and its influence on neighboring Near Eastern cultures would appear to be confirmed linguistically as well. The Persian word for rose (*gul*), which is also the generic term for flower because the rose has always been regarded as the quintessential flowering plant in Persian culture,[11] derives from the Indo-Iranian root *vrd-* ("to grow"), which goes back to the form *⋆wrda-*.[12] The word for rose in the Semitic languages thus has an Iranian origin, appearing as *wered* in Hebrew, *warda* in Aramaic, and *ward* in Arabic.[13] In Egyptian and Coptic, it appears in the form *wrt*, and in Armenian as *vard*.[14] Moreover, the Indo-Iranian root is also related to the Greek *rhodon* (<*wrod-*);[15] from the Greek is derived in turn the Latin *rosa*, whence the English rose.[16]

Rose Cultivation in Iran

According to Persian agricultural and horticultural manuals dating from the thirteenth to the sixteenth centuries, a great number of species of roses were cultivated in Iran in medieval times. The *Irshād al-zirā'a* ("Guidance on Agriculture"), an exhaustive treatise compiled in eastern Iran in 1521, lists besides the many varieties of red rose (*gul-i surkh*), roses named "the five-petalled" (*panj barg*), "the two-colored" or "two-faced" (*dau rang, dau rū*), "the musk rose" (*mushkīn*), "the dappled rose" (*abrash*), and "the Baghdad rose" (*baghdādī*), among others.[17] The following observation made by the Englishman, Sir Robert Ker Porter, who visited Iran at the beginning of the nineteenth century, underscores the pervasiveness of rose cultivation and of the culture of the rose in Iran in the early modern period:

> In no country of the world does the rose grow in such perfection as in Persia; in no country is it so cultivated, and prized by the natives. Their gardens and courts are crowded with its plants, their rooms ornamented with vases,

filled with its gathered bunches and every bath strewed with the full-blown flowers, plucked from their ever-replenished stems.[18]

No species of rose was more important economically or culturally more significant in Iran than the "hundred-petalled rose" (*gul-i ṣad barg*) or *Rosa centifolia*. Characterized by its densely packed petals, it was highly valued for the sweetness of its scent.[19] The thirteenth-century agricultural and horticultural manual, *Āṣār va aḥyā'*, mentions varieties with one hundred and even two hundred petals,[20] and the *Irshād al-zirā'a* refers to both yellow and red varieties, such as the "fiery centifolia of Mashhad" (*ātishī mashhadī*).[21] Commmonly referred to in the West as the Cabbage rose, it was introduced into Europe via the Netherlands in the late sixteenth or early seventeenth century, either directly from Iran during the reign of Shah 'Abbās I, a period of vigorous trade relations and cultural exchange between the Safavid state and Holland, or else through the Ottoman empire during the time of Sultan Süleyman the Magnificent, whose court evinced an unprecedented interest in the culture of flowers, and in roses and rose oil in particular.[22]

A species closely allied to the centifolia, the Damask rose (*Rosa damascena*), which was known in Persian as the Muhammadan rose (*gul-i Muḥammadī*), represented the chief source of rose water (*gulāb*) and rose oil ('*aṭr* > Eng. attar) in Iran.[23] This highly fragrant, rose-pink species apparently obtained its name from the Islamic legendary belief that the rose was created from the drops of perspiration that fell from the forehead of the prophet Muhammad during his miraculous nocturnal ascent (*mi'rāj*) through the seven heavens to the Throne of God.[24] This legend represented an imaginative extrapolation from the traditional Islamic accounts that held that the Prophet exhibited a characteristic feature of sanctity in that his body exuded a fragrant odor.[25] The association of the rose with Islam's prophet was expressed in many spiritually and artistically creative ways, one of which was the depiction of the Muhammadan rose with the ninety-nine epithets of Muhammad inscribed on its petals.[26]

According to medieval sources, the finest quality rose water in the Near East was produced in the region of Shiraz in the southern Iranian province of Fars, whence it was exported to all parts of the world, including Egypt, India, and China, and eventually also to Europe.[27] So closely was the production of rose water identified with Iran that this rose species was often referred to simply as the "Persian rose" (*gul-i fārsī*).[28] In the high caliphal period (circa ninth century), the province of Fars was known to have sent thirty thousand flasks of rose water and one thousand measures of rose honey to the 'Abbasid treasury in Baghdad as part of its annual tribute.[29]

Rose water was a precious commodity in medieval times, primarily because of its medicinal uses as a cardiac drug and as a major ingredient in potions and ointments to relieve skin abrasions, rheumatism, eye strain, and headaches.[30] The *Irshād al-zirā'a* contains a separate chapter devoted to the different methods of distilling rose water and rose oil, as well as to the culinary uses of rose petals and rose water.[31] As a scent with astringent properties, rose water was sometimes used in connection with religious rites, such as ritual ablutions, particularly before the Muslim sabbath prayer; the washing of corpses in preparation for burial; and before handling a copy of the Qur'ān.[32] To anticipate the discussion of the symbolical significance of the rose in ancient Iranian religion, perhaps most significant was the practice, still current in Iran today, of sprinkling rose water over the graves of the deceased at funerals and visitations, as well as at the shrines of Muslim saints and mystics.[33]

The Symbolical Rose

In medieval Perso-Islamic culture, and in poetry in particular, which is the finest expression of the Persian creative genius, the image of the rose was employed as a vehicle for a variety of concepts. It became an especially powerful symbol in the mystical trend that, from the twelfth century onward, permeated Persian religious thought and literary culture. Because the understanding of any symbolic representation presupposes a deep familiarity with the cultural context in which it is operative, an investigation of the metaphorical and symbolical applications of the image of the rose, and the way in which it was interpreted semiotically on the literary level, would contribute to an understanding of certain key aspects of Persian culture, such as the nexus between spiritual experience and poetic sensibility.[34]

Assuming that a symbol is the only possible expression of that which is being symbolized, we may posit, following those scholars who have examined the problem of symbolism and symbolical representation in the context of religious thought, that there is something inherent in the symbol itself that does not merely point to, but actually participates in the transcendent reality it is supposed to symbolize. Hence, a symbol is nothing less than the concrete form that is assumed by a reality beyond the reach of words, which is expressible only through it, and which, to cite Henry Corbin, "*is* the reality."[35]

In the case of the rose *qua* symbol, the qualities that were regarded from the Persian viewpoint as being intrinsic to it were, first and foremost, its beauty. Frequently referred to as "the king of beauty" (*pādshāh-i ḥusn*), the red rose in particular became a symbol for the perfection of the beauty of the beloved, and it was frequently paired with the nightingale in a classic poetic image inspired by the garden and captured in the rhyming formula, *gul u bulbul*.[36] This ubiquitous motif is attested in countless Persian lyrics and depicted in every conceivable artistic medium, from miniature painting and architectural decoration to textile design and lacquerwork.[37] Symbolically depicting the lover, the nightingale pours out in melodious song his unquenchable yearning for union with the rose, who represents the unattainably perfect and perfectly unattainable beloved. The twelfth-century Persian poet, ʿAṭṭār, describes their relationship from the nightingale's perspective in the following way in his *Manṭiq al-ṭayr* ("Discourse of the Birds"):

> No one perceives the secret of the nightingale,
>
> The only one who understands is the rose.
>
> I am so drowned in my love for the rose,
>
> That my own existence has become effaced.
>
> On account of my love for the rose, my head is full of melancholy,
>
> Hence, the only thing I seek [as a cure] is the beautiful rose.[38]

On the mystical plane, the nightingale's love for the rose was interpreted as the mystic's spiritual yearning for the Divine, and the relationship between mystic lover and God as the beloved was construed in erotic terms.[39] Based on an allusion to a late prophetic Tradition according to which Muhammad reportedly stated that, "The red rose is part of the Glory (*bahā'*) of God Most High," many Persian mystical authors utilized the image of the rose as a symbolical depiction of the Divinity.[40] Some of the most striking instances of this usage are to be found in the writings of Rūzbihān Baqlī (d. 1209), a Sufi master from Shiraz (in the region of southern Iran mentioned earlier in connection with rose cultivation). In his diary, entitled *Kashf al-asrār* ("The Unveiling of Secrets"), which records his nocturnal visionary experiences, he often states that the divine Glory appeared to him in the form of a resplendent red rose, thus calling to mind Dante's vision of the Mystical rose.[41] The tension between the seemingly contradictory notions of divine transcendence, which characterizes the Islamic conception of monotheism, on the one hand, and divine self-manifestation in forms perceptible by man's imaginative faculty, on the other, is captured with frank humility by Rūzbihān in a passage that merits citation in full:

After midnight I saw Him, the Transcendent One, as though He appeared in a thousand kinds of beauty, among which I saw a glory of lofty likeness, and "His is the loftiest likeness [in the heavens and the earth] for He is the Mighty, the perfectly Wise" (Qur'ān 30:27).[42] It was as though He were like the glory of the red rose. This is a likeness, but God forbid that He should have a likeness, for "There is nothing whatever like unto Him" (Qur'ān 42:11). Yet I cannot describe except by an expression, and this description is from the perspective of my weakness and incapacity and my lack of comprehension of the qualities of eternity.[43]

As the perfect manifestation of divine Beauty, the rose was regarded by some mystic authors as being mirrored by feminine beauty.[44] As the poet, Jāmī (d. 1492) explains, God made mirrors of the atoms of this world and thereby cast a reflection of His beauty on everything, the radiance of His Face falling first on the rose, and then on the face of Laylā, symbol of the beloved who drove her lover, Majnūn, mad.[45] In the eyes of some mystics, superseding contemplation of the beauty of the feminine face was the controversial notion of contemplation of the beardless youth whose beauty was regarded as a "witness" (shāhid) to God's beauty. In accordance with yet another Sufi aphorism which advised, "Whoever desires to gaze at God's glory, let him contemplate the red rose,"[46] the Persian mystic, Aḥmad Ghazālī (d. 1126), reputedly placed a rose between himself and the handsome youth with whom he sat, now gazing at the one and now at the other, the beautiful face mirroring the beauty of the rose which, in turn, reflected the beauty of the Divine.[47]

The Mystical Rose

The mystical poets underscored the aptness of the rose as a symbolical depiction of the divine Reality by employing a visual pun on the words gul ("rose") and kull ("all"), which are written the same way in the Arabo-Persian script. In Islamic mystical thought, the term kull refers to the concept of God as the ontic source of all being, who is manifest in the multiplicity of the created universe and who at the same time comprises all of creation within Himself, the All deriving from the One and the One encompassing the All. It also contains an unmistakable allusion to the Neoplatonic philosophical notion of the Universal Intellect ('aql-i kull), which was adapted to a monotheistic frame of reference by the Muslim theosophists to refer to the divine Spirit that animates all of creation, which was conceived of as an emanation of the Godhead, from which it was not ontically distinct, and which was often portrayed in angelic form.[48]

In his explanation of the technical theosophical term, sukr ("spiritual intoxication"), the poet ʿAṭṭār illustrated the relationship between the phenomenal world of multiplicity and the divine One who is at the same time the All, by means of an analogy he drew with the rose and its thorns:[49]

> What is sukr (i.e., spiritual intoxication)?
> To imagine the rose (gul) from the thorn,
> To envision the invisible All (kull) from the part.[50]

Possessing spiritual insight and inner vision, the mystic is capable of perceiving from the thorn of the phenomenal world the rose of the Divine, which is immanent in all Creation, yet invisible, as it belongs to the world of the Unseen.[51] The rose thus represents the hidden Reality embedded in the sensible world which the mystics viewed as illusory, often depicting it by means of such negative images as a dung heap, a privy, carrion, an old hag, or a bed of thorns.[52] Because he has discerned the secret of divine Unicity in the plurality of the phenomenal world, the gnostic who is "the possessor of two eyes" (dhū al-ʿaynayn) in Ibn ʿArabī's telling phrase, no longer distinguishes between rose and thorn.[53]

But, paradoxically, just as there is no rose without a thorn (except perhaps in some modern hybrid varieties), so too is it impossible to envision the spiritual Whole save through the medium of the phenomenal part, as the poet, Jalāl al-Dīn Rūmī (d. 1273), states in a comparison he makes between the rose and the theophany at Sinai when God appeared to Moses in the form of a burning bush:

> This [worldly] path is entirely the rose, although externally it may appear as thorns,
>
> [Just as] in the bush of Moses the [divine] Light appeared as fire.[54]

In other words, it is only through the medium of the burning bush that Moses was able to perceive the divine Light that would otherwise be invisible, because it transcends phenomenal representation.

In 'Aṭṭār's view, the apprehension of the imaginal vision of the rose that is All from the thorn of the material world renders the mystic spiritually intoxicated, a notion confirmed by many Sufi authors who describe the state of rapture experienced in the course of visionary contemplation in terms of drunkenness or inebriation. 'Aṭṭār's contemporary, Rūzbihān Baqlī, said of such spiritual visionaries that, "When they see the suprasensible correspondences of the world of the Unseen (*ghayb*), the arcana of the unseen of the Unseen (*ghayb-i ghayb*), and the secrets of Sublimity, they are involuntarily overcome with spiritual intoxication (*mastī*)."[55] And he defined the term *sukr* as denoting "the intoxication (*mastī*) of the spirit from the dew of imaginal envisioning (*mushāhada*) [of the Divine], the wine of love, the fragrance of [divine] discourse, and the lights of the Pre-eternal."[56]

To the well-attested word play on *gul* ("rose") and *kull* ("All") Rūmī added another on *gul* ("rose") and *qul* ("Say!"), the divine command that prefaces numerous verses in Qur'ānic scripture.[57] Applying his radical hermeneutical technique to the stock image of rose and nightingale, Rūmī reversed their traditional roles by calling on his own "nightingale" to be silent in order that it might hear the "fervent song" of the divine Rose:

> Lo, be silent, so that the King of 'Say!' (*shāh-i qul*) may speak,
>
> Do not play the nightingale with a Rose such as this!
>
> This speaking Rose (*gul-i gūyā*) is bursting with fervent song,
>
> O nightingale, hold your tongue and lend an ear![58]

The words of the seemingly silent nightingale of the poet are thus portrayed as being the product of the divine command "Say!" that proceeds from God Himself, whom Rūmī calls the "King of 'Say!'" thereby implying that, like the verses of the Qur'ān, his own poetry is divinely inspired.[59] By integrating the symbolical depiction of God as a red rose with the Qur'ānic notion of divine revelation, Rūmī transforms the "King of 'Say!'" (*qul*) into the Rose (*gul*) of the Divine that speaks through the nightingale of the mutely eloquent poet.[60]

The Ineffability of Scent

Another quality of the rose, apart from its beauty, that was particularly apposite in the economy of the Persian mystical imagination was its scent.[61] The more poetically sensitive authors perceived in the scent of the rose, which has universally been recognized as possessing the most complex and ineffable fragrance of any plant or flower, an apt symbolical allusion to the ineffability of the divine Essence.[62] The scent of the rose thus became the perfect concrete embodiment of a conception too ineffable and too mysterious to be adequately expressed, yet whose very nature could be intimated by means of the phenomenal symbol of mystery and ineffability itself.[63]

The concept of ineffability would seem to lie beyond the limits of language. Yet, paradoxically, it was through language that the mystics sought to express the inexpressible and to depict in words a noumenal reality that is essentially inaccessible.[64] Consonant with the Heideggerian philosophical notion that poetic language, in its subtle interplay between revelation and concealment, makes possible "the seeing of the invisible in the saying of the unsayable," the Persian mystical poets sought to convey the notion of the ineffability of the metaphysically Real by having recourse to the evanescence and impalpability of fragrance.[65] Underscoring its allusive potential, Rūmī states that poetic language is like "scent" that can provide a hint of the Divinity that is otherwise invisible and inaccessible:

> Be content with the scent (*būy*) of the [divine] Friend, for He cannot be seen,
>
> The pre-eternal moon is His face (*rūy*), and poems and verses are His scent (*būy*),
>
> Scent (*būy*) is the lot of him who is not permitted to see His face (*dīdār*).[66]

Thus, just as the scent of the rose (*būy-i gul*) "points the way" (*dalīl*) to the rose bush because it is ontically inseparable from it, so, too, in Rūmī's view does the language of poetry capture the essential nature of that which eludes description, because the poet's inspiration stems from the same divine ontic source.[67]

Expanding on the symbolical congruence between the rose and the notion of the divine All, which is rhetorically supported by the pun referred to earlier on the words *gul* and *kull*, Rūmī states that the fragrance of the rose provides a hint of the mystery of the divine Reality that underlies all things:

> Every rose (*gul*) which is sweet-scented (*būyā*) within,
>
> Conveys something of the secrets of the All (*kull*).[68]

Moreover, he exhorts the mystic to abandon his carnal self in order that he might himself become like the scent of the rose that guides others to the divine Rose Garden.[69] This rose-bodied (*gul-andām*) spiritual master who has surmounted the thorn of fleshly pleasures and opened himself up to the indwelling presence of God in his soul is the embodiment of the rose that is "without why," to cite the enigmatic saying of the seventeenth-century German mystic, Angelus Silesius.[70] Relying on an equivoque (Per. *īhām*) on the metathesized form of the word for "rose garden" (*gulsitān*), which also may be interpreted as "the one who gathers roses" (*gul-sitān*), Rūmī implies that it is to God as the divine Gardener that the scent leads:

> So that you might become like the scent of the rose (*būy-i gul*) for other mystic lovers,
>
> And their leader and guide to the [divine] Rose garden (*gul-sitān*) (or: the divine Gatherer of roses).[71]

"And what," Rūmī asks, "is the scent of the rose?" To which he replies that it is "the breath of intelligence and reason" (*dam-i ʿaql va khirad*), which is man's guide to the Eternal kingdom.[72] By reason and intelligence he does not intend logical reasoning or intellection, but rather intuitive knowledge, the divine inspirational breath (*dam*), which he compares to fragrance, that subtly and involuntarily infuses the mystic at the moment of God's self-disclosure in his heart.[73] Making a characteristically folksy analogy, Rūmī states that just as thorns are "sweetmeats" for the fire, so too is the scent of the rose (*būy-i gul*) "nourishment" for the palate/brain of the intoxicated mystic.[74]

But, in the final analysis, even the scent of the rose falls short of providing anything more than a hint of the ineffable nature of the divine Essence, and hence Rūmī abandons even this symbolical association, while retaining the reference to the olfactory, when he concedes that the "scent" of the Divine cannot be compared to anything except itself:

What is it that with every breath (*dam*) draws me inexorably to Him (*sūy-i ū*)?

It is neither amber nor musk. It is His scent (*būy*) which is the scent of Him (*būy-i ū*).[75]

The olfactory as a mode of spiritual perception is a topic to which little attention has been devoted in the study of the phenomenology of mysticism, as the modalities that predominate in cognition and consciousness both in the Jewish and Christian religious traditions are those of vision and audition.[76] But in Persian Sufism, which represents the most sophisticated poetical elaboration of mystical discourse in the Islamic tradition, and in the poetry of Rūmī in particular, arguably the most experiential of the medieval Persian mystical poets, it appears to play a paramount role.[77] In order to understand this distinctively Persian cultural construction regarding the operation of the senses in the epistemology of medieval Islamic mysticism, it would be instructive to examine the symbolical roots of the rose and the function of scent within the framework of pre-Islamic Iranian religious beliefs and liturgical practices.

The Rose in Ancient Iranian Religion

The use of flowers and floral symbolism figured prominently in ancient Iranian religions, which developed an elaborate "language of flowers," to borrow the nineteenth-century French formula.[78] In Zoroastrian religion, and especially in Mazdaism, a different flower or herb was associated with each of the deities called *yazata*s, who presided over the days of the month and who were honored in special liturgical ceremonies. These floral associations were based on the moral qualities or characteristics that were believed to inhere in the various *yazata*s. In this sacred garden landscape, flowers were construed as the symbolic receptacles for the spiritual energies of the various deities, affording unlimited possibilities for the contemplation and imaginal envisioning of the celestial forms they embodied.[79] Whereas pungent herbs such as myrtle, marjoram and basil were associated with the highest male deities, including the supreme deity, Ahura Mazda ("the Wise Lord"), the rose was associated, appropriately enough, with Daena, one of the female *yazata*s, who was the deity of religion.[80] Moreover, her species of rose was even specified in the Pahlavi texts as the *gul-e sad varg* (New Per. *gul-i ṣad barg*), the "hundred-petalled rose," that is, *Rosa centifolia*, which as already indicated was renowned for its sweet fragrance.[81]

The Daena represented a central concept in Zoroastrian theology, *daēnā* (Middle Per. *dēn*) denoting religion not in the institutional sense but, rather, man's spiritual self, his inner vision, and moral conscience.[82] In view of the importance of the sophianic principle in Zoroastrianism, *daēnā* also referred to innate human wisdom as an emanation of divine wisdom, a quality always associated in Persian thought with the feminine.[83] Henry Corbin discerned in the figure of the Daena the female archetype of wisdom and intuitive vision who represents the secret presence of the Eternal feminine in man, and who, necessarily construed as an angelophany, is the "Angel of his incarnate soul," his heavenly guide and celestial counterpart (see Fig. 1).[84]

According to ancient Iranian eschatological beliefs, the Daena met the soul of the departed three days after death, on the path to the Chinvat bridge that leads to the next world. If a man's soul was righteous, she would appear to him in the form of a luminous maiden, the personification of his transcendent, celestial self, and she would lead him across the bridge to paradise. If a man's actions in life were wicked, however, she would assume the form of an ugly hag, and the bridge would become for him as narrow as a razor blade, causing him to fall into the hellish pit below.[85] Of particular relevance to our topic is the fact that the Daena's appearance was heralded by her beautiful scent, which could be discerned from afar by the soul of the righteous as she wafted toward him like a perfumed breeze.[86] This scent, which stemmed from her association with the centifolia rose, was interpreted as the scent of the soul itself, a notion supported by her reply when asked about her identity: "I am none other than you yourself . . . your own *daēnā*, who has been made beautiful by your own nature."[87]

To appreciate what is implied in the association of scent with the concept of *daēnā*, it is necessary to dwell philologically on the Middle Persian (Pahlavi) term *boy* (also *bod*; New Per. *bū* or *būy*), which was used to denote "scent" or "fragrance" in the Zoroastrian writings. In Avestan texts, the word appears in the form *baodah* (from the root *baod-*, "to perceive"), which has the primary meaning of perception in the sense of spiritual consciousness.[88] Its use in connection with the apprehension of the presence of the Daena, the visionary soul, underscores the epistemic function of the olfactory as a modality of spiritual perception in ancient Iranian religious

Fig. 1. Angel holding a hundred-petalled rose, *Rosa centifolia*. Iran, 1575–1600. Courtesy of the Arthur M. Sackler Gallery, Smithsonian Institution, Washington, D.C., Art and History Collection, LTS1995.2.72.

thought.[89] Probably the most striking instance of scent functioning as an epistemic mode in Zoroastrian literature occurs in the Pahlavi epic, *Ayādgār-i Zarērān*, where the sage and seer, Jāmāspa, is described as having received knowledge of all the sciences, as well as the ability to foretell future events, through the fragrance of a consecrated flower which had been given to him by the prophet Zoroaster.[90]

Scent and Synaesthesia

It has been noted that scent is the most subjective of the human senses, and scientific evidence has revealed the existence of strong links between scent and memory.[91] Odor memory is exceptionally strong in humans, and as suggested by such nineteenth-century French literary authors as Proust, Baudelaire, and Rimbaud, it is perhaps stronger than other sensory experiences.[92] According to widely held views in the field of aromatherapy, fragrance enhances powers of perception and induces various psychic and mental states.[93] Researchers who have investigated the role of olfaction in religious and creative rituals have even suggested that scent may aid in the creation of "an emotional, ecstatic state of consciousness that would render individuals more susceptible to religious experience."[94]

These ideas appear to be confirmed in the poetry of Rūmī for whom scent is the medium of reminiscence, evoking the memory of the divine Beloved, who is often referred to simply as "the Friend" in Persian mysticoerotic poetry and symbolically identified in Sufi poetics with the rose. Imitating a poem by the tenth-century eastern Iranian poet, Rūdakī, Rūmī infuses the famous opening line, "The scent (*būy*) of the Muliyan stream continually comes to me; The remembrance (*yād*) of the gracious friend continually comes to me," with mystical meaning, radically transforming its original panegyrical intent into a metaphorical elucidation of the role of scent as a mode of spiritual cognition:

> The scent (*būy*) of the [divine] Rose garden (or: the [divine] Gatherer of roses) continually comes to me,
>
> The scent (*būy*) of the gracious Friend (*yār-i mihrbān*) continually comes to me.[95]

Rūmī's replacement of the word *yād* ("remembrance") in Rūdakī's original by the term *būy* ("scent"), which is repeated twice in the same line, causes the verse to function like a poetical palimpsest which still bears traces of the notion of reminiscence beneath the overlay of the olfactory references.[96] Also preserved in Rūmī's interpretation is what I believe to be an allusion to the association made in some Middle Persian texts that reflect a courtly, nonreligious setting, between the scent of the rose and the beloved. In the Pahlavi narrative, *Khusraw ut rētak*, which records an interview between the Sasanian king, Khusrau, and a royal page whom he tests in his knowledge of Persian tastes and customs, the king inquires about the significance of the scents of various herbs and flowers. Whereas myrtle and basil are described as being redolent of sovereignty and kingship, the scent of the rose is characterized as being reminiscent of the beloved (*mihrbān*).[97] It is noteworthy that both Rūmī and Rūdakī apply this same ancient Persian term to the beloved "friend" in their respective poems.[98]

In Rūmī's understanding of the concept of sensory correspondences, the five senses are linked to one another because they all stem from a single ontic root (*aṣl*), each acting as a "cupbearer" (*sāqī*), as it were, to the rest.[99] By the term "sense" (*ḥiss*) Rūmī intends not the physical or "external" senses, but rather the spiritual or "inner senses," which he calls the "senses that perceive the Light" (*ḥavāss-i nūr-bīn*), and to which he often refers by means of such metaphorical locutions as "the ear of the soul" (*gūsh-i jān*) and "the eye of the soul" (*chashm-i jān*).[100] When one of the senses is stimulated, all the senses become interconnected to produce a unitary mystical experience. Thus, ocular vision stimulates speech; speech stimulates visual perception; this perception then becomes the means of awakening every sense, so that all the senses "taste" (*zauq*) the rapture of spiritual vision:[101]

> When, in the process, one sense loosens the bonds,
>
> The rest of the senses become transformed.
>
> When one sense has perceived things that are not objects of sense-perception,
>
> That which is of the Unseen world (*ghaybī*) becomes apparent to all the senses.[102]

Rūmī appears, on the one hand, to be echoing the idea of sensory synaesthesia expressed in the writings of such early Christian authors as Origen, and, on the other, to anticipate the theory of sensory correspondences poetically articulated in the nineteenth century by Baudelaire in whose view the sensory modalities correspond to each other on the phenomenal level, while at the same time reflecting a single underlying noumenal reality.[103]

The phenomenon of sensory synaesthesia involves a complex interplay among the different sense modalities, whereby one modality may be expressed in terms of another, or several modalities may converge to act as one to create an altered state of consciousness.[104] The cross-modal relationship between vision and audition, for example, where visual imagery is expressed in auditory terms, or hearing is expressed in terms of seeing, has been discussed in the phenomenology of various mystical traditions.[105] In the epistemology of Islamic mysticism, the concept of the fundamental unity of the senses signifies not just that preference cannot be accorded to one over another, but also that visionary experience does not necessarily imply ocular perception.[106] Imaginal vision (ru'yā) or "witnessing" (mushāhada), as it is sometimes referred to in the technical terminology of Sufi theosophy, ensues when one of the sense modalities stimulates a synaesthetic response in the others. Thus, Rūmī states that for a gnostic who has suppressed his physical or "external" senses, even "the ear and the nose can become an eye."[107]

The Olfactory as a Mode of Spiritual Perception

For Rūmī, the lead modality of sensory perception appears more often than not to be the olfactory. In a particularly pointed passage that pivots on an equivoque on the word, bīnī, which in Persian means both "nose" and "seeing," Rūmī explains that, since scent provides an intimation of the world of the Unseen, whoever does not have a "nose" to perceive this scent is unable to "see" spiritually:[108]

> The nose (bīnī) is that which perceives a scent (būyī),
> And scent (būy) is that which leads us to the abode (kūy) [of the Beloved].[109]
> Whoever does not possess this scent is without a nose (bī-bīnī) (or: lacks vision),
> [Since] by 'scent' (būy) I mean the scent that is religious (dīnī).[110]

It is possible to discern in Rūmī's rather unusual use of the word dīnī, which in this particular context appears to refer primarily to Islamic religious practice (from Ar. dīn), a punning allusion to the Daena, the Middle Persian form of whose name, Den, was rendered into Arabic as dīn, thus contributing to the confusion between two very different conceptions of "religion"—the Semitic and the ancient Iranian.[111] Hence, the word dīnī may also be interpreted as meaning, "pertaining to the Daena" (dēnī), thereby yielding the alternative reading: "By 'scent' I mean the scent that is associated with the Daena (i.e., the visionary soul)." Rūmī thus combines in his definition of scent as a mode of spiritual perception both the domain of religious law and praxis (dīn), and the visionary aspect of faith symbolized by the feminine soul (dēn). Support for my contention that both conceptions of religion are to be understood by the word dīnī in the abovementioned verse, is Rūmī's notorious penchant for expressing profoundly mystical ideas in the language of pun and equivoque, as well as other allusions in his poetry to the meeting of the soul after death with the Daena, the luminous maiden who embodied man's good actions in life.[112]

A particularly evocative illustration of the role of the olfactory and the way in which it relates to spiritual vision is Rūmī's extended esoteric commentary on the Qur'ānic verse, "Cast [my shirt] upon my father's face, and he will come to see" (Qur'ān 12:93), which occurs in the story of Joseph and his brothers. Just as the scent of Joseph's shirt miraculously restored blind Jacob's sight when it was placed on his face, so too is the mystic's inner vision illuminated when he perceives the scent (bū) of the Divine:[113]

> In order that that scent (bū) may attract your soul,
> And in order that that scent (bū) may become the light (nūr) of your eyes.

It was for the sake of [that] scent (*bū*) that Joseph, the son of the prophet Jacob, said:

'Cast [my shirt] upon my father's face.'[114]

Accentuating the significance of the Qur'ānic motif of Joseph's shirt is the implicit comparison made by the Persian poets between the rose and Joseph's shirt, which according to Islamic legend was woven in the Garden of Paradise, and hence imbued with its scent.[115] Moreover, on account of his physical beauty, Joseph was sometimes himself likened to a rose.[116] Conflating these seemingly disparate images, Rūmī states that, "It was through the scent (*būy*) of that Rose [Joseph] that Jacob's eyes were opened."[117] Given that, in the poetics of Persian Sufism, Joseph symbolized the soul not just on account of his beauty but also because of his sexual purity, the implied metaphor in this verse may further be interpreted as alluding to the scent of the soul itself, thus again harkening back to the notion of the Daena, the luminous maiden who represented the visionary soul and who was associated with the fragrance of the rose.[118]

The metaphorical association of Joseph's shirt with the rose was also underscored by the aetiology the Persian poets provided for the natural phenomenon of a rose in full bloom.[119] Based on the perceived similarity in appearance between a rose whose petals have opened to disperse their fragrance, and Joseph's shirt which, according to the Qur'ānic account, was torn by Potiphar's wife (named Zulaykhā in the Islamic legendary accounts) when she attempted to seduce him, the rose was described as "rending the collar of its shirt," a traditional gesture of grief as well as of ecstatic rapture.[120] Rūmī deftly weaves together the many connotations of the motif of Joseph's shirt in a complex image which he relates to the mystic himself who, like a fragrant rose opening its petals in springtime, rends the collar of his own shirt when, Jacob-like, he experiences the rapture of mystical insight:

Like the red rose, rend your collar in rapture,

For it is that time when [Joseph's] shirt comes to Jacob.[121]

Perhaps the most compelling connection between scent and spiritual vision is to be found in an analogy Rūmī draws between Jacob's regaining his sight on perceiving the scent of Joseph's shirt, and the prophet Muhammad's experience of visionary prayer. Citing the last part of an oft-quoted prophetic Tradition according to which Muhammad reportedly said that women, fragrance, and prayer were the three things he held dear in this world, Rūmī states:[122]

It was on account of this [very] scent (*bū*) that Muhammad always used to say:

'The delight of my eyes is in prayer' (*qurratu 'aynī fī al-ṣalāt*).[123]

By playing on the literal meaning of the idiomatic Arabic expression, *qurratu 'aynī* ("the delight of my eyes"), with its obvious allusion to sight,[124] and at the same time implying by the reference to "scent" (*bū*) that the second of the three items listed in the Tradition (i.e., fragrance) served as the catalyst for spiritual vision, Rūmī is suggesting that the visionary experience of God that the Prophet of Islam was accorded through his performance of ritual prayer was the result of his having perceived the same "spiritual" scent emanating from the realm of the Divine that had rendered the prophet, Jacob, clairvoyant.[125]

In sum, the marked preference that may be discerned in Rūmī's poetry for the olfactory as an epistemic mode of spiritual perception appears to have reflected, on the one hand, the horticultural tradition of Iran that viewed scent as an all-important criterion in rose cultivation, and, on the other, the Sufi notion of sensory correspondences, which, as I have sought to demonstrate, was reinforced by the role accorded to scent in ancient Iranian religious conceptions relating to the Daena, the representative of the visionary soul, whose associated flower was the centifolia rose.

Horticultural Continuity and Poetical Innovation

In view of the unbroken continuity of rose cultivation in Iran from deep antiquity, it would be futile to look for changes in medieval Persian culture resulting from innovations in horticultural practice. It seems that, in the case of Iran, it was the long-standing culture of flowers that molded particular cultural responses, the most imaginative being attested in the poetics of Persian mysticism. Applying their innovative hermeneutical strategies to the image of the rose, the Persian mystical poets transformed this mythopoeic signifier of beauty and mystery, which was redolent of associations with ancient Iranian deities and Qur'ānic prophets, into the symbol of the ineffability of the divine Essence, by focusing on the ineffable nature of its scent. In the poetic imagination of Rūmī in particular, the scent of the rose became the sensory catalyst for imaginal vision of the Divine and for spiritual perception of the Unseen, pointing the way to the rose bush of the Real and the numinous beauty of the soul itself.

To gauge cultural changes resulting from innovations in rose cultivation, one would have to look sooner to Europe. Although initially debased under the influence of early Christian ascetic ideals that associated it with the hedonism and voluptuousness of the pagan Roman empire, the culture of the rose "returned" to Europe in the twelfth century. Only during the Renaissance, however, was it truly revived.[126] It may be surmised that the introduction of the *Rosa centifolia* into Europe in the early modern period, most probably from Persia, and the subsequent establishment of the perfume industry, contributed along with other social and economic factors connected with the rise of the bourgeoisie, to an unprecedented appreciation for floral fragrance in nineteenth-century France, and to the creation of a new olfactory aesthetic that ascribed to scent a psychospiritual dimension for which eroticism provided the top note.[127] This cultural change, which found its most dramatic expression in Parisian intellectual and literary trends of an esoteric and quasi-cabalistic nature, may even have been inspired indirectly by Iranian religious ideas that revolved around the sacred status of flowers.[128]

Acknowledgements:
I would like to express my gratitude to James R. Russell for many illuminating exchanges about roses and scent in ancient Iranian languages and religions, as well as to Jamsheed Choksy, Oktor Skjaervø, Willem Floor, Mehdi Baghi, and Neguine Salehinia for their generous assistance at various stages of my research.

NOTES

1. He cast his spell by reciting an incantation over the rose and then blowing upon it, which was a standard practice in medieval Islamic magic. See Constant Hamès, "L'usage talismanique du Coran," *Revue de l'histoire des religions* 218, no. 1 (2001): 86.

2. Niẓāmī Ganjeh'ī, *Kulliyāt-i Khamseh* (Tehran, 1351/1972), 89–90; for a discussion of the tale, see Edward G. Browne, *Arabian Medicine* (Cambridge: Cambridge University Press, 1921), 89–90; for an illustration of the scene from a sixteenth-century manuscript of the *Makhzan al-asrār*, see Laurence Binyon, *The Poems of Nizami* (London: The Studio Limited, 1928), pl. 5.

3. The tale is not far-fetched by medieval standards. According to Arabic works on toxicology, certain scents were believed to be lethal—see Martin Levey, *Medieval Arabic Toxicology: The "Book on Poisons" of Ibn Waḥshīya and Its Relation to Early Indian and Greek Texts*, Transactions of the American Philosophical Society, n.s., vol. 56, pt. 7 (Philadelphia: The American Philosophical Society, 1966), 39ff.

4. Abū al-Qāsim Firdausī, *Shāh-nāmeh*, ed. Jalāl Khāliqī-Muṭlaq, 6 vols. to date (New York: Bibliotheca Persica, 1988–1997), 1: 92, line 43; also *Naurūz-nāmeh,* attributed to the eleventh-century mathematician and philosopher, 'Umar Khayyām. See 'Omar Khaiiām, *Traktaty,* ed. and trans. B.A. Rozenfel'd (Moscow: Izdatel'stvo vostochnoi literatury, 1961), 191 (Russ. text), 81b (Per. text) where it is stated that Farīdūn also introduced the violet (*benafsheh*), narcissus (*nargis*), and water lily (*nīlūfar*), among other flowers. See also Aḥmad Tafażżolī, "Ferēdūn," *Encyclopaedia Iranica*, ed. Ehsan Yarshater, 11 vols. to date (London–Boston–Henley: Routledge and Kegan Paul, 1985-), 9: 533. Another mythical Iranian king, Jamshīd, was credited with the discovery of aromatics, including the method of distilling rose water. See Firdausī, *Shāh-nāmeh*, 1: 43, line 42; and Reuben Levy, trans., *The Epic of the Kings: Shah-Nama, the National Epic of Persia by Ferdowsi*, rev. ed. Amin Banani, Persian Heritage Series, no. 2 (London–Henley–Boston: Routledge and Kegan Paul, 1967; repr. ed., 1977), 10.

5. See Jack Goody, *The Logic of Writing and the Organization of Society* (Cambridge: Cambridge University Press, 1986; repr. ed., 1996), 84–86; also Jack Goody, *The Culture of Flowers* (Cambridge: Cambridge University Press, 1993), 18–27, and 102–105.

6. See, for example, Germain Bazin, *Paradeisos: The Art of the Garden* (Boston–Toronto–London: Bulfinch Press, and Little, Brown and Company, 1990), 12–13. For an extensive discussion of the topic and bibliographical references, see Maria E. Subtelny, *Le monde est un jardin: Aspects de l'histoire culturelle de l'Iran médiéval*, Cahiers de Studia Iranica, 28 (Paris: Association pour l'Avancement des Études Iraniennes, 2002), 103–106.

7. On this point, see Goody, *Culture of Flowers*, 102ff., and esp. 110–11. For the gardens of Islamic Spain as inheritors of the Persian tradition of garden design, see Stefano Bianca, *Hofhaus und Paradiesgarten: Architektur und Lebensformen in der islamischen Welt*, 2nd ed. (Munich: C.H. Beck, 2001), 118. For the continuity of Persian ideas in Islamic civilization, see Ehsan Yarshater, "The Persian Presence in the Islamic World," in *The Persian Presence in the Islamic World*, eds. Richard G. Hovannisian and Georges Sabagh, Thirteenth Giorgio Levi Della Vida Conference (Cambridge: Cambridge University Press, 1998), 4–125.

8. For the importance of arboriculture in imperial Achaemenid ideology, and the concept of the king as gardener, see Pierre Briant, *Histoire de l'Empire perse: De Cyrus à Alexandre* (Paris: Fayard, 1996), 244–46.

9. For the Persian model of kingship, see Subtelny, *Le monde est un jardin*, 69.

10. See Maria E. Subtelny, "The Tale of the Four Sages Who Entered the *Pardes*: A Talmudic Enigma from a Persian Perspective," *Jewish Studies Quarterly* 11, nos. 1–2 (2004): 15–21.

11. Thus according to Kaempfer (d. 1716), physician and secretary to the Swedish embassy in Iran: "Nomen Rosa proprium non habet apud Persas, nisi generis *gul* [in Persian script] id est, *Floris*, quo ob excellentiam indigatatur." See Engelbert Kaempfer, *Amoenitatum exoticarum politico-physico-medicarum*, fasc. 5: *Variae relationes, observationes & descriptiones rerum persicarum & ulterioris Asiae* (London, 1712), 374. The term *gul* generally denoted cultivated flowers as opposed to wildflowers, the generic term for which appears to have been *lāleh*. See I. Mélikoff, "La fleur de la souffrance: Recherche sur le sens symbolique de *lāle* dans la poésie mystique turco-iranienne," *Journal asiatique* 255 (1967): 345–46, and 357–58.

12. The root *vrd-* is attested in Avestan in the form *vareda-*, meaning plant or rose. See Christian Bartholomae, *Altiranisches Wörterbuch* (1904; repr. ed., Berlin and New York: Walter de Gruyter, 1979), 1368–69; Henrik Samuel Nyberg, *A Manual of Pahlavi*, pt. 2: *Glossary* (Wiesbaden: Otto Harrassowitz, 1974), 86; Jean Kellens, *Les noms-racines de l'Avesta* (Wiesbaden: Dr. Ludwig Reichert Verlag, 1974), 65; and Hušang A'lam, "Gol," *Encyclopaedia Iranica*, 11: 46.

13. See Ludwig Koehler and Walter Baumgartner, *The Hebrew and Aramaic Lexicon of the Old Testament*, 2 vols. (repr. ed., Leiden–Boston–Cologne: Brill, 2001), 1: 260; Marcus Jastrow, *A Dictionary of the Targumim, the Talmud Babli and Yerushalmi, and the Midrashic Literature* (repr. ed., New York: Judaica Press, 1996), 375; Ernest Klein, *A Comprehensive Etymological Dictionary of the Hebrew Language for Readers of English* (New York: Macmillan Publishing Co., 1987), 192; and especially Immanuel Löw, *Die Flora der Juden*, 4 vols. (1924–1934; repr. ed., Hildesheim: Georg Olms Verlag, 1967), 3: 194–95. The contention of some comparative Semiticists that the word for rose in the Semitic languages belongs to the category of "old culture words" whose origins cannot be ascribed to any particular language, fails to take into account its Indo-European etymology. Thus Edward Lipiński, *Semitic Languages: Outline of a Comparative Grammar*, 2nd ed., Orientalia Lovaniensia Analecta, 80 (Louvain: Peeters, 2001), 573–74. The word *shoshanah*, which was taken to mean rose in medieval Hebrew, actually denoted the lily. See Löw, *Die Flora der Juden*, 2: 168–69, and 3: 195–96.

14. Egyptian *wrt* is attested only in Demotic texts from the Ptolemaic (i.e., Persian) period, as the rose was unknown in Pharaonic Egypt. See Renate Germer, *Flora des pharaonischen Ägypten* (Mainz am Rhein: Verlag Philipp von Zabern, 1985), 64; and Victor Loret, *La flore pharaonique d'après les documents hiéroglyphiques et les spécimens découverts dans les tombes* (1892; repr. ed., Hildesheim and New York: Georg Olms Verlag, 1975), 82. The word must have come into Egyptian during the Ptolemaic period along with the design of the Persian architectural garden, or *paradeisos*. See also J. Černý, *Coptic Etymological Dictionary* (Cambridge: Cambridge University Press, 1976), 215 (also for the reference to Armenian).

15. For early linguistic contacts between Greeks and Iranians, see Walter Burkert, *The Orientalizing Revolution: Near Eastern Influence in the Early Archaic Age*, trans. Margaret E. Pinder and Walter Burkert (Cambridge, Mass.: Harvard University Press, 1995), 33ff. Roses were abundantly cultivated on the island of Rhodes (whence the name), and the flower is thought to have been depicted on its coinage. See Stuart Mechlin and Janet Browne, eds., *The Rose* (New York: Mayflower Books, 1979), 15. Although poetically evocative, the derivation of the Latin *rosa* from *ros*, meaning "dew," is a popular etymology that is not supported linguistically. See, for example, Alexander Roob, *The Hermetic Museum: Alchemy and Mysticism* (Cologne: Taschen, 1996), 377.

16. See *The American Heritage Dictionary of Indo-European Roots,* 2nd rev. ed. (Boston and New York: Houghton Mifflin Company, 2000), 102, s.v. *wrod-*.

17. See Qāsim b. Yūsuf Abū Naṣrī, *Irshād al-zirāʻa*, ed. Muḥammad Mushīrī (Tehran, 1346/1968), 202–206. For a scientific catalogue of the species cultivated in Iran in modern times, see Karl Heinz Rechinger, ed., *Flora Iranica: Flora des iranischen Hochlandes und der umrahmenden Gebirge* (Graz: Akademische Druck- u. Verlagsanstalt, 1963–), vol. 152 (1982): J. Zieliński, *Rosaceae II–Rosa.*

18. Sir Robert Ker Porter, *Travels in Georgia, Persia, Armenia, Ancient Babylonia. . . . during the Years 1817, 1818, 1819, and 1820* (London, 1821–1822), cited in Donald Newton Wilber, *Persian Gardens and Garden Pavilions*, 2nd ed. (Washington, D.C.: Dumbarton Oaks, 1979), 71.

19. The scent of the rose is carried generally in the petals, hence the greater the number of petals, the larger the volume of scent. Scent is released when the essential oil, known as attar (< Persian *ʻaṭr*) is dispersed as the flowers open. This occurs when atmospheric conditions are favorable. Rose scents have been divided into several categories by rose specialists. The category that has the most potent and luxurious scent is referred to as the "Old rose fragrance," that is, the extremely complex fragrance associated with the centifolia roses. See David Austin, *David Austin's English Roses* (London: Conran Octopus, 1996; repr. ed., 2002), 41–43.

20. Rashīd al-Dīn Fażlullāh Hamadānī, *Āṣār va aḥyā'*, eds. Manūchihr Sutūdeh and Īraj Afshār (Tehran, 1368/1989), 63.

21. Qāsim b. Yūsuf, *Irshād al-zirāʻa*, 204–205; see also Aʻlam, "Gol," 49.

22. Most handbooks on roses state that the origins of the *Rosa centifolia* are unknown. Thus, for example, Charles and Brigid Quest-Ritson, *The American Rose Society Encyclopedia of Roses* (New York: DK Publishing, 2003), 9–11, and 85; and Mechlin and Browne, *The Rose*, 50ff. The *Rosa centifolia* appears to have been a complex hybrid (*R. gallica* x *R. moschata* x *R. canina*), and it may therefore be presumed to have had a garden origin. Although it was long believed to be the oldest of the Old European roses, according to the authoritative rose scholar, C.C. Hurst, it was actually the "youngest of all our old Roses," having evolved only from the end of the sixteenth century, and Hurst corroborates its introduction into Europe via the Netherlands (although he does not specify where it was introduced from). See C.C. Hurst, "Notes on the Origin and Evolution of Our Garden Roses," in Graham Stuart Thomas, *The Old Shrub Roses,* 4th rev. ed. (London: J.M. Dent and Sons Ltd., 1979; repr. ed., 1986), 70–71. Attesting to the role played by the Dutch is its old botanical name, *Rosa centifolia batavica*, which was changed to *Rosa centifolia L.* in the eighteenth century by the Swedish botanist, Linnaeus. See C. Linnaeus, *Species Plantarum*, 2 vols. (Stockholm, 1753), 1: 491–92. It was often depicted in seventeenth- and eighteenth-century Dutch floral paintings. For the possibility of its introduction via Ottoman Turkey, see Nurhan Atasoy, *A Garden for the Sultan: Gardens and Flowers in the Ottoman Culture* ([Istanbul]: Aygaz, 2002), 128–30, and 164.

23. Although it shares a similar genealogy with the *Rosa centifolia* (*R. gallica* x *R. moschata*), it appears to have been a distinct species. See Hurst, "Notes on the Origin of the Moss Rose," in Thomas, *Old Shrub Roses,* 102; also Hurst, "Notes on the Origin and Evolution of Our Garden Roses," 63. For the *Rosa damascena,* see Peter Beales, *Classic Roses: An Illustrated Encyclopaedia and Grower's Manual of Old Roses, Shrub Roses and Climbers* (London: Collins Harvill, 1985), 184ff. I am indebted to Holly Shimizu, Director of the United States Botanic Garden, for elucidating for me the differences between these rose species. For references to the *gul-i Muḥammadī* in Persian sources, see Ḥabībullāh Ṣābitī, *Dirakhtān va dirakhtchehhā-yi Īrān* [Trees and Shrubs of Iran] (Tehran, 1344/1966), 323; Aʻlam, "Gol," 47; and Hušang Aʻlam, "Golāb," *Encyclopaedia Iranica*, 11: 58. See also Austin, *David Austin's English Roses*, 13; Penelope Hobhouse, *Gardens of Persia* ([Carlsbad, Calif.]: Kales Press, 2004), 34; and F. Aubaile-Sallenave, "'Aṭr," *Encyclopaedia Iranica*, 3: 14–16.

24. See Annemarie Schimmel, *And Muhammad Is His Messenger: The Veneration of the Prophet in Islamic Piety* (Chapel Hill and London: University of North Carolina Press, 1985), 34–35; and Annemarie Schimmel, "The Celestial Garden in Islam," in *The Islamic Garden*, eds. Elisabeth B. Macdougall and Richard Ettinghausen, Dumbarton Oaks Colloquium on the History of Landscape Architecture, 4 (Washington, D.C.: Dumbarton Oaks, 1976), 31. The poppy/tulip (*lāleh*), by contrast, was believed to have been produced from a tear that fell from Muhammad's right eye, and it was generally regarded as the poetic symbol of martyrdom and suffering in love. See Mélikoff, "La fleur de la souffrance," 349.

25. The women of his circle were even said to have collected his perspiration to use as a perfume. See Schimmel, *And Muhammad Is His Messenger*, 34. On this point see Annick Le Guérer, *Scent: The Mysterious and Essential Powers of Smell,* trans. Richard Miller (New York: Turtle Bay Books and Random House, 1992), 120–27 (with a discussion of the possible metabolic causes); and Peter Brown, *The Cult of the Saints: Its Rise and Function in Latin Christianity* (Chicago: The University of Chicago Press, 1981; paperback ed., 1982), 76 (for Christian parallels).

26. For a late medieval depiction, see Schimmel, *And Muhammad Is His Messenger*, 111. One would expect the number of his names to correspond to the hundred petals of the centifolia, but in fact by denoting something incomplete, the number ninety-nine points to the comprehensive unity represented by the totality of the names, just as in the case of the ninety-nine divine Names. See Annemarie Schimmel, *The Mystery of Numbers* (New York and Oxford: Oxford University Press, 1993), 271. For the comparison of the prophet Muhammad, regarded as the perfect man in the Islamic tradition, to a red rose, see note 41 below.

27. See G. Le Strange, *The Lands of the Eastern Caliphate: Mesopotamia, Persia, and Central Asia from the Moslem Conquest to the Time of Timur* (Cambridge: Cambridge University Press, 1905; repr. ed., 1930), 293; [Thaʻālibī], *The Laṭāʼif al-maʻārif of Thaʻālibī: The Book of Curious and Entertaining Information*, trans. C.E. Bosworth (Edinburgh: Edinburgh University Press, 1968), 127; and [Ḥamdullāh Mustaufī Qazvīnī], *The Geographical Part of the Nuzhat-al-Qulūb Composed by Ḥamd-Allāh Mustaufī of Qazwīn in 740 (1340)*, trans. G. Le Strange, E.J.W. Gibb Memorial Series, 23.2 (Leiden: E.J. Brill, and London: Luzac and Co., 1919), 117. For rose water production in the Safavid period, see Willem Floor, *The Economy of Safavid Persia* (Wiesbaden: Reichert Verlag, 2000), 165–66; and Kaempfer, *Amoenitatum exoticarum*, 373: "Rosam uti Persia ex omnibus mundi partibus maxime copiosam ac suave olentem gignit, ita *Sjirasum* [i.e., Shiraz] ejusve pagus praelaudatus, praeceteris Persiae provinciis, fert copiosissimam ac fragrantissimam: cujus aqua destillata per omnem Indiam, atque ipsius quoque Persiae provincias laboriose divehitur."

28. Thus Rashīd al-Dīn, *Ās̱ār va aḥyāʼ*, 63–64; see also A.K.S. Lambton, "The *Āt̲h̲ār wa Aḥyāʼ* of Rashīd al-Dīn Faḍl Allāh Hamadānī and His Contribution as an Agronomist, Arboriculturist and Horticulturalist," in *The Mongol Empire and Its Legacy*, eds. Reuven Amitai-Preiss and David O. Morgan (Leiden–Boston–Cologne: Brill, 1999), 144.

29. See Thaʻālibī, *Laṭāʼif al-maʻārif*, 127–28. The unit of measure for the latter was the Iraqi *riṭl*, which weighed approximately four hundred grams.

30. See Qāsim b. Yūsuf, *Irshād al-zirāʻa*, 203–204; and Bess Allen Donaldson, *The Wild Rue: A Study of Muhammadan Magic and Folklore in Iran* (London: Luzac and Co., 1938), 145–46. See also Hakeem Abdul Hameed, ed., *Avicenna's Tract on Cardiac Drugs [Risāla al-adwiya al-qalbiyya] and Essays on Arab Cardiotherapy* (Karachi: Hamdard Foundation Press, 1983), 45–46; and the discussion of the function of exhilirating fragrances in Islamic medicine in Ali Akbar Husain, *Scent in the Islamic Garden: A Study of Deccani Urdu Literary Sources* (Karachi: Oxford University Press, 2000), 124ff.

31. See Qāsim b. Yūsuf, *Irshād al-zirāʻa*, 250–52; also Expiración García-Sánchez, "Les techniques de distillation de l'eau de rose à Al-Andalus," in *Parfums d'Orient*, ed. Rika Gyselen, Res Orientales, 11 (Bures-sur-Yvette: Groupe pour l'Étude de la Civilisation du Moyen-Orient, 1998), 125–40. Rose water is mentioned as an ingredient in one of the dishes prepared by the Devil, who disguised himself as a cook in order to entice the mythical Iranian king, Żaḥḥāk—see Levy, *Epic of the Kings*, 14.

32. On the admissibility of using rose water in connection with various religious rites, see Ibn Ḥajar al-ʻAsqalānī, *Fatḥ al-bārī bi-sharḥ Ṣaḥīḥ al-Bukhārī*, 28 vols. (Cairo: 1398/1978), 1–2: §258 and §280; 5–6: §879, etc.; also Aʻlam, "Golāb," 58. The prophet Muhammad was apparently favorably disposed toward scent. See Schimmel, *And Muhammad Is His Messenger*, 51; for the prophetic Tradition to this effect, see note 122 below.

33. Both the medicinal and funerary uses of rose water are illustrated in a tale told by the twelfth-century Persian poet, ʻAṭṭār (whose name means "perfumer" or "druggist"), about a sick miser for whom he prepared a potion of rose water. When he tried to administer it, he refused, considering it too precious to be wasted on himself. The man died soon thereafter (perhaps as a result of not having taken the potion), and ʻAṭṭār used the same rose water to sprinkle on his grave. See Hellmut Ritter, *The Ocean of the Soul: Man, the World and God in the Stories of Farīd al-Dīn ʻAṭṭār*, trans. John O'Kane, ed. Bernd Radtke, Handbook of Oriental Studies, section 1, vol. 69 (Leiden and Boston: Brill, 2003), 99.

34. We might well ask, paraphasing Michel Conan, what is left for the study of the rose when the horticultural reality is removed, and only the symbol remains? See Michel Conan, "From Vernacular Gardens to a Social Anthropology of Gardening," in *Perspectives on Garden Histories*, ed. Michel Conan, Dumbarton Oaks Colloquium on the History of Landscape Architecture, 21 (Washington, D.C.: Dumbarton Oaks, 1999), 181–82.

35. I have found the definition of Henry Corbin to be the most appropriate to the Perso-Islamic cultural context, and particularly to the domain of Persian mystical literature, with which he was thoroughly familiar. See Henry Corbin, *Temple and Contemplation*, trans. Philip Sherrard and Liadain Sherrard (London and New York: KPI and Islamic Publications, 1986), 304, and 13; see also the comments of Sebastian Brock regarding the (originally Persian) term *raza* (meaning both "mystery" and "symbol") in Syriac mystical literature, in St. Ephrem, *Hymns on Paradise*, trans. Sebastian Brock (Crestwood, N.Y.: St. Vladimir's Seminary Press, 1998), 42.

36. A stock image employed by the poets was the comparison of the cheeks of the beloved to the petals of a red rose. See C.-H. de Fouchécour, *La description de la nature dans la poésie lyrique persane du XIe siècle: Inventaire et analyse des thèmes* (Paris: Librairie C. Klincksieck, 1969), 68–72.

37. For the depiction of flowers, trees and gardens in Persian painting, see Norah M. Titley, *Plants and Gardens in Persian, Mughal and Turkish Art* (London: The British Library, 1979). For the rose and nightingale motif as a decorative theme in Persian art, especially in late Safavid and Qajar lacquer bookbindings and penboxes, see Layla S. Diba, "Gol o Bolbol," *Encyclopaedia Iranica*, 11: 52–57.

38. Farīd al-Dīn ʻAṭṭār, *Manṭiq al-ṭayr*, ed. Sayyid Ṣādiq Gauharīn (Tehran, 1366/1988), 42–43, lines 763–65. The last line contains an allusion to the medicinal uses of rose water, which was believed to alleviate headaches and to reestablish the equilibrium of the bodily humours. The term *saudā* (which I have translated as "melancholy") referred to black bile, which denoted the melancholy temperament, one of the four humours in the Galenic system of medicine.

39. It is interesting to note in this connection that the Persian image of rose and nightingale was taken up by the medieval Armenian Christian devotional poets who applied it to Christ and John the Baptist. See James R. Russell, "On Armeno-Iranian Interaction in the Medieval Period," in *Au carrefour des religions: Mélanges offerts à Philippe Gignoux*, ed. Rika Gyselen, Res Orientales, 7 (Bures-sur-Yvette: Groupe pour l'Étude de la Civilisation du Moyen-Orient, 1995), 237.

40. For this prophetic Tradition (*ḥadīth*), which does not occur in the canonical compilations, see Annemarie Schimmel, *Mystical Dimensions of Islam* (Chapel Hill: University of North Carolina Press, 1975), 472.

41. See Ruzbihan Baqli, *The Unveiling of Secrets: Diary of a Sufi Master*, trans. Carl W. Ernst (Chapel Hill, N.C.: Parvardigar Press, 1997), 48–49, §71; and Rūzbehân Baqlî Shîrâzî, *Le jasmin des fidèles d'amour (Kitâb-e ʻAbhar al-ʻâshiqîn)*, trans. Henry Corbin (Paris: Verdier, 1991), 87, §76–77. See also Annemarie Schimmel, *As through a Veil: Mystical Poetry in Islam* (New York: Columbia University Press, 1982), 76. Dante's celestial rose was not red, however, but white (or gold). See Dante Alighieri, *The Divine Comedy*, vol. 3: *Paradise*, trans. Mark Musa (New York: Penguin Books, 1986), Cantos 30–31, and commentary, 369. Rūzbihān sometimes also likens the prophet Muhammad to a red rose, thereby supporting the later nomenclature of the "Muhammadan" rose. See Ruzbihan Baqli, *Unveiling of Secrets*, 61, §94, and 84, §134.

42. Midnight was considered the most propitious time for devotional and contemplative activity by Muslim mystics and Jewish Kabbalists alike.

43. Ruzbihan Baqli, *Unveiling of Secrets*, 57, §87 (translation slightly altered). Note also the passages where God is described as scattering an infinite quantity of red roses. See Ruzbihan Baqli, *Unveiling of Secrets*, 48–49, §71, and 91, §145.

44. For the idea of feminine beauty as a manifestation of divine Reality, as elaborated by the great Spanish Arab theosopher, Ibn 'Arabī (d. 1240), based on a well-known prophetic Tradition (for which see note 122 below), see Ibn al'Arabi, *The Bezels of Wisdom*, trans. R.W.J. Austin (Mahwah, N.J.: Paulist Press, 1980), 275: "Contemplation of the Reality without formal support is not possible, since God in His Essence, is far beyond all need of the Cosmos. Since, therefore, some form of support is necessary, the best and most perfect kind is the contemplation of God in women." See also Sachiko Murata, "Witnessing the Rose: Ya'qūb Ṣarfī on the Vision of God in Women," in *Gott ist schön und Er liebt die Schönheit / God Is Beautiful and He Loves Beauty*, eds. Alma Giese and J. Christoph Bürgel (Bern: Peter Lang, 1994), 352ff.; and Annemarie Schimmel, *My Soul Is a Woman: The Feminine in Islam*, trans. Susan H. Ray (New York and London: Continuum, 1997), 98ff.

45. Nūr al-Dīn 'Abd al-Raḥmān Jāmī, *Yūsuf va Zulaykhā*, in *Ma<u>s</u>navī-i Haft aurang*, ed. Āqā Murtaẓā-Mudarris Gīlānī (Tehran, 1337/1958), 592; see also Ritter, *Ocean of the Soul*, 495. The star-crossed lovers, Laylā and Majnūn (whose name means "madman"), are the counterparts of Romeo and Juliet in the Islamic romantic tradition.

46. The aphorism was based on the prophetic Tradition mentioned above in connection with Rūzbihān Baqlī. See Ruzbehan Baqli Shirazi, *Commentaire sur Les paradoxes des soufis (Sharh-e Shathíyât)*, ed. Henry Corbin, 4th rev. ed., ed. Mohammad-Ali Amir-Moezzi, Bibliothèque Iranienne, vol. 12 (Tehran: Institut Français de Recherche en Iran, 2003), 145, §265.

47. For the account, see Ritter, *Ocean of the Soul*, 488.

48. For the concept of the Universal Intellect, also referred to as the First Intellect (*al-'aql al-awwal*), which was frequently equated with the Spirit of divine Revelation, hence with the Angel Gabriel, and contrasted with the Partial Intellect (*al-'aql al-juz'ī*), see William C. Chittick, *The Self-Disclosure of God: Principles of Ibn al-'Arabī's Cosmology* (Albany: State University of New York Press, 1998), xxix; Henry Corbin, *L'imagination créatrice dans le soufisme d'Ibn 'Arabī*, 2nd ed. (Paris: Aubier, 1993), 133; also Henry Corbin, *Avicenne et le récit visionnaire* (Paris: Verdier, 1999), 77.

49. I have opted for the variant reading, *sukr* ("intoxication"), instead of the *shukr* ("gratitude") of the edited text. Not only is it the *lectio difficilior*, but it also follows shortly after the definition of the term *ṣaḥv* ("sobriety"), and conforms better to the idea that 'Attār is articulating. Textual support for this reading is provided by Rūzbihān Baqlī—see p. 18 above. On the definition of the term in the compilative works on Sufi technical terminology, see Michael A. Sells, *Early Islamic Mysticism: Sufi, Qur'an, Mi'raj, Poetic and Theological Writings* (New York and Mahwah: Paulist Press, 1996), 124 (citing al-Qushayrī's *al-Risāla al-Qushayriyya*).

50. Farīd al-Dīn 'Aṭṭār Nīshāpūrī, *Muṣībat-nāmeh*, ed. Nūrānī Viṣāl (Tehran, 1338/1959), 41. A great many examples of the application of this pun may be adduced. The poet, Rūmī, echoes 'Aṭṭār's words when he states: "See the rose (*gul*) in the thorn . . . ; See the Whole (*kull*) in the part . . . ," and "Because of you, my thorn has become a rose (*gul*), and the particulars (*juzvhā*) have all become the Universal (*kull*)"—Jalāl al-Dīn Rūmī, *Kulliyāt-i Shams yā Dīvān-i kabīr* (henceforth *Dīvān*), ed. Badī' al-Zamān Furūzānfar, 10 vols. in 9 (repr. ed., Tehran, 1363/1984) (Note: citations are to the volume of the edition, and the number and line of the poem), 1: 326/3,547, and 1: 326/3,546. See also [Jalāl al-Dīn Rūmī], *The Mathnawí of Jalálu'ddín Rúmí* [*Ma<u>s</u>navī-i ma'navī*] (henceforth *Ma<u>s</u>navī*), ed., trans. and commentary by Reynold A. Nicholson, 8 vols., E.J.W. Gibb Memorial Series, n.s., 4.1–4.8 (London: Luzac and Co., 1925–1940; repr. ed., vols. 1–6, 1985) (Note: citations are to the 6 books of the *Ma<u>s</u>navī* and the line number, unless indicated otherwise), 1: 763:

    The faces of the particulars (*juzvhā*) are all set towards the Universal (*kull*),

    [Just as] the task of nightingales is to play the game of love with the rose (*gul*).

51. Similar comparisons made by the mystical poets include perceiving the wine in the grape, the spark in the flint, and so on. See Schimmel, "Celestial Garden," 30.

52. Rūmī, for example, refers to the phenomenal world as an "abode of deception" (*dār al-ghurūr*). See *Ma<u>s</u>navī*, 4: 1,366 (the allusion is to Qur'ān 57:20).

53. For the phrase, see William C. Chittick, *The Sufi Path of Knowledge: Ibn al-'Arabi's Metaphysics of Imagination* (Albany: State University of New York Press, 1989), 361–62. Compare Rūmī, *Dīvān*, 2: 729/7,652: "Thorn and rose are one for him . . ." (referring to the great Sufi master, Shams-i Tabrīzī). Alternatively, rose and thorn represent two aspects of a single Reality, the former representing God's beauty (*jamāl*), the latter His awesome majesty (*jalāl*). See Annemarie Schimmel, *The Triumphal Sun: A Study of the Works of Jalāloddin Rumi* (London: Fine Books, and The Hague: East-West Publications, 1978), 92.

54. Rūmī, *Dīvān*, 2: 859/8,971.

55. Ruzbehan Baqli, *Commentaire*, 81, §89, and French text, 12.

56. Ruzbehan Baqli, *Commentaire*, 403, §1,062. It is noteworthy that, in his commentary on al-Ḥallāj, Rūzbihān does not even include the term *shukr* ("gratitude") in his long list of Sufi *termini technici*. For the variant reading I have adopted, see note 49 above; for other corroborating examples, see Sayyid Ja'far Sajjādī, *Farhang-i iṣṭilāḥāt va ta'bīrāt-i 'irfānī* (Tehran, 1375/1997), 468–72, s.v. *sukr*.

57. See, for example, Qur'ān 5:60, 5:68, 5:76, 5:77, 5:100, etc.

58. Rūmī, *Ma<u>s</u>navī*, 6: 1,815–16.

59. Nicholson interprets *shāh-i qul* to mean the saint or spiritual man who alone is worthy of uttering the Word of God. See *The Mathnawí of Jalálu'ddín Rúmí*, vol. 8: 205 (Commentary to *Ma<u>s</u>navī*, 4: 3,259). But it seems to me that Rūmī's figurative expression contains an unmistakable allusion to the Qur'ānic command.

60. Rūmī's expression, *shāh-i qul* ("the king of 'Say!'"), also alludes to the rose's traditional epithet, "the king of beauty" (*shāh-i ḥusn*), which it parallels in form but outdoes in content.

[61] The early Persian poets never ignored the aspect of scent in their descriptions of the rose, but they did not accord it the primacy that the mystical poets did. See, for example, de Fouchécour, *La description de la nature*, 68–73.

[62] For the discovery of the chemical compound in rose oil that accounts for its mysterious scent, see Günther Ohloff, *Scent and Fragrances: The Fascination of Odors and Their Chemical Perspectives*, trans. Wilhelm Pickenhagen and Brian M. Lawrence (Berlin–Heidelberg–New York: Springer-Verlag, 1994), 154–58.

[63] Contributing to the evanescent quality of the rose was the fact that most Old roses, such as the centifolia, usually bloomed only once a season. The feature of continuous flowering was not introduced into Europe until circa 1800 with the China roses. See Hurst, "Notes on the Origin and Evolution of Our Garden Roses," 59–60.

[64] The literature on the concept of ineffability is extensive, and in my discussion I largely follow Ben-Ami Scharfstein, *Ineffability: The Failure of Words in Philosophy and Religion* (Albany: State University of New York Press, 1993), esp. 67–84, and 193–200; Steven T. Katz, "Mystical Speech and Mystical Meaning," in *Mysticism and Language*, ed. Steven T. Katz (New York and Oxford: Oxford University Press, 1992), 3–41; and Michael A. Sells, *Mystical Languages of Unsaying* (Chicago and London: University of Chicago Press, 1994), 1–13.

[65] For this succinct formulation of a central theme in Heidegger's thinking about the function of poetic language, and its application to the poetics of Jewish mysticism, see Elliot R. Wolfson, *Pathwings: Philosophic and Poetic Reflections on the Hermeneutics of Time and Language* (Barrytown, N.Y.: Barrytown/Station Hill Press, 2004), 230–31. See Martin Heidegger, *On the Way to Language*, trans. Peter D. Hertz (New York: Harper San Francisco, 1982), 159ff., and especially 188, where he states: "Poetry's spoken words shelter the poetic statement as that which by its essential nature remains unspoken."

[66] Rūmī, *Dīvān*, 1: 469/4,976. I have sought in my translation to convey the word play on *dīdār*, which means both sight/vision and face, by combining the two meanings. The concept of imaginal vision of God *(ru'yā)* is probably the most recondite doctrine in Islamic mysticism. Rūmī refers to the mystic who is privileged to "see" God as *maḥram*, meaning one who has access to what is normally forbidden *(ḥarām)*.

[67] Rūmī, *Masnavī*, 4: 2,215. For Rūmī's views on poetry and inspiration, see my "La langue des oiseaux: L'inspiration et le langage chez Rumi," in *L'inspiration: Le souffle créateur dans les arts, littératures et mystiques du Moyen Âge européen et proche-oriental*, eds. Claire Kappler and Roger Grozelier (Paris: L'Harmattan, 2006), 363–75.

[68] Rūmī, *Masnavī*, 1: 2,022. The lines are reminiscent of the following idea expressed by 'Aṭṭār:

> Whoever arrives at this place loses his head,
> He can no longer find the door to the four boundaries of his self.
> But he who finds his way here,
> Finds in an instant the supreme secret of the All *(sirr-i kull)*.

—see 'Aṭṭār, *Manṭiq al-ṭayr*, 217, lines 3,905–06.

[69] Rūmī, *Masnavī*, 5: 3,349.

[70] This saying of Angelus Silesius, who was influenced by Rūmī's European near-contemporary, the thirteenth-century German mystic, Meister Eckhart, resonates in the thought of the twentieth-century philosopher, Martin Heidegger, who states:

> What is unsaid in the saying—and everything depends on this—is rather that man, in the most hidden ground of his
> essence, truly is for the first time, when he is in his own way like the rose—without why.

—see John D. Caputo, *The Mystical Element in Heidegger's Thought* (New York: Fordham University Press, 1986), 97ff., and esp. 140. Rūmī appears to be using the phrase, "without why" *(bī-chūn)*, in the same way, as in the following verse: "Mystic lovers concern themselves with the beauty of love, without why"—see Rūmī, *Dīvān*, 1: 132/1,521.

[71] For the reference, see note 69 above. The word *gulistān*, meaning rose garden, usually occurs in the metathesized form, *gulsitān*, only when required by the poetic metre, but Rūmī uses it often as the basis for a favorite word play, as in the following verse in which he clearly differentiates between the two meanings:

> You were an intoxicated nightingale in the midst of the owls [of this world],
> When you perceived the scent of the [divine] Rose garden *(gulistān)*, you made your way to the [divine] Gatherer of roses
> *(gul-sitān)*.

—Rūmī, *Dīvān*, 6: 3,051/32,463. For the image of God as divine Gardener in Rūmī's poetry, see my *Le monde est un jardin*, 131.

[72] Rūmī, *Masnavī*, 5: 3,350.

[73] Elsewhere Rūmī exhorts the Sufi adept to "Strive to become old in (or: a master of) intelligence *('aql)* and religion *(dīn)*; So that like the Universal Intellect *('aql-i kull)*, you may become one who perceives the hidden Reality *(bāṭin)*"—Rūmī, *Masnavī*, 4: 2,178. The idea of *ham-damī* ("sympathy" or "conspiration," as it was rendered brilliantly into French by Corbin), which denotes the correspondence of the spiritual and the sensible, figures prominently in Rūmī's theology. See Corbin, *L'imagination créatrice*, 131–32. For the equation of the heart with the mind in Sufi epistemology, see Subtelny, *Le monde est un jardin*, 139.

[74] Rūmī, *Masnavī*, 6: 30. Yet another allusion to the efficaciousness of rose scent and rose water for curing headaches and hangovers. Rūmī's formulation would appear to attest to an uncanny appreciation for the fact that scent travels directly to the brain without mediation. See note 92 below.

[75] Rūmī, *Dīvān*, 5: 2,147/22,726.

[76] See Elliot R. Wolfson, *Through a Speculum That Shines: Vision and Imagination in Medieval Jewish Mysticism* (Princeton, N.J.: Princeton University Press, 1994), Chapters 3–4; David Chidester, *Word and Light: Seeing, Hearing, and Religious Discourse* (Urbana and Chicago: University of Illinois Press, 1992), 43ff., and 59 where he notes that St. Augustine dismissed taste, smell, and touch as modalities through which one could experience religious truth. For the

various modes of vision in Islamic mysticism, see Henry Corbin, *Le paradoxe du monothéisme* (Paris: L'Herne, 1981), 16–17.

[77.] This aspect of Rūmī's poetry has never been subjected to deeper scholarly scrutiny. See the comments of Schimmel, *As through a Veil,* 98; and Annemarie Schimmel, "Yūsuf in Mawlānā Rumi's Poetry," in *The Heritage of Sufism,* vol. 2: *The Legacy of Medieval Persian Sufism (1150–1500),* ed. Leonard Lewisohn (Oxford: Oneworld, 1999; repr. ed., 2003), 46 (although she associates Rūmī's focus on the olfactory only with the Qur'ānic story of Joseph). For a useful discussion of the role of scent in Urdu literature of the nonmystical variety, see Husain, *Scent in the Islamic Garden,* esp. 154ff.

[78.] For the esoteric "langage des fleurs" and the Parisian intellectual and cultural trend it represented, see Beverly Seaton, "A Nineteenth-Century Metalanguage: *Le Langage des Fleurs,*" *Semiotica* 57, nos. 1–2 (1985): 73–86; also Goody, *Culture of Flowers,* 232–53. For the application of the French formula to this aspect of the Zoroastrian religious tradition, see Henry Corbin, *Corps spirituel et Terre céleste: De l'Iran mazdéen à l'Iran shî'ite,* 2nd rev. ed. (Paris: Buchet/Chastel, 1979), 54. See also Jivanji Jamshedji Modi, *The Religious Ceremonies and Customs of the Parsees* (Bombay, 1922; repr. ed., New York and London: Garland Publishing Inc., 1979), 396 (although without the French cultural allusion).

[79.] See the sensitive treatment of this point by Corbin, *Corps spirituel,* 32, and 54.

[80.] See *Zand-Ākāsīh: Iranian or Greater Bundahišn,* translit. and trans. Behramgore Tehmuras Anklesaria (Bombay, 1956), Chapter 16, 152–53; also Modi, *Religious Ceremonies and Customs,* 396–98. The *Bundahishn* ("Creation") was also known as *Zand-Āgāhīh* ("Knowledge of the Zand"), that is, the commentary on the Avesta, the Zoroastrian scripture. The Daena was connected with the twenty-fourth day of the month, which does not appear to have been significant in itself. The highest Amesha Spentas, Ahura Mazda (Ohrmazd) and Shaherivar, who presided over the sovereignty of kings, were associated with myrtle (*mūrd-yāsmīn*) and basil (*shāh-sparem*), respectively. See *Zand-Ākāsīh* [*Bundahishn*], 151–52. Anklesaria's English translation of *môṛt yâsmin* as two separate flowers, "the myrtle and the jasmine," is to be corrected, as the variety of myrtle cultivated in Iran was in fact characterized by its white jasmine-like flowers. See Valīullāh Muẓaffariyān, *Farhang-i nāmhā-yi giyāhān-i Īrān—Lātīnī, Inglīsī, Fārsī* (Tehran, 1377/1998), 357, no. 4942. It may well be a sprig of the sacred myrtle that is depicted in the well-known scene of the investiture of the Sasanian king, Ardashīr I, by Ahura Mazda in the rock relief at Naqsh-i Rustam in southern Iran, an interpretation that, to the best of my knowledge, has never been proposed. See Roman Ghirshman, *Iran: Parthes et Sassanides* (Paris: Gallimard, 1962), 133, and pl. 168.

[81.] See p. 15 above.

[82.] For the many related translations and interpretations of the term, which apparently derives from the root *day–* ("to see"), see Bartholomae, *Altiranisches Wörterbuch,* 662–66 (where he notes that it is "öfters kaum übertragbar"); Mary Boyce, *Zoroastrians: Their Religious Beliefs and Practices* (London and New York: Routledge, 1979; repr. ed., 2001), 71; Gherardo Gnoli, *Zoroaster's Time and Homeland: A Study on the Origins of Mazdeism and Related Problems* (Naples: Istituto Universitario Orientale, 1980), 195, note 70; Geo Widengren, "La rencontre avec la *daēnā,* qui représente les actions de l'homme," *Orientalia Romana* 5 (1983), 47; Werner Sundermann, "Die Jungfrau der guten Taten," in *Recurrent Patterns in Iranian Religions: From Mazdaism to Sufism,* ed. Philippe Gignoux, Cahiers de Studia Iranica, 11 (Paris: Association pour l'Avancement des Études Iraniennes, 1992), 160–61; also Mansour Shaki, "Dēn," *Encyclopaedia Iranica,* 7: 279–80. Although Zoroastrianism itself was referred to as *Veh Dēn,* "the Good Religion," the term *dēn* related to the psychology of religion rather than to its institutional aspect. Moreover, the Middle Persian term, *dēn* (from the Indo-Iranian root *day-* "to see"), is not to be confused linguistically with the Arabic word, *dēn,* which is derived from the Semitic root DYN (meaning "to judge" in Hebrew), and which was customarily used for the Islamic religion. For the confusion between the two in early Arabic translations of Pahlavi texts, see Shaul Shaked, "From Iran to Islam: Notes on Some Themes in Transmission," in Shaul Shaked, *From Zoroastrian Iran to Islam: Studies in Religious History and Intercultural Contacts* (Aldershot, England: Variorum, 1995), VI: 36.

[83.] See, for example, Aturpāt-i Ēmētān, *The Wisdom of the Sasanian Sages (Dēnkard VI),* trans. Shaul Shaked, Persian Heritage Series, no. 34 (Boulder, Colo.: Westview Press, 1979), 103, §262. See also Carsten Colpe, "Daēnā, Lichtjungfrau, zweite Gestalt: Verbindungen und Unterschiede zwischen zarathustrischer und manichäischer Selbst-Anschauung," in *Studies in Gnosticism and Hellenistic Religions: Presented to Gilles Quispel on the Occasion of His 65th Birthday,* eds. R. van den Broek and M.J. Vermaseren (Leiden: E.J. Brill, 1981), 65–67; Sundermann, "Die Jungfrau," 159–65; and Jamsheed K. Choksy, *Evil, Good, and Gender: Facets of the Feminine in Zoroastrian Religious History,* Toronto Studies in Religion, vol. 28 (New York–Washington–Baltimore–Bern: Peter Lang, 2002), 68–73, and 70 (fig. 6) for a depiction of the encounter of the soul with the Daena from a seventeenth-century manuscript of the *Ardā Vīrāz.*

[84.] See Corbin, *Corps spirituel,* 58ff.; Henry Corbin, *L'homme et son Ange: Initiation et chevalerie spirituelle* (Paris: Fayard, 1983), 61; also Corbin, *L'imagination créatrice,* 210.

[85.] For accounts of the meeting with the Daena in the *Hadokht Nask, Ardā Vīrāz,* and other Pahlavi texts, see Widengren, "La rencontre avec la *daēnā,*" 41ff.; Colpe, "Daēnā, Lichtjungfrau," 68; *The Book of Arda Viraf,* trans. Martin Haug and Edward William West (1872; repr. ed., Amsterdam: Oriental Press, 1971), 154–55; and *Le livre d'Ardā Vīrāz: Translittération, transcription et traduction du texte pehlevi,* ed. and trans. Philippe Gignoux, Éditions Recherche sur les Civilisations, cahier no. 14 (Paris: Éditions Recherche sur les Civilisations, 1984), 156–57. For the appearance of the Daena in the vision of the Zoroastrian high priest Kirdir, see Prods Oktor Skjaervø, "'Kirdir's Vision': Translation and Analysis," *Archaeologische Mitteilungen aus Iran* 16 (1983): 281–86, and 295; Philippe Gignoux, "Apocalypses et voyages extra-terrestres dans l'Iran mazdéen," in *Apocalypses et voyages dans l'au-delà,* eds. Claude Kappler et al. (Paris: Éditions du Cerf, 1987), 368–70; and Philippe Gignoux, ed. and trans., *Les quatre inscriptions du mage Kirdīr: Textes et concordances,* Cahiers de Studia Iranica, 9 (Paris: Association pour l'Avancement des Études Iraniennes, 1991), 96–97. See also A.D.H. Bivar, "The Iranian Tradition concerning Visits to the 'Afterlife'," Appendix B to his *The Personalities of Mithra in Archaeology and Literature* (New York: Bibliotheca Persica Press, 1998), 90–91.

[86.] Widengren, "La rencontre avec la *daēnā,*" 41–42, 44–45, and 78; Gignoux, *Le livre d'Ardā Vīrāz,* 156. See also Antonio Panaino, "'Smell,' 'Good Smell,' 'Stench' and 'Stinking' in Avestan—Some Reflections on a Semantic Area," in *Parfums d'Orient,* ed. Rika Gyselen, Res Orientales, 11 (Bures-sur-Yvette: Groupe pour l'Étude de la Civilisation du Moyen-Orient, 1998), 167–68, and 172. The gentle breeze perfumed by the scent of the rose was a stock image in medieval Persian poetry even of the nonmystical variety—see de Fouchécour, *La description de la nature*, 69.

[87.] Widengren, "La rencontre avec la *daēnā*," 45–46, and 41–42.

[88.] *Baodah* was considered one of the five spiritual faculties. See Shaki, "Dēn," 279; Colpe, "Daēnā, Lichtjungfrau," 68; and Panaino, "'Smell,' 'Good Smell'," 167ff. For the derivation of the word, see Bartholomae, *Altiranisches Wörterbuch*, 917–19, s.v. *baod-*, and *baoδah-*; also Jean Kellens, *Liste du verbe avestique* (Wiesbaden: Dr. Ludwig Reichert Verlag, 1995), 39, s.v. *BUD*.

[89.] In the present context, the translation of the term *daena* as "visionary soul" (Schauseele) is especially felicitous. See Sundermann, "Die Jungfrau," 172.

[90.] See James R. Russell, "The Sage in Ancient Iranian Literature," in *The Sage in Israel and the Ancient Near East*, eds. John G. Gammie and Leo G. Perdue (Winona Lake: Eisenbrauns, 1990), 84–85 (citing *Jāmāspi*, ed. and trans. J.J. Modi [Bombay, 1903]); also Boyce, *Zoroastrians*, 31, who notes that Jāmāspa, chief counsellor to the eastern Iranian king, Vīshtāspa, who established Zoroastrianism in his Bactrian kingdom, was proverbial for his wisdom.

[91.] Apparently, certain odors stimulate the right cerebral hemisphere, and their "emotive" capacity is attributed to the effect they have on the limbic system, which plays a central role in memory and emotion. See R. Tisserand, "Essential Oils as Psychotherapeutic Agents," in *Perfumery: The Psychology and Biology of Fragrance*, eds. Steve Van Toller and George H. Dodd (London and New York: Chapman and Hall, 1988), 168; and especially Dwight E. Hines, "Olfaction and the Right Cerebral Hemisphere," *Journal of Altered States of Consciousness* 31, no. 1 (1977): 51–52.

[92.] Trygg Engen, *The Perception of Odors* (New York and London: Academic Press, 1982), 14–15; and Richard E. Cytowic, *Synesthesia: A Union of the Senses*, 2nd ed. (Cambridge, Mass.: MIT Press, 2002), 130–32, and 196. For the difference between the way in which sensory input is processed in the case of vision and audition on the one hand, and olfaction on the other, see Hines, "Olfaction and the Right Cerebral Hemisphere," 48. For an excellent treatment of the subject of odor memory in European literary sources, see Hans J. Rindisbacher, *The Smell of Books: A Cultural-Historical Study of Olfactory Perception in Literature* (Ann Arbor: University of Michigan Press, 1992).

[93.] Tisserand, "Essential Oils," 175; see also Abdul Hameed, *Avicenna's Tract on Cardiac Drugs*, 27–28.

[94.] See Hines, "Olfaction and the Right Cerebral Hemisphere," 54–55.

[95.] Rūmī, *Dīvān*, 6: 2,897/30,781. For Rūmī's favorite equivoque on the word *gulsitān*, which means both "rose garden" and "one who gathers roses," see p. 19, and note 71 in this chapter. The scent of the divine Friend is identified with the scent of the soul, as he states later in the same poem (line 30,785):
> From the kitchens of the soul, the smell of bread continually comes to me.

[96.] This is an unusual strategy, as medieval Persian rhetorical norms dictated that a word should never be repeated in the same line (*bayt*) unless used to convey a different meaning. The quotation translated here actually represents two hemistichs of a single line of poetry in the Persian original.

[97.] See Davoud Monchi-Zadeh, "*Xusrōv i Kavātān ut Rētak*: Pahlavi Text, Transcription and Translation," in *Monumentum Georg Morgenstierne II*, Acta Iranica, 22 (Leiden: E.J. Brill, 1982), 78–81; Arthur Christensen, *L'Iran sous les Sassanides*, 2nd ed. (Copenhagen: Ejnar Munksgaard, 1944), 477; and Albert de Jong, "Sub Specie Maiestatis: Reflections on Sasanian Court Rituals," in *Zoroastrian Rituals in Context*, ed. Michael Stausberg (Leiden and Boston: Brill, 2004), 348–49. The correspondences provided in this passage generally conform to those recorded in the *Bundahishn* for the various Zoroastrian deities.

[98.] It is not surprising that Rūmī, who identified himself as a native of Balkh (ancient Bactria, the putative homeland of Zoroastrianism), should have been familiar with ancient Iranian ideas and motifs.

[99.] Rūmī, *Masnavī*, 2: 3,236–37.

[100.] See, Rūmī, *Masnavī*, 1: 3,180 and 1,462. The Muslim theosophers defined *ḥiss* as "the manner in which an attribute of the soul becomes manifest." See Ruzbehan Baqli, *Commentaire*, 413, §1,107. The Persian mystical author, Shihāb al-Dīn Yaḥyā Suhravardī (d. 1191), explained the notion of the inner senses in the following manner:
> When the inner eye (*dīdeh-i andarūnī*) is opened, the outer eye should be sealed to everything, the lips shut to everything, and the five external senses (*panj ḥiss-i ẓāhir*) should cease to be used and the internal senses (*ḥavāss-i bāṭin*) set in motion so that when the [spiritual] 'patient' grasps something, he grasps it with his inner hand, and when he sees, he sees with his inner eye, and when he hears, he hears with his inner ear, and when he smells, he smells with his inner nose, and his taste (*zauq*) is from the palate of the soul. When this is comprehended, he will continually contemplate the secret of the heavens and always be aware of the world of the Unseen (*'ālam-i ghayb*).

—see Shihabuddin Yahya Suhrawardi, *The Philosophical Allegories and Mystical Treatises*, ed. and trans. Wheeler M. Thackston, Jr., Bibliotheca Iranica, Intellectual Traditions Series, no. 2 (Costa Mesa, Calif.: Mazda Publishers, 1999), 40. For the various classifications of the postsensational faculties, or internal senses, in Islamic philosophy, see Harry Austryn Wolfson, "The Internal Senses in Latin, Arabic, and Hebrew Philosophic Texts," *Harvard Theological Review* 28, no. 2 (1935): 69–71, and 77; see also *The Mathnawí of Jalálu'ddín Rúmí*, vol. 7: 352 (Commentary to *Masnavī*, 2: 3,236).

[101.] See Rūmī, *Masnavī*, 2: 3,238–39. "Taste" or "tasting" (Per. *zauq*; Ar. *dhauq*) is a technical term denoting spiritual rapture or the mystical experience in general. See Schimmel, *Mystical Dimensions*, 253; and Sells, *Early Islamic Mysticism*, 126–27 (citing al-Qushayrī, *al-Risāla al-Qushayriyya*).

[102.] Rūmī, *Masnavī*, 2: 3,240–41.

[103.] See Chidester, *Word and Light*, 70, citing Origen's statement that, "There are many forms of this [interior] sense. . . . So also there is a sense of smell which smells spiritual things, as Paul speaks of 'a sweet savour of Christ unto God' (2 Cor. 2:15)." On this central theme in Baudelaire's poetic philosophy, see Lawrence E. Marks, *The Unity of the Senses: Interrelations among the Modalities* (New York–San Francisco–London: Academic Press, 1978), 224–29. Probably the most famous expression of this idea in Baudelaire's poetry is to be found in his poem, "Correspondances":
> Comme de longs échos qui de loin se confondent
> Dans une ténébreuse et profonde unité,
> Vaste comme la nuit et comme la clarté,
> Les parfums, les couleurs et les sons se répondent.

—see Baudelaire, *Les fleurs du mal* (Paris: Éditions Garnier Frères, 1961), 13.

[104.] See Marks, *Unity of the Senses*, 83ff.; Cytowic, *Synesthesia*, 2; Scharfstein, *Ineffability*, 31ff.; and Chidester, *Word and Light*, 69–72.

[105.] For the phenomenon of synaesthesia in Jewish mysticism, see Wolfson, *Through a Speculum*, 287–88; for the interplay of modalities in early Christianity, see Chidester, *Word and Light*, 30ff., and 139–43.

[106.] See the interesting observation by Cytowic that vision is a "psychophysical phenomenon"—Cytowic, *Synesthesia*, 3–4.

[107.] Rūmī, *Maṣnavī*, 4: 2,400. See also Rūmī, *Maṣnavī*, 4: 2,401, where he states, paraphrasing Bāyazīd Basṭāmī, that "Every hair of the gnostic becomes an eye"; and *The Mathnawí of Jalálu'ddín Rúmí,* vol. 8: 186 (Commentary to *Maṣnavī*, 4: 2,401).

[108.] Compare Suhravardī's reference to the "inner nose" (*bīnī-i bāṭin*), in note 100.

[109.] Rūmī is playing here on the words *būy* ("scent") and *kūy*. In Persian mysticoerotic poetry, the latter denoted the narrow lane where the beloved person lived. It is always described as being filled with his/her scent, especially the scent of the hair, which was customarily perfumed with strong fragrances such as musk, ambergris, and camphor.

[110.] Rūmī, *Maṣnavī*, 1: 440–41.

[111.] See note 82 above. The verse under discussion occurs in the tale about the Jewish king who persecuted Christians, which is replete with references to other religions, including Christianity and Zoroastrianism—see Rūmī, *Maṣnavī*, 1: 324ff.

[112.] Thus, for example, Rūmī states, "Your good deeds will run to meet you after your death; Just like moon-faced maidens, your attributes will walk elegantly [before you]"—see Rūmī, *Dīvān*, 1: 385/4,099. On this point, see also Schimmel, *Triumphal Sun*, 261.

[113.] The motif of Joseph's shirt occurs three times in the Qur'ānic version of the biblical narrative, the third occurrence being the episode under discussion, according to which Joseph, after revealing his identity to his brothers in Egypt, instructs them to take his shirt to their father, Jacob. The esoteric interpretation of the restoration of Jacob's sight would appear to be anticipated already in the Qur'ānic narrative, when Jacob says to his sons, "Did I not tell you, I know from God what you do not?" (Qur'ān 12:96). For the cloak-motif in the biblical account, where it does not figure as prominently or have the same esoteric significance, see Donald B. Redford, *A Study of the Biblical Story of Joseph (Genesis 37–50),* Vetus Testamentum, suppl. 20 (Leiden: E.J. Brill, 1970), 87ff., and 139. According to the Islamic legendary account, Jacob went blind on account of old age and weeping for Joseph—see Abū Isḥāq Aḥmad al-Thaʻlabī, *ʻArāʼis al-majālis fī qiṣaṣ al-anbiyāʼ or "Lives of the Prophets,"* trans. William M. Brinner (Leiden–Boston–Cologne: Brill, 2002), 228–29; thus also in Jewish legend—see Louis Ginzberg, *Legends of the Jews*, trans. Henrietta Szold and Paul Radin, 2nd ed., 2 vols. (Philadelphia: The Jewish Publication Society, 2003), 1: 377. Elaborating on the Qur'ānic verse in which Jacob is reported as saying, "I perceive the scent (*rīḥ*) of Joseph" (Qur'ān 12:94), the Islamic legendary accounts explain that God commanded the east wind to carry Joseph's scent to Jacob even before the messenger brought him the shirt. See al-Thaʻlabī, *ʻArāʼis al-majālis*, 228; and Muḥammad ibn ʻAbd Allāh al-Kisāʼī, *Tales of the Prophets* (*Qiṣaṣ al-anbiyāʼ*), trans. Wheeler M. Thackston, Jr. ([Chicago]: Great Books of the Islamic World, Inc., 1997), 188–89. It may be noted that, in the Islamic mystical lexicon, the term *baṣīr*, which is used in the Qur'ānic verse to indicate Jacob's restored vision, denotes "one endowed with intuitive insight."

[114.] Rūmī, *Maṣnavī*, 2: 3,233–34. Compare Rūmī, *Maṣnavī*, 1: 125: "At this moment, my Soul tucked up the hem of its skirt (i.e., in preparation); For it had caught the scent of Joseph's shirt" (with a play on the words for "hem" and "shirt").

[115.] According to Qur'ānic legend, Joseph's shirt had originally been brought from Paradise by the angel Gabriel to Abraham in order to protect him from the fire, after which it was inherited by Isaac, and then passed on to Jacob. The latter placed the shirt in an amulet around Joseph's neck to ward off the evil eye. When Joseph was in the well into which he had been cast by his brothers, Gabriel removed the shirt from the amulet and dressed him in it. See al-Thaʻlabī, *ʻArāʼis al-majālis*, 190; and al-Kisāʼī, *Tales of the Prophets*, 170. Thus also in the Jewish legendary accounts—see Ginzberg, *Legends of the Jews*, 1: 336, and 364 (where it is indicated that Joseph wore this same shirt when he appeared before Pharoah, and even when he became ruler of Egypt), and 1: 405 (where the shirt is mentioned as originally having been woven for Adam).

[116.] According to a Judaeo-Islamic legendary account, Joseph's younger brother, Benjamin, named one of his sons Ward ("Rose") in remembrance of Joseph's beauty. See al-Thaʻlabī, *ʻArāʼis al-majālis*, 217–18; thus also Ginzberg, *Legends of the Jews*, 1: 380 (where the reading of his name as Ard is to be amended to Ward, the Arabic equivalent of Hebrew Wered, meaning "rose").

[117.] Rūmī, *Dīvān*, 3: 1,153/12,240 (with an equivoque on the word *gul*, which may also be read as *kull*, as explained earlier; and with an equivoque on the word "opened," *begushād*, which also means "bloomed"). The second part of the verse, which reads, "[So] do not look down upon the breeze (i.e., scent) of the Joseph of our self (i.e., our soul) which [emanates] from our shirt," contains a) an auditory pun on the word *khvār* ("lowly") in the idiomatic expression, *khvār magīr* ("do not look down upon"), which in Persian is pronounced the same as *khār*, "thorn," thereby echoing the previous line which ends in *khār magīr* ("don't take the thorn"), and b) a visual pun on the word *kurteh* ("shirt" or "tunic"), which also can be read as *kirteh* ("thorn").

[118.] See, for example, the extended metaphor of "the Joseph of the soul" in ʻAṭṭār, *Manṭiq al-ṭayr*, 234, lines 4,224ff.; see also Schimmel, "Yūsuf," 48ff.

[119.] In doing so they employed the peculiar Persian rhetorical figure called *ḥusn-i taʻlīl*, usually translated as "fantastic aetiology," for which see Schimmel, *As through a Veil*, 58–59; and Annemarie Schimmel, *A Two-Colored Brocade: The Imagery of Persian Poetry* (Chapel Hill and London: University of North Carolina Press, 1992), 41. This is not just a metaphor but the ascription of a fanciful explanation, often in terms of human characteristics, to phenomena in nature, especially trees and flowers. Thus, rose water was explained as being "the tears of the rose" (*sirishk-i gul*)—today the name of a popular brand of rose water bottled in Shiraz. For other examples of the application of this rhetorical device to the rose, see Ritter, *Ocean of the Soul*, 11, 50, and 415.

[120.] Zulaykhā's name is mentioned only in the Islamic legendary accounts, according to which Joseph even married her in the end, and she became the mother of his sons, Ephraim and Manasseh. See al-Kisāʼī, *Tales of the Prophets*, 179–80; al-Thaʻlabī, *ʻArāʼis al-majālis*, 212 (where she is called Rāʻīl); also Ginzberg, *Legends of the Jews*, 1: 352–59. In contrast to the account in Genesis, which simply states that Joseph left his shirt behind when Potiphar's wife caught him by it (Gen. 39:12), the Qur'ān mentions several times that Joseph's shirt had been torn from behind, as proof of his innocence (Qur'ān

12:25–28).

[121.] Rūmī, *Dīvān*, 4: 2,003/21,179. The verse is based on a rhetorical congruence (Per. *tanāsub*) between the words for collar (*girībān*) and shirt (*pīrāhan*).

[122.] The prophetic Tradition reads: "Three things have been made dear to me from this world of yours: women (*nisā'*), fragrance (*ṭīb*), and the delight of my eyes is in [the performance of] the ritual prayer (*ṣalāt*)"—see Badī' al-Zamān Furūzānfar, *Aḥādīs̱-i Mas̱navī* (Tehran, 1334/1956), no. 182.

[123.] Rūmī, *Mas̱navī*, 2: 3,235.

[124.] Literally, "my eye is cooled/refreshed."

[125.] It does not appear that Rūmī was influenced by the highly idiosyncratic esoteric commentary on this Tradition by Ibn 'Arabī, who interpreted the placement of the word "fragrance" between "women" and "prayer" in the list as representing a masculine entity (*ṭīb*) between two feminine ones (*nisā'*, *ṣalāt*), and hence as being analogous to the ontological position of Adam between the divine Essence, the source of all existence, and Eve, whose existence stems from him. See Ibn al'Arabi, *Bezels of Wisdom*, 277–79.

[126.] See Goody, *Culture of Flowers*, 88–89, 120ff., and 166ff.; and Le Guérer, *Scent*, 148–54. For the use and importance of fragrances, especially that of the rose, in Roman culture, see Andrew Dalby, *Empire of Pleasures: Luxury and Indulgence in the Roman World* (London and New York: Routledge, 2000), 244–46.

[127.] See Alain Corbin, *Le miasme et la jonquille: L'odorat et l'imaginaire social XVIIIe–XIXe siècles* (Paris: Flammarion, 1986), 85–101. My thanks to Michel Conan for acquainting me with this fascinating study.

[128.] See Goody, *Culture of Flowers*, 245–46, in which he notes that many influential French writers on the topic made the connection to ancient Iranian religion.

*Botanical Progress,
Horticultural Innovations
and Cultural Changes*

*Botanical Progress,
Horticultural Innovations
and Cultural Changes*

*Botanical Progress,
Horticultural Innovations
and Cultural Changes*

*Botanical Progress,
Horticultural Innovations
and Cultural Changes*

Alain Touwaide

The Countryside into Cities

According to the Latin encyclopedist Pliny (23/4–79 A.D.), first-century Roman citizens brought *the countryside into cities*:[1]

> . . . Nowadays indeed under the name of gardens people possess the luxury of regular farms and country houses actually within the city . . .

In other words, Pliny's contemporaries built houses with gardens that contained a variety of utilitary and decorative plants in a way that allowed them to enjoy many, if not all of the advantages of the countryside without having to leave the capital, its active and delightful life, its political contacts, and its sources of revenues.

Gardens with probably a lavish and colorful vegetation dramatically contrast with Cato's simple backyard. A bellicose and indefatigable adversary of Carthage who punctuated all his discourses at Rome's senate with the formule *Cartago delenda est* (*Carthage must be destroyed*), and a virulent partisan of old Roman traditions, Cato (234–149 B.C.) strongly defended and illustrated the cultivation of the only cabbage in Roman gardens in his *Manual of agriculture.*[2] According to him, indeed, it was not only an excellent nutrient and an all-healing medicine, but also a source of aesthetic pleasure!

The transformation or, probably better, the revolution of Roman gardens from Cato to Pliny has been traditionally attributed to the seduction and influence exerted by Greek culture on Rome, as Pliny himself already did:[3] In a moralistic way not rare in his work,[4] the Latin encyclopedist attributes the responsibility of such transformation to an emblematic figure: Epicurus (341–270 B.C.), the best representative of hedonistic philosophy. This practice was first introduced at Athens by that connoisseur of luxurious ease, Epicurus; down to his day the custom has not existed of having country dwellings in town. Whatever Pliny's motivation might have been, it is true that Greek civilization deeply influenced Rome, particularly after the battle of Pydna (168 B.C.), to which we shall return. Horace (65–27 B.C.) perfectly expressed the fact in his well-known and somewhat paradoxal verses:[5]

> . . . Captured Greece captured her savage victor and brought the arts into the agricultural Latium . . .[6]

Archeology of Roman houses confirms such a Greek influence on Roman gardens. At Pompeii, for example, the traditional structure was transformed from a simple building with a small garden at the back (Fig. 1), to a larger construction that included a central court with a colonnade, that is, the peristyle of Greek houses (Figs. 2 and 3).[7] Roman gardens were embellished with sculptures in the Greek style.[8] Characteristically, the adoption of the Greek peristyle-structure went along with an

1. The structure of a traditional Roman house (from Maureen Carroll, *Earthly Paradises: Ancient Gardens in History and Archaeology*. London: British Museum, 2003).

2. A reconstruction of a Greek house in Delos, with its peristyle (from M. Carroll-Spillecke, ed., *Der Garten von der Antike bis zum Mittelalter* [Kulturgeschichte der antiken Welt 57]. Mainz: Philipp von Zabern, 1992).

3. The structure of a Pompeian house with a peristyle-garden (from Maureen Carroll, *Earthly Paradises: Ancient Gardens in History and Archaeology*. London: British Museum, 2003).

adaptation/transformation: the peristyle was no longer a court under the open sky, and a source of light and air; it became a garden. This is interpreted as a Roman innovation that would have introduced the *country in town*.[9]

Such a history has been summarized and explained by the creator of Roman garden archeology, Wilhelmina Jashemski:[10]

> . . . Only at Pompeii can domestic architecture be traced for a period of almost four hundred years. Pompeii preserves examples of the early Italic house with the garden at the rear, houses that date back to the late fourth or early third century B.C. They remind us that the *hortus* is old; it formed a significant part of the primitive *heredium* ("hereditary estate"). The *hortus* was primarily a kitchen garden, but I suspect that even so the ancient gardener tucked in a few flowers amid the herbs and vegetables . . . Next at Pompeii come the elegant houses built by the Samnites during the second century B.C. . . . the Hellenistic peristyle was added to the Italic house . . . a living, breathing garden, instead of the paved courtyard found in Hellenistic houses . . . A love of beauty and gardens was a basic part of their lives [i.e. of Romans]; the desire for a bit of green, a few herbs, and flowers appears to have been an integral part of their character. The garden was intimately related to so many aspects of Roman life . . .

This explanation, of a romantic nature based on a supposed Roman character, can be complemented by economic history. In Pliny's time, indeed, Rome was at its zenith. During the first century B.C. it extended its domination to the entire Mediterranean basin. In 31 B.C. Octavius (63 B.C.–14 A.D.) defeated the float of Antony (ca. 83–30 B.C.) off the coast of Actium, and put in this way an end to almost fifty years of civil war. Then, in 27 B.C., he transformed the old Republic into an Empire, inaugurating a long period of peace and prosperity. He received the title of *Augustus,* under which he is better known. Roman ships sailed the Mediterranean without other risks than intemperies, and delivered to Rome goods of all kinds from all circum-Mediterranean regions, their hinterland, and even well beyond. Plants were included in this profusion of luxury that supposedly contributed to Rome's decadence, if it did not provoke it. This is all the more true because, in Nero's time (b. 15; emp. 54–68 A.D.), imperial politics aimed at creating an economic circuit by encouraging the circulation of wealth and provoking a demand for goods. Gardens, be they public or private, were part of this new economy.[11]

4. The garden fresco from the so-called Casa di Livia in Rome (40–ca. 20 B.C.) (from Salvatore Settis. *Le pareti ingannevoli: La villa di Livia e la pittura di giardino.* Milan: Electa, 2002).

5. A garden fresco from Pompeii, Casa del Bracciale d'Oro (from Annamaria Ciarallo. *Il giardino pompeiano: Le piante, l'orto, I segreti della cucina.* Naples: Electa, 2002).

Visualization and Perception of Gardens

If archeological findings confirm a transformation of Roman houses and gardens, and, in the best excavated ones, make it possible to know the plant species and their arrangement, they do not tell much on the appearance of such gardens or on the principle(s) that guided the selection and arrangements of their plants.[12] Some frescos of the late republican and early empire period with garden representations have been preserved, however, in Rome, Pompeii, and Oplontis.

In Rome, such frescos decorated the so-called *Casa di Livia,* that is, the *House of Livia* (58 B.C.–29 A.D.), the wife of August (63 B.C.–14 A.D.). This residence, which was also known in ancient sources as *ad gallinas albas* (*the villa of the white hens*) because of a mythological tale, was the empress' summer residence, and not her house on the Palatine.[13] It was located on the heights of the *via Flaminia* at the outskirts of Rome, close to present Prima Porta. These frescos, preserved since 1951 in Rome, at the Museo Nazionale Romano, Palazzo Massimo alle Terme,[14] have been differently dated. Ranuccio Bianchi Bandinelli suggests that they might be posterior to Livia's death[15] and date back to the first century A.D.[16] Recent publications have proposed earlier periods: 20–10 B.C.,[17] 30–20 B.C.,[18] and, more recently, 40–circa 20 B.C.[19] Whatever their period, these frescos decorated a large (5.90 x 11.70 m.) *tricilinum* (dining room) that was underground so as to keep fresh during the summer's heat. Its walls were painted with a representation of a garden so as to suggest that such a subterranean room was an open space in a garden with a lavish vegetation (Figure 4).[20]

Similar frescos (although not so well preserved) can be found in Rome, between present via Merulana and via Leopardi, in the so-called *Auditorium of Maecenas.* The building was located in the gardens of Maecenas (d. 8 B.C.), mostly known as a patron of letters and the arts under August, and has been considered without proof as an auditorium.[21] Its frescos are stylistically so close to those of the *Casa di Livia* that they have been considered to be by the same hand and dated to the first half of the first century, even though the construction of the building was estimated to be of the years 40–35 B.C.[22]

At Pompeii, frescos with garden representations adorned some houses such as the so-called *Casa del Bracciale d'oro,* the *Casa del Frutteto,* and the *Casa dei cubiculi floreali,* which, like the entire city, were all buried under the pietra pomice, the ashes and all the material from the eruption of the Vesuvius in 79 A.D. (Figure 5).[23]

Close to Pompeii, the villa of Oplontis, supposedly of Poppaea Sabina, Nero's second wife, contains a similar decoration.[24]

The most representative from our viewpoint is probably the *Casa del Bracciale d'oro* (Pompeii, Reg. VI, 17 [Ins. Occid.], 42 = Inv. 40690–40693), which was unpublished until recently.[25] This kind of fresco, which is structured in the same way as those of the *Casa di Livia,* is dated to the first half of the first century A.D.

The visual effect suggested by such frescos is echoed in the literary description of a garden in the pastoral romance *Daphnis and Chloe* by Longus, a Greek author of unknown epoch, maybe the second century A.D.(4.1–3):[26]

> . . . the garden he trimmed with great care and diligence, that all might be pleasant, fresh, and fair. And that garden indeed was a most beautiful and goodly thing, . . . Trees it had of all kinds, the apple, the pear, the myrtle, the pomegranate, the fig, and the olive; and to these on the one side there grew a rare and taller sort of vines, that bended over and reclined their ripening bunches of grapes among the apples and pomegranates . . . To these were not wanting the cypress, the laurel, the platan, and the pine. And towards them, instead of the vine, the ivy leaned . . . Within were kept, as in a garrison, trees of lower growth that bore fruit. Without stood the barren trees, enfolding all, much like a fort or some strong wall that had been built by the hand of art . . . The roses, hyacinths, and lilies were set and planted by the hand, the violet, the daffodil, and pimpernel the earth gave up of her own good will. In the summer there was shade, in the spring the beauty and fragrancy of flowers, in the autumn the pleasantness of the fruits . . .

On the basis of building remains, such frescos as previously mentioned, and Longus's passage, it seems that the selection of plants and their arrangement in early empire gardens was rather simple and conventional: at the forefront, small colored flowers constituted a sort of low garland; distinguishable fruit trees occupied the center of the compositions, with colored spots representing their fruits; the back was constituted of higher trees that created an indistinct green wall as a fence. If this pictorial structure corresponded to reality, the principle for plant selection and ordering was mainly their size (with three types: low, middle, and higher) and their color (with the colored spots of flowers, those of the fruits, and then just green).

Be that as it may, the interpretation of such gardens has varied. Ranucio Bianchi Badinelli compared these gardens to the parks of Iranian origin known in Rome through their Greek version, the so-called *paradeisos,*[27] whereas Wilhelmina Jashemski attributed the creation of gardens in Pompeian private houses to the typically Roman genius.[28] Salvatore Settis has renewed the analysis, by introducing a chain from the imaginary to the actual garden and, beyond, to oriental *paradeisos* artificially recreated in a private house, which made it possible for wealthy urban Romans to transfer the countryside in the cities.[29]

In the context of an analysis of the possible impact of cultural changes on horticultural practices, I wish to go beyond such an interpretation. As a historian of ancient botany, I would like to introduce into the debate an element not necessarily taken into consideration so far: plant cataloguing and classification. I shall argue that early imperial Roman private gardens, sometimes described as *botanical catalogues,* result in fact from an accurate selection and arrangement of plants that reflect—and rely on—a determined botanical system. It is my purpose here to highlight the very existence of this system, which has not been noticed yet by historians of botany, to detail its contents, and to explain its function, as well as to investigate its origins and application in Roman private gardens of the period under consideration. In so doing, I shall also suggest that such a botanical system closely corresponded to the new geopolitical and cultural situation of the Mediterranean world resulting from the Roman conquest, with the awareness of, and opening to new—that is, different—realities among the population under Roman rule, and the ordering of the single elements of this vast ensemble in an all-encompassing synthesis, which was all together the Roman political world, the structure of society, the literary genre of the encyclopedia, and the garden.

Background: Theophrastus and the Alexandrian School

The history of scientific plant catalogue and classification starts with Theophrastus (372–370 to 288–286 B.C.), a student of Aristotle (384–322 B.C.) and his successor at the direction of his school, the Lyceum.[30] Theophrastus analyzed the plant world in his *Enquiry into plants.*[31] The title of the work is significant: in Greek *peri futôn istorias,* exactly rendered into Latin by *Historia plantarum.* It does not refer to *history* in the current meaning of the word, but to a *research,* an *enquiry.*[32] Such *research* reflects Aristotle's scientific method, particularly in biology, which proceeded in two major steps: first, a collection of currently available data and their analysis—this is the *historia*—and second, on the basis of such material, the construction of an interpretative system.[33] The collection of data in Theophrastus's *Historia plantarum* was large, all the more because it included material brought from Africa, Persia, India, and Arabia by the scientists who accompanied the troops of Alexander the Great (356–323 B.C.) in his military expedition to Egypt and India.[34] In describing such material, Theophrastus not only created—or, most probably, codified—the Greek botanical lexicon, but also followed the method of description of his teacher Aristotle in his works of natural history, with two complementary descriptors:[35] similarities and differences.[36] Similarities made it possible to regroup individuals with common features (whatever their nature, animals in Aristotle, plants in Theophrastus), and to constitute coherent groups, and differences made it possible to distinguish individuals within such groups. Such grouping principles gradually led Theophrastus to the highest division of the *regnum vegetale* (or *Plant Kingdom*), with four ultimate categories:[37] trees, shrubs, under-shrubs, and herbs. If we had to stop here, there would be quite a gap from Theophrastus to Rome, both chronologically and conceptually. This would induce one to think that plant arrangement in Roman gardens—whatever its nature—was an original Roman creation. Such a view would not be correct, however, because it would not take into consideration a piece that has been forgotten in previous research and needs to be introduced here: Alexandrian science and its further developments in the Hellenistic and early Roman worlds.

The School of Alexandria was founded by the first Greek king of Egypt, Ptolemy I (b. 367/6 B.C.; king of Egypt 304–283/2). No explicit element on the botanical research done in the school has survived. Only later works relying on Alexandrian research such as those by Dioscorides and Pliny have been preserved. However, they allow a hypothetical reconstruction of earlier Alexandrian activity.[38]

Dioscorides, who is considered without any good reason to have been a military physician in the Roman army under the emperors Claudius (born in 10 B.C., emp. 41–54 A.D.) or Nero (born 15 A.D.; emp. 54–68 A.D.), has supposedly achieved his work sometime around the completion of *Naturalis historia,* dated to 77 A.D.[39] His treatise is entitled Περὶ ὕλης ἰατρικῆς (*peri ulês iatrikês*), which is exactly translated by *De materia medica,* that is, *On the natural substances used to prepare medicines.*[40] Although such substances were of plant, animal or mineral origin, their great majority were plants (almost 70 percent); hence, the frequent—and wrong—affirmation according to which *De materia medica* is an herbal. Each substance is dealt with in a monographic chapter that mainly includes its description, the therapeutic properties of its part(s) to be used as therapeutic agent(s), and the medical conditions for the treatment of which such substance(s) could be used.

Before studying the plant classification in *De materia medica,* I must briefly deal with their inventory and description. Inventory aimed to be exhaustive. Dioscorides explicitly mentioned his intention to be so in the preface of the work[41] and sometimes in the course of *De materia medica,* he included substances that were not used medicinally, but were correlated in one way or another to other substances. He did so because, he says, he did not want to be exposed to the possible objection to have forgotten some substance, even if not particularly useful.[42] As for the description of plants, it proceeded by comparisons and differences according to the Aristotelian method. In *De materia medica,* however, this method was slightly modified. Comparisons dealt very often with only one element of the plants (the leaf), and referred to a limited number of plants: ivy, olive tree, garden rocket, and rue, which have typical shapes:[43]

> ivy: polygonal
>
> olive tree: oblong and entire
>
> garden rocket: oblong and cut
>
> rue: small

Recurrent references to the leaves of these plants seem to point to a system for plant classification in which the leaf was the major element, and plants were classified in major groups according to their leaf form. In other words: a classification proceeding by major morphological types defined on the basis of only one part of plants. This thus suggests that Theophrastus's analytical and classificatory method was not only systematically applied to all plants, but also—if not above all—transformed so as to become a simple and productive system for plant description and classification.

In Pliny's *Naturalis historia,* Books 12 to 19 deal with botany, and Books 20 to 27 with medicinal plants. Plant descriptions not only provide data similar or identical to those of *De materia medica* by Dioscorides, but also proceed according to the same system as *De materia medica,* that is, by comparisons and differences. Dioscorides is not quoted in Pliny's encyclopedia, neither in the list of sources in the beginning of the entire work and of each book, nor in the text of any single book. The similarities between *De materia medica* and *Naturalis Historia* do not result from the fact that the latter reproduced data from the former, but from the fact that both reproduced information from the same source(s). Such a source(s), currently lost or not yet brought to light, was certainly Greek, and most probably originated in the eastern Mediterranean. Because it prolonged and expanded

Theophrastus's method and applied it, among others, to the plants brought from the East by the scientists who accompanied Alexander's troops, it originated most probably in the context of the School of Alexandria. Such work(s) resulted in all probability from the research activity of the first generation of scholars in the School, particularly because the institution declined during the third century B.C. and was reactivated only later, during the Roman period (first century B.C./A.D.). If so, the time between these Alexandrian scholars, on the one hand, and Dioscorides and Pliny, on the other, corresponded to three centuries, and probably to more than only one work by one author. In short, there were probably several works from the Alexandrian scholars to our two scientists.

Plant Classification and Classificatory Models of Life in the Greek World

For the classification of plants, let's return to *De materia medica.* According to a classical interpretation, the work is divided into five books, each of which deals with a specific topic. A count of their lines reveals, however, that they all are approximately of the same length: almost twenty-five hundred lines each. This fact suggests that *De materia medica* was not necessarily composed by its author in five books defined as thematic entities, but rather that it was cut into five units, the length of which was determined by the maximum length of the medium of the book in Dioscorides' epoch, the papyrus roll.[44] If so, *De materia medica* constituted a sequence of almost one thousand chapters without any divisions. It surely could not have been used by a practitioner in the daily exercise of therapy.

This is not the case, however. Chapters in *De materia medica* are classified according to a system organized in two levels that allowed easily finding the required information. On the first level, plants and other materia medica are gathered in coherent groups according to such changing parameters as their structure, their supposed therapeutic property (or properties), or their smell and taste. On the second level, these very groups are organized according to a scale of properties, which requires some explanation. The first group contains the plants used for the preparation of perfumes, starting with the iris. The last group contains the matters used for the treatment of dermatological pathologies, mainly minerals. These two groups are in opposition from all points of view: perfumed plants are warm, while minerals are cold; as such they are light and heavy respectively; and, to quote just a few, perfumed and warm plants treat the excess of humidity in the body, whereas minerals, as cold, were efficacious against excessive warmth. The very first and very last groups are thus opposed by all their elements, and create a bipolar axis within *De materia medica.* To these supposed objective properties of such groups were superimposed subjective values, linked with the perception of the qualities of the two groups. Warm and perfumed matters were seen as positive, whereas cold and heavy minerals were negatively connoted in ancient Greek culture. Also, the very first and very last matters of the whole work are totally opposed in their colors: the iris presents all the colors of the rainbow as Dioscorides himself states—and we need not to forget that in ancient Greece the word *iris* designed both the plant *Iris* spp. and the rainbow, for the flower contains all the colors of the rainbow—and the last matter is *the black substance with which we write,* as Dioscorides says, that is, soot.[45] Yet in the chromatic system of ancient Greece, colors were not defined by their nature, but by their perception, their luminosity. In this context, black was not a color, but the absence of any color. As such, it was opposed to the rainbow, which contained all possible colors. In other words: from the iris to soot, there is a shift from all the colors to the total absence of colors, from all the positive qualities to their absence, from one thing to its opposite.

On this basis, we can return to the other groups of matters dealt with in *De materia medica,* that is, all the groups included between the first and last, between iris and soot. If we bear in mind that these two matters and the groups they open and conclude are opposed by the subjective values they are credited with, we discover that all the groups in between are ordered according to the degree of their supposed positive or negative value, that is, the degree of their warmth or cold property or any other property that takes place between these two fundamental ones. Such a classification relying on the intensity of positive or negative qualities constitutes a scale, with a gradual reduction of positive qualities (warm, perfumed, colored, etc.) and a parallel increase of negative qualities (cold, unscented, and black, etc.)—in other words, a *scala mundi* or *naturae.* [46]

An important point in such classification is that it implicitly contains a theory on evolution, which is sometimes explicitly mentioned in *De materia medica.* The shift from positively connoted substances to their opposite seems to refer, indeed, to a theory of evolution dominated by entropy, that is, by the loss of characteristics, rather than by the acquisition of new ones. Such an interpretation is confirmed by some passages where Dioscorides mentions that wild plant species result from the degeneration of other ones, which are normally qualified not with the adjective "domesticated" (since it refers to a process of acquisition of properties), but *kêpaios,* which means "*of*" or "*from the garden,*" or, in other words, from the human environment. [47] We shall return to this theory, because it will give us a key on the origin of Dioscorides's system.

On this basis, we can return to Roman gardens. The selection and arrangement of plants in the Roman and Pompeian frescos seem to reflect the classification of plants that we have identified in *De materia medica.* At the forefront, we have small, perfumed and colored plants; then, fruit trees, with their production clearly in evidence; and at the very end taller trees, which are not a pictorial convention to close the space of the picture, but the trees in general, exactly as in Dioscorides's *De materia medica.*

Similarly, the garden described by Longus might result from the same ordering principle. Somebody entering the garden saw first the low perfumed flowering plants with a great variety of colors, such as roses, hyacinths, lilies, violet, daffodils, and pimpernel. And they grow almost spontaneously. Then came fruit trees, which were higher than the flowering plants, but smaller than the trees coming next: apple, pear, myrtle, pomegranate, fig and olive. Intertwined in their branches were, on one side, the grape, and, on the other, ivy. Then there were taller trees, which made a wall closing the garden: cypress, laurel, plane tree, and pine (Figure 6). We thus have an organization in three concentric circles with not only a different and gradual increase in the height of plants, but also a selection of plants with a gradual reduction of positive qualities, with first all the perfumes and colors, second sweet fruits, and third no specific property. Such a selection and organization of plants closely corresponds to that of *De materia medica*'s first section, where we have, successively, the plants used in the production of perfumes, the fruit trees, and the trees in general.

This parallelism suggests that a theory similar, if not identical, to that one I have detected in Dioscorides's treatise might have underpinned the design of the early empire private Roman gardens preserved and excavated so far in Rome and Pompeii.

If so, it would be interesting to identify the origin of such theory. Yet the evolution process we have highlighted—by entropy and not by acquisition of new qualities—does not appear in Pliny's *Natural History.* Here, evolution proceeds by progress. According to Pliny, indeed, the first age of humankind was of a rather wild nature: men got their subsistence from trees, without cultivating any land. Human curiosity pushed them to gradually discover the world, to use its resources and, by Pliny's time, to live in a luxurious way, in total contradiction with the austere model Cato mentioned earlier. [48]

Dioscorides's decline model, instead, resembles the story of the *Five Ages of Humankind* as told by the seventh-century Greek poet Hesiod in *Works and Days*.[49] According to this tale, the history of humankind is divided into five successive generations, each of which diminished in degree of positive qualities, and corresponded accordingly to a metal: the first generation, of gold, was of "mortal men . . . [who] lived like gods without sorrow . . ."; the second generation, of silver, "was less noble by far . . . they lived only a little"; the third, of bronze, "was terrible and strong . . . they loved war and deeds of violence; they ate no bread . . . they were destroyed by their own hands . . ."; finally, after an intermediary generation, there was the fifth, of iron, which was that of Hesiod and remains the current one on Earth.

The concept underlying this history of humankind corresponds pretty well to the structure I have identified in *De materia medica*. The distance between Hesiod and his age (seventh century B.C.) and Dioscorides (most probably first century A.D.) is not an objection: the thinking system reflected by Hesiod remained alive until a late epoch as the case of the first century B.C. poet Lucretius (ca. 94–55 B.C.) shows. In his work *On the nature of things,* he describes the history of humankind in very similar terms.[50] This was a theme of ancient philosophical literature, probably

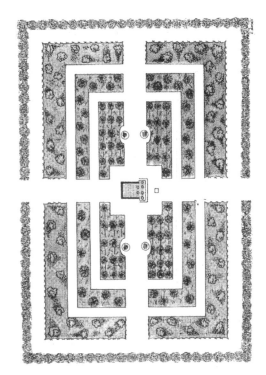

6. A reconstitution of the garden described by Longus (from Salvatore Settis. *Le pareti ingannevoli. La villa di Livia e la pittura di giardino.* Milan: Electa, 2002).

dating back, in the case of Lucretius, to his model, the Greek Epicurus (341–270 B.C.). Such a theme was also exploited during the same period by Virgil (70–19 B.C.) in his famous fourth *Eclogue,* where he announced the birth of a child who will herald the return of the Golden Age.[51] This sort of prophecy has been interpreted as a premonition of the birth of Christ. More probably it is a homage to August. In this same homage, Virgil was wishing August to have a descendant who would continue the golden age of domestic peace begun by his father.

Although probably created in the context of Greek culture rather than in the Roman one, this model, with its variant of the return of the golden age, was largely diffused in the first century B.C./A.D., and thus could have contributed to Roman gardens, all the more because, at that time, they underwent a transformation.

Science and Politics

In a period of restoration of the ideals of ancient Roman morals and discipline, the adoption of a Greek model of decline could be surprising. At first glance, it seems indeed that a model of progress would have been more appropriate. A closer analysis reveals, however, that the plant classification system we have identified in *De materia medica,* which relies on an exhaustive inventory of the elements of the world and classifies them, that is, assigns its right place to each of them, corresponds to a conception of the ongoing history of that time.

The Greek historian Polybius (ca. 200–118 B.C.) provides us with significant insights on that point. A native Greek, he was involved in the political life of his country until the forces of Perseus (born 212/3; king 179–168 B.C.), the king of Macedonia, were defeated by the Roman troops at Pydna in 168. Polybius was then deported to Italy for further investigation with a thousand other Greek notables, and detained there for ten years without trial. On the basis of his experience in Rome, he started writing the history of the rise of the Roman world for, he mentioned:[52]

> . . . the Romans have subjected to their rule not portions, but nearly the whole of the world . . . since [then] history has been an organic whole and the affairs of Italy and Libya have been interlinked with those of Greece and Asia, all leading up to one end . . . Special histories . . . contribute very little to the knowledge of the whole . . . It is only indeed by study of the interconnexion of all the particulars, their resemblances and differences, that we are enabled at least to make a general survey . . .

What Polybius described in this way was not only the contemporary ongoing historical process of unification of the Mediterranean world under the Roman rule and the project of his *Histories* aimed at reflecting the state of things at that time, but also a program for a possible encyclopedic project that would exhaustively describe the contemporary world, with all its elements and their inter-relationships. A similar project underpins Dioscorides' *De materia medica* and, to a lesser extent, Pliny's *Naturalis Historia,* in which, however, no structure ordering the single elements is present.

It thus seems that such an encyclopedia as *De materia medica* not only resulted from a mere scientific project but also reflected the geopolitical state of the world, and the self-consciousness of a culture shared by all the world known at that time. As a result, it aimed at exhaustively inventorying all relevant botanical data, and organizing them at the same time in a structure where each one was assigned a place in function of its nature and its contribution to the whole. Such a theoretical system, which was not created by Dioscorides but came almost certainly from earlier sources, was also applied to horticulture and contributed to shaping the Roman gardens by guiding both the selection and the arrangement of plants.

Gardens, Science, and Politics

Private Roman gardens of the first century B.C./A.D. were very different from the piece of land cultivated by the austere Cato. Although certainly influenced by Greek culture as already suggested in previous literature, such gardens of a new kind were not just the countryside introduced to the city that Pliny criticized. They probably also included a more theoretical component for they relied on a selection and arrangement of plants based on a scientific theory, which in turn reflected the order of the world of that time. As a consequence, Roman gardens of the first century B.C./A.D. were a miniature image of the action of Rome, which included all inhabitants of the universe, and assigned to each group a well-determined place in its geographical space, in its social organization, and in its economic system. From Cato to such gardens, there had been a shift from the individual and his particular needs, mainly alimentation and medicinal plants, to the collectivity, defined not only as a sum of elements but also as an organizing principle.

NOTES

[1] *Natural History*, 19.50. See Pliny, *Natural History*, Books 17–19. With an English translation by H. Rackham (Loeb Classical Library, 371) (Cambridge, Mass. and London: Harvard University Press, 1950), 452–53:

> . . . *iam quidem hortorum nomine in ipsa urbe delicias agros villasque possident* . . .

See also the English translation (with notes of commentary) of Pliny's book *19* by John Henderson, *The Roman book of Gardening* (London and New York: Routledge, 2004), 67–101. The title that Henderson gave to this chapter is significant: *Nature's Miracles*.

[2] Marcus Porcius *Cato, On Agriculture*. With an English translation by William Davis Hooper revised by Harrison Boyd Ash (Loeb Classical Library, 283). (Cambridge, Mass.: Harvard University Press; London: William Heinemann, 1934), 140–51 (chapter 156).

[3] *Natural History*, 19.51. In the edition by Rackham, see 452–53:

> . . . *primus hoc instituit Athenis Epicurus otii magister; usque ad eum moris non fuerat in oppidis habitari rura.*

[4] On this point, see, for example, Sandra Citroni Marchetti, *Plinio il Vecchio e la tradizione del moralismo romano* (Biblioteca di Materiali e discussioni per l'analisi dei testi classici, 9) (Pisa: Giardini, 1991).

[5] *Epistles*, 2.1.156–157:

> . . . *Graecia capta, ferum victorem cepit et artis*
> *intulit agresti Latio* . . .

Latin text with English translation in *Horace, Satires, Epistles and Ars Poetica*. With an English translation by H. Rushton Fairclough (Loeb Classical Library, 194). (Cambridge, Mass.: Harvard University Press; London: William Heinemann, 1929), 408–409.

[6] On this question of garden hellenization, see for example Pierre Grimal, *Les Jardins Romains* (Paris: De Boccard, 1944), 63, and, more recently, Andrew Wallace-Hadrill. "*Horti* and hellenization." In Maddalena Cima and Eugenio La Rocca, eds. *Horti romani. Atti del Convegno Internazionale*, Roma 4–6 maggio 1995 (Bullettino della Commissione Archeologica Comunale di Roma, Supplementi 6) (Rome: L'Erma di Bretschneider, 1998), 1–12.

[7] On the introduction of this structure in Campanian architecture, see Wilhelmina Jashemski. "The Campanian peristyle garden." In Elisabeth B. Macdoughal and Wilhelmina Jashemski, eds., *Ancient Roman Gardens* (Dumbarton Oaks Colloquium on the History of Landscape Architecture, 7) (Washington, D.C.: Dumbarton Oaks, 1981), 29–48.

[8] On this point, see Brunilde Sismondo Ridgway. "Greek antecedents of garden sculpture." In Macdougal and Jashemski, eds., 7–28.

[9] For this expression and a study of the relationship between country and town in Roman gardens, see Nicholas Purcell. "Town in country and country in town." In Elisabeth Blair Macdougall, ed. *Ancient Roman Villa Gardens* (Dumbarton Oaks Colloquium on the History of Landscape Architecture, 10). (Washington, D.C.: Dumbarton Oaks, 1987), 185–203.

[10] Wilhelmina Jashemski. "Introduction." In Macdougall and Jashemski, eds., 3–6 (see 3–4).

[11] For such a view, see Marisa Mastroroberto. "Una visita di Nerone a Pompei: la *deversoriae tabernae* di Moregine." In Pier Giovanni Guzzo and Marisa Mastroroberto, eds., *Pompei: Le stanze dipinte* (Milan: Electa, 2002), 34–87 (see particularly 73–74, the paragraph entitled *Il modello neroniano di politica economica*).

[12] The literature on Roman gardens is vast, from the classical work of Grimal with several reeditions) to the more recent archeological work and syntheses by Wilhelmina Jashemski (later).

For an inventory and description of gardens in Rome and the Roman world, see Wilhelmina Jashemski, ed., *Gardens in the Roman Empire*, forthcoming.

For an analysis of different components of Roman gardens, see the proceedings of the 1995 conference edited by Cima and La Rocca.

For Pompeii gardens, see Wilhelmina Jashemski. *The gardens of Pompeii, Herculanum and the Villas destroyed by Vesuvius* (New Rochelle, N.Y.: Caratzas, 1979). For a synthesis on the "natural history" of Pompeii, see Wilhelmina Feemster Jashemski and Frederick G. Meyer, eds., *The natural history of Pompeii* (Cambridge: Cambridge University Press, 2002).

On the so-called "Sallust' gardens" in Rome, see: Emilia Talamo, "Gli *horti* di Sallustio a Porta Collina." In Cima and La Rocca, 113–69, and the recent monograph by Kim J. Hartswick, *The Gardens of Sallust: A Changing Landscape* (Austin: University of Texas Press, 2004).

See also (in chronological order) the volumes edited by Macdoughal and Jashemski, and Macdougall; Mariette de Vos. "La casa, la villa, il giardino. Tipologia, decorazione, arredi." In Salvatore Setti, ed., *Civiltà dei Romani. Il rito e la vita privata* (Milan: Electa, 1992), 140–54; M. Carroll-Spillecke, ed., *Der Garten von der Antike bis zum Mittelalter* (Kulturgeschichte der antiken Welt, 57) (Mainz: Philipp von Zabern, 1992); Linda Farrar, *Ancient Roman gardens* (Stoud: Sutton Publishing, 1998); Annamaria Ciarallo, *Gardens of Pompeii* (Los Angeles: the J. Paul Getty Museum, 2001); Annamaria Ciarallo, *Il giardino pompeiano. Le piante, l'orto, I segreti della cucina* (Naples: Electa, 2002); Maureen Carroll, *Earthly Paradises. Ancient gardens in history and archaeology* (London: British Museum, 2003); Maureen Carroll-Spillecke, "Gardens." In *Brill's New Pauly. Encyclopedia of the Ancient World*. Vol. 5. (Leiden and Boston: Brill, 2004), cols. 692–97.

[13] On Livia's Palatine house, see Lawrence Richardson, *A new topographical dictionary of Ancient Rome* (Baltimore and London: Johns Hopkins University Press, 1992), 73–74, with bibliography.

[14] See the catalogue of the Museum: Adriano La Regina , ed., *Museo Nazionale Romano. Palazzo Massimo alle Terme* (Milan: Electa, 1998), 208–13.

[15] See, for example, Ranuccio Bianchi Bandinelli, *Rome: Le centre du pouvoir* (Paris: NRF-Gallimard, 1969), 126.

[16] See 403–404 in the *Iconographic documentation*.

[17] Marina Sapelli, *The National Roman Museum: Palazzo Massimo alle Terme*. (Milan: Electa, 1998) 50–53.

[18] See La Regina, ed., 213.

[19] Salvattore Settis. "Le pareti ingannevoli. Immaginazione e spazio nella pittura romana di giardino." *Fondamenti* 11, (1988) 3–39, which is a preliminary version of Salvatore Settis. *Le pareti ingannevoli. La villa di Livia e la pittura di giardino* (Milan: Electa, 2002).

[20] On these frescos, see the earlier bibliography, particularly the works by Settis 1988 and 2002. For an analysis, see also Ida Baldassarre, Angela Pontrandolfo, Agnes Rouveret and Monica Salvadori, *Pittura romana: Dall'ellenismo al tardo-antico* (Milan: Federico Motta, 2002), 151–54.

[21] On this construction, see Ernest Nash, *Pictorial Dictionary of Ancient Rome*. 2 vols. London: (Thames and Hudson, 1968), vol. 1, 160–62, and, more recently: Richardson, 44–45, both with bibliography. See also Baldassarre et al., 183–84.

[22] Bianchi Bandinelli, 126.

[23] On the garden frescos in Pompeii houses, see Angela Donati, ed. *Romana pictura. La pittura romana dalle origini all'età bizantina* (Milan: Electa, 1998), 130–32 and 275–76; Baldassarre et al., 192–98.

[24] For an excavation report, see Wihelmina F. Jashemski, "Recently excavated gardens and cultivated land of the Fillas at Boscoreale and Oplontis." In Macdougal, ed., 31–75.

[25] See *Riscoprire Pompei. Musei Capitolini, Palazzo dei Conservatori, 13 novembre 1993–12 febbraio 1994.* (Rome: L'Erma di Bretschneider, 1993), 325–35, with the identification of the plants represented in the fresco. See also Ciarallo 2002.

[26] 4.1–2 (*passim*). For the Greek text and an English translation, see Longus, Daphnis & Chloe. With the English translation of George Thornley revised and augmented by J. M. Edmonds (Loeb Classical Library, 69) (Cambridge, Mass.: Harvard University Press; London: William Heinemann, 1916), 188–93:

[27] See Bianchi Bandinelli, p. 125.

[28] See above. Such interpretation is repeated in *Riscoprire Pompei*, 329–31.

[29] Settis 2002, 39–41.

[30] On Theophrastus, see Suzanne Amigues, "Les traités botaniques de Théophraste." In Georg Wöhrle, ed., *Geschichte der Mathematik un der Naturwissenschaften in der Antike: 1 Biologie.* (Stuttgart: Franz Steiner, 1999), 124–54. See also the several studies in *Theophrastus of Eresus: On his Life and Work.* Edited by William W. Forthebaugh together with Pamela M. Huby and Anthony A. Long (Rutgers University Studies in Classical Humanities) (New Brunswick and Oxford: Transaction Books, 1985).

1. ... τὸν παράδεισον ἐθεράπευεν ὡς ὀφθείη καλός. 2. Ἦν δὲ ὁ παράδεισος πάγκαλόν τι χρῆμα καὶ κατὰ τοὺς βασιλικούς ... εἶχε δὲ πάντα δένδρα, μηλέας, μυρρίνας, ὄχνας καὶ ῥοιὰς καὶ συκῶν καὶ ἐλαίας. ἑτέρωθι ἄμπελος ὑψηλὴ ἐπέκειτο ταῖς μηλέαις καὶ ταῖς ὄχναις περκάζουσα ... ἦσαν δὲ καὶ κυπάριττοι καὶ δάφναι καὶ πλάτανοι καὶ πίτυς · ταύταις πάσαις ἀντὶ τῆς ἀμπέλου κιττὸς ἐπέκειτο ... ἔνδον ἦν τὰ καρποφόρα φυτά, καθάπερ φρουρούμενα, ἔξωθεν περιειστήκει τὰ ἄκαρπα, καθάπερ θριγκὸςχειροποίητος ... ῥοδωνιὰ καὶ ὑάκινθοι καὶ κρίνα χειρὸς ἔργα, ἰωνιὰς καὶ ναρκίσσους καὶ ἀναγαλλίδας ἔφερεν ἡ γῆ · σκιά τε ἦν θέρους καὶ ἦρος ἄνθη καὶ μετοπώρου ὀπώρα ...

[31] Greek text (with English translation): Theophrastus, *Enquiry into plants*. With an English translation by Arthur Hort. 2 vols. (Loeb Classical Library, 70 & 79). (London: William Heinemann; Cambridge, Mass.: Harvard University Press, 1916). A new edition of the Greek text (with French translation and original identifications of plants) is currently in preparation by Suzanne Amigues. Four volumes have been published so far (books I-VIII): *Théophraste. Recherche sur les plantes.* Texte établi et traduit par Suzanne Amigues. 4 vols. (Paris: Les Belles Lettres, 1988–2003).

[32] On the meaning of this term, see Amigues, vol. 1, p. XVI-XVIII.

[33] On Aristotle's method, see Michael Boylan. *Method and practice in Aristotle's biology.* (Lanham, Md. and London: University of America Press, 1983), and the essays in *Biologie, logique et métaphyique chez Aristote.* Daniel Devereux and Pierre Pellegrin, eds. (Paris: Editions du CNRS, 1990).

[34] On the botanical aspects of Alexander's expedition, see (in chronological order): V. Ball. "On the identification of the animals and plants of India which were known to early Greek authors" *Proceedings of the Royal Irish Academy. Polite literature and antiquities.* Ser. II, vol. II, no. 6 (January 1885), 302–46; Charles Joret, *La flore de l'Inde d'après les écrivains grecs.* (Paris: Librairie Emile Bouillon, 1901); Hugo Bretzl, *Botanische Forschungen des Alexanderzuges.* (Leipzig: Teubner, 1903); Paul Pedech, (Paris: Les Belles Lettres, 1984), and, more recently, Klaus Kartunnen, *India and the Hellenistic World* (Studia Orientalia edited by the Finnish Oriental Society 83) (Helsinki: The Finnish Oriental Society, 1997).

[35] On Aristotle's method, see principally: Pierre Pellegrin, *La classification des animaux chez Aristote. Statut de la biologie et unité de l'aristotélisme.* (Paris: Les Belles Lettres, 1982).

[36] For Theophrastus's method, see principally: Georg Wöhrle, *Theophrasts Methode in seinen botanischen Schriften* (Studien zur antiken Philosophie 13). (Amsterdam: B.R. Grüner, 1985); Luciana Repici, *Uomini capovolti. Le piante nel pensiero greco* (Biblioteca ci cultura moderna 1152). (Rome and Bari: Laterza, 2000).

[37] On Theophrastus's system for plant classification, see Wöhrle 1985. See also Helen S. Lang, *The order of nature in Aristotle's physics* (Cambridge: Cambridge University Press, 1998).

[38] A study of Alexandrian research activity can be found in Lucio Russo, *La rivoluzione dimenticata. Il pensiero scientifico greco e la scienza moderna* (Milan: Feltrinelli, 1996). An anthology of scientific texts of the Hellenistic period has been published by Georgia L. Irby-Massie and Paul T. Keyser, *Greek science of the Hellenistic era* (London and New York: Routledge, 2002). For some specialistic studies, see (among others and in chronological order): Gabriele Giannantoni and Mario Vegetti, eds., *La scienza ellenistica. Atti delle tre giornate di studio tenutesi a Pavia dal 14 al 16 aprile 1982* (Elenchos 9) (Naples: Bibliopolis, 1984); Gilbert Argoud and Jean-Yves Guillauin, eds., *Sciences exactes et sciences appliquées à Alexandrie (IIIe siècle av. J.-C.–Ier siècle ap. J.-C.)* (Centre Jean-Palerne, Memoires, 16) (Saint-Etienne: Publications de l'Université de Saint-Etienne, 1994).

[39] On Dioscorides, see principally John Marion Riddle, *Dioscorides on pharmacy and medicine* (History of science series 3) (Austin: University of Texas Press, 1985).

[40] Critical edition of the Greek text: Pedanii Dioscuridi Anazarbei. *De materia medica libri quinque.* Edidit Max Wellmann. 3 vols. Berlin: Weidmann, 1906–

1914 (reprinted in 1958 and 2002. An English translation has been recently published by Lilly S. Beck, Pedanius Dioscorides of Anazarbus, De materia medica (Altertumswissenschaft Texte und Studien, 38) (Hildesheim, Zürich and New York: Olms-Weidmann). It makes obsolete the English translation published in 1934 (*The Greek herbal of Dioscorides illustrated by a Byzantine A.D. 512, Englished by John Goodyer A.D. 1655*. Edited and first printed A.D. 1933 by Robert Gunther. Oxford: Oxford University Press, 1934, with several reprints), as well as its revision by Tess A. Osbaldeston and Robert P. Wood under the title: *Dioscorides. De materia medica being an herbal with many other medicinal materials written in Greek in the first century of the common era*. Johannesburg: IBIDIS Press, 2000.

[41] See *praef.*, 1 (vol. 1, p. 1 in Wellmann's edition), where Dioscorides criticized his predecessors because they did not exhaustively treat the topic.

[42] See, for example, 5.10 (vol. 3, p. 19 in Wellmann's edition), where Dioscorides declares:

> . . . it is not useless, we think, to describe the preparation of the most different wines, so that the work will be complete for those interested in medicine. I do so, not because they are often used or are indispensable, but not to give the impression to have ommitted them at all . . .

> ... οὐκ ἄχρηστον δὲ ὑπογράψαι νομίζομεν πρὸς τὸ πλήρη τὴν ἱστορίαν τοῖς φιλιατροῦσι
> γενέσθαι καὶ τὴν τῶν ποικιλωτέρων οἴνων σκευασίαν, οὐχ ὅτι πολλή ἐστιν ἢ ἀναγκαία ἡ χρῆσις
> αὐτῶν ἀλλ' ἵνα κατὰ μηδὲν αὐτῶν ἐλλείπειν δοκῶμεν ...

[43] On this system, see Alain Touwaide, "La botanique entre science et culture au Ier siècle de notre ère.." In Wöhrle, ed., 219–252.

[44] See the pioneering remarks of Theodor Birt, *Das antike Buchwesen in seinem Verhältnisse zur Literatur, mit Beiträgen zur Textgeschichte des Theokrit, Catull, Properz, und anderer Autoren.* (Berlin: W. Hertz, 1882), 332, and, more recently, Touwaide in Wöhrle , ed., 248–249.

[45] Dioscorides, *De materia medica*, 1.1 and 5.162, respectively (vol. 1, 5–7, and vol. 3, 108 in the edition by Wellmann).

[46] On this notion, see the classical work by Arthur O. Lovejoy, *The great chain of being.* (Cambridge, Mass. and London: Harvard University Press, 1936). For the medieval developments of such theory, see: E.P. Mahoney, "Metaphysical foundations of the hierarchy of being according to some late medieval and Renaissance philosophers." In P. Morewedge , ed., *Philosophies of existence. Ancient and medieval* (New York: Fordham University Press, 1982), 165–257; D. Buschinger and A. Crepin , eds., *Les quatres éléments et la culture médiévale. Actes du Colloque d'Amiens, 25–27 mars 1982* (Göttingen: Kümmerle Verlag, 1983).

[47] On this point, see Touwaide in Wöhrle, ed., 241–42, with examples.

[48] Among many passages, see *Naturalis Historia* 12.1, 16.1, 17.1. See also Citroni Marchetti, 55–75.

[49] *Work and days*, 110–69. See Hesiod, the *Homeric Hymns and Homerica*. With an English translation by Hugh G. Everlyn-White (Loeb Classical Library, 67) (Cambridge, Mass.: Havard University Press; London: William Heinemann, 1914), 10–15.

[50] 4.910–1050. Latin text with an English translation: Lucretius, *De natura rerum*. With an English translationby W.H.D. Rouse. Revised with new text, introduction, notes and index by Martin Ferguson Smith (Loeb Classical Library, 181) (Cambridge, Mass.: Harvard University Press; London: William Heinemann, 1975), 346–59. For a more recent English translation, see: Lucretius: *On the nature of things. De natura rerum*. Edited and translated by Anthony M. Esolen. (Baltimore and London: Johns Hopkins University Press, 1995 (for vv. 4.910–1050, see 147–51). For a study of some themes in Lucretius's biological thinking: P.H. Schrijvers. Lucrèce et les sciences de la vie (Mnemosyne, Supplementum 186) (Leiden, Boston, and Köln: Brill, 1999).

[51] *Virgil*. With an English translation by H. Rushton Fairclough. *I Eclogues, Georgics, Aeneid I-VI*. Revised edition (Loeb Classical Library, 63) (Cambridge, Mass.: Harvard University Press; London: William Heinemann, 1978), 28-33.

[52] *Polybius, Histories*, 1.2.7, 1.3.4, 1.4.10–11. Greek text in: *Polybius, The histories*. With an English translation by W.R. Platon (Loeb Classical Library, 128) (Cambridge, Mass.: Harvard University Press; London: William Heinemann, 1922), 6-13:

> ... Ῥωμαῖοί γε μὴν οὐ τινὰ μέρη, σχεδὸν δὲ πᾶσαν πεποιημένοι τὴν οἰκουμένην ὑπήκοον αὐτοῖς
> ... ἀπὸ δὲ τούτων τῶν καιρῶν οἷον εἰ σωματοειδῆ συμβαίνει γίνεσθαι τὴν ἱστορίαν,
> συμπλέκεσθαί τε τὰς Ἰταλικὰς καὶ Λιβυκὰς πράξεις ταῖς τε κατὰ τὴν Ἀσίαν καὶ ταῖς
> Ἑλληνικαῖς καὶ πρὸς ἓν γίνεσθαι τέλος τὴν ἀναφορὰν ἁπάντων ... διὸ παντελῶς βραχύ τι
> νομιστέον συμβάλλεσθαι τὴν κατὰ μέρος ἱστορίαν πρὸς τὴν τῶν ὅλων ἐμπειρίαν ... ἐκ μέντοι γε
> τῆς ἁπάντων πρὸς ἄλληλα συμπλοκῆς καὶ παραθέσεως, ἔτι δ' ὁμοιότητος καὶ διαφορᾶς, μόνως
> ἄν τι ἐφίκοιτο καὶ δυνηθείη κατοπτεύσας ...

Botanical Progress, Horticultural Innovations and Cultural Changes

Botanical Progress, Horticultural Innovations and Cultural Changes

Botanical Progress, Horticultural Innovations and Cultural Changes

Botanical Progress, Horticultural Innovations and Cultural Changes

Rose of Eros and the Duplicity of the Feminine in Zoharic Kabbalah

Elliot R. Wolfson

Rose! O rareness of the revealed!

Dance! Dance out the truth of being, as act and existence!

Dance up the linear measurement of time to its quintessence!

Beat it up into synthesis!

Under the storm of your heels

Pound to perfection!

Make it be!

Establish it in your movement,

The perdurable moment that cannot die!

Incarnate it!

Divinize it in your fierce unquenchable essence!

Confer beatitude upon it!

Immortalize it in your glance,

Your look and your laughter!

By the kiss of your lips,

By the ecstasy of your breath,

Re-create it anew!

O clear expression of beauty!

O melody of existence!

Rose!

Reality Unfolded!

> *William Everson*, "The Raging of the Rose"

The ubiquity of the utilization of the rose as an image to depict the allure of eros in manifold cultural settings has been well documented in scholarly literature. Focusing our attention on the three monotheistic faiths, it can be said that the rose assumed special significance in the religious symbolism cultivated by Jews, Christians, and Muslims in the high Middle Ages. In this study, I will limit my reflections to one of the Abrahamic faiths, Judaism, and even with respect to this particular choice I am concerned primarily with analyzing the symbol of the rose in the major compendium of medieval Jewish esotericism, *Sefer ha-Zohar,* the "Book of Splendor," an anthology of kabbalistic teachings that began to circulate in Castile in the late thirteenth and early fourteenth centuries, although it would take several centuries before the literary contours of the work assumed a relatively stable form.[1]

In her study *The Symbolic Rose,* Barbara Seward suggested that in the medieval world, which she reconstructed exclusively from Latin sources and/or vernacular texts written in Christian Europe, the rose exemplified the "allegorical habit of mind as shaped by Christianity and expressed in religious literature."[2] More specifically, Steward delineates four major metaphorical applications of the rose in medieval sources: Paradise, martyrdom, the Virgin Mary, and Christ's Passion and Resurrection.[3] What is most noteworthy for the purposes of the ensuing discussion is Seward's observation that symbolic meanings accorded the rose did not efface the older identification of this flower as a token of erotic passion. Indeed, the ideal of spiritual eros, which evolves most conspicuously in the twelfth century, although its conceptual roots are much older, is well captured by the image of the rose precisely because the latter still conveyed the concrete sense of carnality, mediated through the visual, tactile, and olfactory senses. The ancient identification of the rose as the flower of Venus, the goddess of earthly love, was easily appropriated in the medieval Christian setting and applied to Mary, queen of the heavenly realm, to the point that in Dante's *Paradiso* the concrete figure of courtly love was identified with the mystical symbol of the soul's marriage to God.[4]

In spite of fundamental theological differences that separated medieval Jewish and Christian liturgical communities, one of the common threads tying them together was the confluence of the spiritual and somatic, the aetheral and sexual. The commonly held (mis)perception notwithstanding, the body as such was not denigrated in the ascetic culture of the late Middle Ages. What was denounced was the corporeality of corruptible flesh, the body subsumed by brazen passion for things ephemeral. By contrast, the holy body, variously described as the luminous or angelic body, was affirmed as an eschatological state of beatitude that could be attained to some degree even when the soul remained imprisoned in the world of generation and decay. The rose emerged, therefore, as an exemplary icon of the erotic underpinnings of the religious quest. The thorns stood as a steadfast reminder that one beholden to the rose of spiritual eros is always at risk of being lured by carnal lust and temptation. It comes as no surprise that figures such as Bonaventura and Albertus Magnus depicted Mary as a thornless rose, a perfectly apt image to convey the perfection of eros divested of sexual gratification.[5] The sensuality of the flower was thereby transformed into an ideal of sexual purity and abstinence.

Turning to the image of the rose in medieval kabbalah, I commence with a passage that appears in the beginning of the literary unit that is printed under the title "introduction" in the Mantua edition of *Zohar* (1558), and all subsequent editions based thereon.[6] *Shekhinah,* referred to as *kenesset yisra'el,* "Community of Israel," the last of the ten *sefirot,* the emanations that constitute the divine pleroma, is portrayed in the scriptural image *shoshannah ben ha-ḥoḥim,* "rose amongst the thorns" (Song 2:2), for just as the rose "is red and white, so the Community of Israel has judgment and mercy."[7] In a second passage, this symbolic interpretation is reiterated: "Come and see: It is written 'as a rose amongst the thorns so is my beloved amidst the daughters' (Song 2:2). 'As a rose,' this is the Community of Israel, for she is in the company of her hosts like the rose amidst its thorns."[8] Building on older rabbinic sources wherein the rose mentioned in the Song is explicated as a reference to the people of Israel, the medieval kabbalists understand the allegorical intent as a symbolical allusion to *Shekhinah,* the counterpart to the

Jewish people in the divine pleroma,[9] and just as the rose can be either red or white, so *Shekhinah* has the dual nature of being merciful and judgmental, and just as a rose sits atop its thorns, so *Shekhinah* is positioned above the angelic hosts that correspond to the twelve tribes of Israel in the mundane sphere.[10]

But why is *Shekhinah* singled out in this way? Surely, the author of these passages would have assented to the view articulated by kabbalists before the zoharic period that each of the divine attributes displays the androgynous character of overflowing and withholding, the former valenced as masculine mercy and the latter as feminine judgment. The principle is articulated clearly by the thirteenth-century Catalonian kabbalist Jacob ben Sheshet in a passage in *Sefer ha-Emunah we-ha-Bittaḥon* where he explains why the divine names can be both masculine and feminine:

> The names at times are masculine and at other times feminine, and there is no doubt about this because it is the truth. . . . And the reason for this matter is that even though with respect to the names there is a difference between the attribute of judgment and the attribute of mercy, whatever is in the one is in the other [*kol mah she-yesh ba-zeh yesh ba-zeh*], and on account of this the attribute of judgment changes [*mithappekhet*] into the attribute of mercy and the attribute of mercy into the attribute of judgment . . . for if the attributes were not all contained one within the other [*kullan kolelot zo et zo*] it would be impossible for one to change into the other.[11]

Utilizing contemporary jargon, we might say the *sefirot* manifest a "gender hybridity," attributes aligned with the masculine possess the potential for judgment and attributes aligned with the feminine possess the potential for mercy.

Insofar as the binary opposition pertains equally to male and female, it would have been tenable logically to speak of the two faces of the male, Samael and *Tif'eret,* the sinister force of dark and the godly force of light. However, as the opening comment in the Mantua recension of the zoharic text attests, the author of the homily focused exclusively on the twin nature of the female conveyed by the scriptural image of the rose perched atop its thorns as well as by the fact that it can be either red or white, the former signifying judgment and the latter mercy. Fluctuation of *Shekhinah,* her "unstable quality" according to Tishby,[12] or as Scholem put it, the "ambivalence" manifest in her "alternating phases,"[13] is measured by the fact that it is one and the same energy-force that may be divine or demonic.[14] *Shekhinah* and Lilith, the woman of valor (*eshet ḥayyil*) and the estranged woman (*ishshah zarah*), are not autonomous beings but two facets of the self-same reality.[15]

The boldness of this claim is brought into sharp relief by a passage from the stratum of the zoharic anthology called *Sitrei Torah,* "Mysteries of the Law," an exegesis of Jacob's sojourn from Be'er Sheva to Haran (Gen. 28:10). Be'er Sheva stands symbolically for *Shekhinah,* the attribute that comprises the lower seven attributes (the name *Be'er Sheva* is decoded, accordingly, as the "well of seven"), and Haran, the place of anger (the word *ḥaron* denotes rage or fury), stands symbolically for the feminine potency in the domain of the demonic, the "side of the woman of harlotry, the adulterous woman, the mystery of the mystery of mysteries that comes forth from the strength of the lustre of Isaac." The latter, which is described as being "red like a rose," is designated as well as the "serpent" (*naḥash*), "end of all flesh" (*qeṣ kol basar*), and "end of days" (*qeṣ ha-yamim*). Just as on the side of holiness *Shekhinah* is comprised within *Tif'eret,* so on the side of impurity Lilith is comprised within Samael: "Two evil spirits conjoined as one, the spirit of the male is subtle, the spirit of the female spreads out in several ways and paths, but she is conjoined to the spirit of the male. She adorns herself in jewelry like a prostitute, and from a distance she stands at the head of the ways and paths to seduce men."[16] The symbolic meaning of the rose in this context is obvious enough. The red color of the flower is highlighted for it figuratively depicts the temptation of carnal passion, which is understood principally from the vantage point of male kabbalists as the personification of the feminine potency of the demonic in the form of sexual seduction, the enticement of the forbidden other, embodied in the Gentile woman.[17]

Jacob's journey is thus read as an allegorical portrayal of the spiritual struggle on two planes, the psychological and theological, although I hasten to add that there is no justification to treat these as disparate as the assumption of every kabbalist is that the soul of the Jew emanates from and hence is ontically on a par with the divine. From the psychological perspective, the narrative marks the danger of desire that lurks as the soul moves from the pleromatic unity of being to the contingent world of becoming, from the timespace of identity to the spacetime of difference; from the theological perspective, the danger translates into the Jew's struggle against idolatry, the battle to withstand the temptation to render imaginal representations of the formless transcendent iconically in material form.[18] The biblical figure of Jacob initially fends off the advances of Lilith and then victoriously wrestles with Samael at a later point in the journey (Gen. 32:25–31). The change of name from "Jacob" to "Israel" denotes that "he was perfected in his perfection and he ascended to his perfect gradation, and he was called 'Israel,' and then he ascended to the supernal gradation, and he was perfected in everything, and he became the central pillar."[19] The transfiguration of Jacob, his apotheosis, is facilitated by encountering and overcoming the force of unholiness in its dual character as male and female.

In the imagination of the Castilian kabbalists responsible for the zoharic text, the rose depicts both *Shekhinah,* consort of the Holy One, and Lilith, companion of Samael, the archon of Edom, the spiritual force of Christendom. However, as I noted above, these two images are not to be treated as polar opposites; on the contrary, in the zoharic understanding, the chaste bride and wanton whore are two sides of one phenomenon, two forms of dissimilitude by which the spiritual reality is configured in human imagination. The image of the rose, therefore, signifies the dynamism of *Shekhinah,* her ability to change from one aspect to another. Perhaps the most characteristic feature of *Shekhinah* is this quality of metamorphosis, the seemingly unstable nature of the feminine to become one thing and its opposite, the very quality that underlies the texture of eros. One passage in the zoharic anthology is especially pertinent and worthy of full citation. The mythopoetic imagination in this instance is inspired by the verse "I am a rose of Sharon, a lily of the valleys," *ani ḥavaṣṣelet ha-sharon shoshannat ha-omaqim* (Song 2:1).

> Come and see: initially, the rose [*ḥavaṣṣelet*] was green with green leaves, but afterward the lily [*shoshannah*] had two colors, red and white. The lily on six leaves—the lily [*shoshannat*], for it changes its colors [*deshani'at gawwanaha*], and it changes from one color to another [*we-ishtani'at me-gawwana le-gawwana*]. The lily at first was [called] a rose [*ḥavaṣṣelet*], for at the time she desired to unite with the king she is called *ḥavaṣṣelet,* but after she is conjoined to the king by means of the kisses, she is called *shoshannah.* . . . The "lily of the valley" [*shoshannat ha-omaqim*], for she changes her colors, sometimes for good and sometimes for bad, sometimes for judgment and sometimes for mercy.[20]

This passage articulates the vibrancy of eros emblematized in the rose. Before being conjoined to the masculine, the feminine betrays the color green, which in this context means she has not yet blossomed. When does the flower bloom, a transition marked semantically by the change in name from *ḥavaṣṣelet,* which is translated as "rose," to *shoshannah,* which is rendered as "lily"? When she unites with her masculine partner, and she then assumes one of two possible colors, red or white, which, as we have seen, respond respectively to the attributes of judgment and mercy. Underlying the duplicitous nature of the female in the imaginal representations of kabbalists is the archaic priestly approach to the menstrual cycle of the Jewish woman, part of the month deemed pure and permitted to her husband and another part impure and forbidden.[21] The dual portrait of the female body, reflected in laws and customs pertaining to menstruation, alludes to the mystical secret mentioned above: Just as the same woman fluctuates from one state to its opposite, *Shekhinah* and Lilith are two facets of one being rather than two distinct beings. In her judgmental capacity, *Shekhinah* assumes the title *shushan edut,* "rose of testimony" (Ps. 60:1), for she attests to the unique providential role that God exercises in relation to Israel in general and to the spiritual elite in particular. The point is emphasized in the following zoharic interpretation of the idiom "who browses among the lilies," *ha-ro'eh ba-shoshannim* (Song 2:16): "Just as this lily [*shoshan*] is red and its waters are white, so the Holy One, blessed be He, governs his world [*manhig olamo*] from the attribute of judgment to the attribute of mercy, as it is written, 'If your sins are like crimson, they can turn snow-white' (Isa. 1:18)."[22]

One passage that is especially interesting relates this imagery to the act of walking, a central theme in zoharic literature and subsequent kabbalistic texts influenced thereby that denotes the gesture of meandering about on the hermeneutical path, the imaginary physical localities within the confines of the land of Israel corresponding to interpretive zones in the textual landscape.[23] According to the passage in question, the image of the rose is applied to the Jewish people, for just as a rose emits a pleasing fragrance only when its petals open, so Jews are considered praiseworthy when they open their hearts to repent. After this fairly standard moralistic claim is made, the text relates that as the members of the fraternity were sitting an eagle swooped down and plucked a rose, to which they responded, "From here forward we can go on our way."[24] The eagle[25] and the rose signify the providential supervision entrusted to *Shekhinah,*[26] the guardianship that the sages needed before they could resume their excursion. The dramatic act of the eagle inspires R. Phineḥas to expound the verse "For the leader, concerning the rose of testimony [*shushan edut*], an epigram [*mikhtam*] for David, to be taught / when he fought with Aram-Naharaim and Aram-Zobah, and Joab returned and defeated Edom" (Ps. 60:1–2). The expression *shushan edut* in the verse is explained by R. Phineḥas as follows: "This is the rose of testimony that exists here, the stars are in heaven and *Shekhinah* is upon us, and the supernal gradations that are with her, and she provides sacred assistance for [the utterance of] praise. This is the [the meaning of] *shoshan,* in tact [*bi-shelimu*], as is appropriate."[27] It is reasonable to assume that this passage served as the basis for the homiletical elaboration in the following citation: "*Shushan edut,* this is the testimony of *Shekhinah* [*sahaduta di-shekhinta*], for she is the 'rose of witness,' as she attests that she stands over us, and she brings testimony concerning us before the king, and the supernal, holy gradations are with her, and the sacred assistance regarding the liturgical praise."[28]

The exegetical moves employed in this homily are too complex to be disentangled properly in this context. Briefly, it can be said that the presumed experience of the eagle plucking a rose enabled the zoharic author to connect the two images and thereby juxtapose various biblical verses. Especially significant is the wisdom (*ḥokhmata*) elicited from the verses from Psalms 60:1–2, which relates to the providential guidance accorded David and Joab when they went to do battle against the enemies of Israel. The intricate web of exegesis and experience is reinforced when the zoharic author interrupts the narrative by saying that the "great eagle" encircled the heads of the companions of the mystical fraternity as they were going about their way. R. Phineḥas takes this to be an omen: "Surely, now is a propitious time [*iddan re'uta*], and in this moment the gates of mercy are opened for all those who are ill, as this is the time for their healing. Even though they are prisoners of the king, this eagle is a sign of compassion."[29] This comment leads to the citation and explication of the verse "Like an eagle who rouses his nestlings, gliding down to the young, so did he spread his wings and take him, bear him along his pinions" (Deut. 32:11). Once again, the exegetical track crisscrosses with the experiential in the zoharic literary imagination, as the interpretation of the aforementioned verse is interrupted by the comment that the eagle encircled and went before the sages, which prompts R. Phineḥas to inquire of the eagle if it has come on a mission from God or for some other purpose, an inquiry that occasions the ascent of the eagle, which, in turn, compels the sages to sit down.

The zoharic text continues with a comment of R. Ḥiyya that recounts a tradition about Solomon and the great eagle. Of necessity we will again have to bracket many of the technical details of this legend, but what is most important for our purposes is the depiction of the eagle transporting Solomon to the place in the wilderness called Tadmor (based on the masoretic reading of 1 Kgs. 9:17) where Azza and Azzael, the impure forces of the demonic, were bound in iron chains, "a place to which no mortal in the world, and not even the birds of heaven, could enter therein except for Balaam."[30] Perched beneath the wings of the eagle, Solomon wrote a magical formula and tossed it into the place of darkness and was thereby saved from the sinister spirits. Moreover, as the eagle came closer to the spot where Azza and Azzael were incarcerated, Solomon placed a ring inscribed with the holy name (I presume a reference to the Tetragrammaton) into the mouth of the eagle, and the demonic beings became subservient to Solomon.[31] This is the way, we are told, that Solomon learned the wisdom (ḥokhmata), which one must assume refers to the occult wisdom of magic and sorcery.[32] Immediately succeeding this legend, the zoharic text circles back (much like the motion of the eagle) to the experience of the members of the fraternity and reiterates the description that preceded the exegesis of R. Phineḥas that led to the excursus on Solomon proffered by R. Ḥiyya:

> As they were sitting the eagle came toward them, and a rose was in his mouth, and he cast it before them, and he went away. They saw and they were happy. R. Phineḥas said: Did I not tell you that this eagle comes and goes by the agency of our master? This rose is the "rose of witness" [shushan edut] that I mentioned, and the Holy One, blessed be He, sent it to us.[33]

The message of this intricate labyrinth of exegesis and narrative seems to be that the path of interpretation is fraught with danger and hence, before one can proceed, one needs the protection of divine providence, iconically portrayed by the image of the eagle and rose in the case of the members of the imaginary kabbalistic fraternity, and as the eagle and ring upon which the divine name was inscribed in the case of Solomon. Indeed, the concurrence of the legend about Solomon and the story concerning the sages suggests that the rose has the same theurgic potency as the ring. With this in mind, we can better understand the connotation of the locution shushan edut, "rose of witness," that is, the rose attests that the eagle has come at the behest of God to serve as the guide on the path, to protect the sojourners from the potentially deleterious consequences of the demonic power. From the story about Solomon, however, we learn that there is wisdom to be gained from even the side of darkness and impurity, the science associated with magic and sorcery. I think it reasonable to assume that a similar role be assigned to the rose, especially the red rose, which signifies the lure of passion and sexual temptation, qualities that the zoharic authorship associated with the realm of the demonic. Yet, it is precisely by discerning this wisdom that one can apprehend the unity of the divine.[34] This very point is made in the continuation of the zoharic passage where the following teaching of R. Phineḥas is recorded:

> He began [to expound] as before and said: "For the leader, concerning the rose of witness [shushan edut], an epigram [mikhtam] for David, to be taught" (Ps 60:1). This "rose of witness," what testimony does it offer [sahaduta sahid]? This "rose of witness" is testimony for the account of creation, and it is testimony for the Community of

Israel, and it is testimony for the supernal unity, for the rose has thirteen petals, and all of them exist from one root, and it has five strong petals on the outside that surround and protect it. Everything is in the mystery of wisdom [*raza de-ḥokhmata*]: the thirteen petals are the thirteen attributes of mercy that the Community of Israel inherits from above, and all of them adhere to one root, and this is the one covenant [*berit ḥada*], which is in the pattern of the covenant of the foundation of everything [*dugmat di-verit yesoda de-khola*]. The five strong ones that surround her are the fifty gates, the five hundred years in which the Tree of Life extends.[35]

It lies beyond the parameters of this study to elucidate all of the issues intimated in this passage. Suffice it to note that the mystery of wisdom elicited from the image of the rose involves cultivating the ability to discern the underlying identity of the holy and unholy. The holy dimension is symbolized by the thirteen petals, which correspond to the thirteen attributes of mercy[36] located in *Keter*,[37] the highest aspect of the divine,[38] depicted here as well as a "covenant" that is parallel to the "covenant of the foundation of everything," that is the ninth attribute of *Yesod,* which is the phallic potency;[39] the unholy aspect is symbolized by the five strong petals on the outside, which are associated with the fifty gates, a likely allusion to the rabbinic notion of fifty gates of understanding,[40] and with the tradition that the Tree of Life extended a measure of five hundred years.[41] Inasmuch as the rose embodies in its very composition both aspects of existence, it is well suited to assume the role of witness that attests to the act of creation, the *Shekhinah,* and to the supernal unity that is predicated on affirming that both light and dark, good and evil, left and right, stem from one root.[42]

The fluctuating character of the rose affords us an opportunity to ascertain the mysterious nature of eros and the erotic nature of mystery that informed the kabbalistic sensibility. As I have noted, the image of the rose enthroned on its thorns portends that mercy and judgment are entwined on one stem. The two-faced character of *Shekhinah* is expressive of an ontological principle embraced by practitioners of kabbalah from its inception as a literary-historical phenomenon: Androgyny is applicable to each of the divine attributes, that is, not only is it the case that the divine anthropos in its totality comprises masculine and feminine, but every aspect of the anthropomorphic configuration displays the interplay of the erotic, the power to overflow and the will to receive. Although all of the *sefirot* are characterized by this dual potentiality, the documents produced by kabbalists from an early period attest that the principle of androgyny was applied to *Shekhinah* in a distinctive manner, as the instability, unpredictability, and duplicity of the feminine provided adequate metaphorical language to capture something of the inscrutable nature of being.

The quality of indeterminacy linked to the rose bespeaks the nature of time as well, for in kabbalistic gnosis, the essential component of time is linked to the moment, which is portrayed as a double-edged sword, the temporal interval that opens and closes, binds and unbinds.[43] The duplicity is captured, for instance, in the zoharic gloss on the scriptural expression *ḥerev ha-mithappekhet,* "ever-turning sword" (Gen. 3:24), which is applied to *Shekhinah:* "It changes from this side to that side, from good to evil, from mercy to judgment, from peace to war, it changes in everything, good and evil, as it is written, 'the tree of knowledge of good and evil' (ibid., 2:17)."[44] The vacillation between good and evil, which is attributed to the divine presence, is indicative of the incisive quality of time, the fullness of the moment realized in the cut that binds, the fork in the road that splits into a path to the right and a path to the left. The sojourner on the way knows, however, that the two paths are not to be construed dichotomously, as they spring from one and the same font. The rose of eros, ever changing from red to white and from white to red, is this very bridge of time that connects being and becoming, a bridge that swerves this way and that way, and, thereby, facilitates the constant motion that the dance of the erotic dictates.

NOTES

[1] On the history of the composition, redaction, and literary structure of zoharic literature, see Gershom Scholem, *Major Trends in Jewish Mysticism* (New York: Schocken Books, 1956), 156–204; Isaiah Tishby, *The Wisdom of the Zohar,* translated by David Goldstein (Oxford: Oxford University Press, 1989), 1–12; Yehuda Liebes, *Studies in the Zohar,* Translated by Arnold Schwartz, Stephanie Nakache, and Penina Peli (Albany: State University of New York Press, 1993), 85–138; Boaz Huss, "*Sefer ha-Zohar* as a Canonical, Sacred and Holy Text: Changing Perspectives of the Book of Splendor between the Thirteenth and Eighteenth Centuries," *Journal of Jewish Thought and Philosophy* 7 (1998): 257–307; idem, "The Appearance of *Sefer ha-Zohar,*" *Tarbiz* 70 (2001): 507–542 (Hebrew); Charles Mopsik, "Le corpus Zoharique ses titres et ses amplifications," in *La Formation des Canons Scripturaires,* ed. Michel Tardieu (Paris: Cerf, 1993), 75–105; idem, "Moïse de León, le Sheqel ha-Qodesh et la rédaction du Zohar: Une réponse à Yehuda Liebes," *Kabbalah: Journal for the Study of Jewish Mystical Texts* 3 (1998): 117–218; Daniel Abrams, "Critical and Post-Critical Textual Scholarship of Jewish Mystical Literature: Notes on the History and Development of Modern Editing Techniques," *Kabbalah: Journal for the Study of Jewish Mystical Texts* 1 (1996): 17–71, esp. 61–64; Ronit Meroz, "Zoharic Narratives and their Adaptations," *Hispania Judaica Bulletin* 3 (2000): 3–63; Pinchas Giller, *Reading the Zohar: The Sacred Text of Kabbalah* (Oxford and New York: Oxford University Press, 2001).

[2] Barbara Seward, *The Symbolic Rose* (New York: Columbia University Press, 1960), 20.

[3] Ibid., 20–24.

[4] Ibid., 43–44. On the relationship between the images of Venus, Eve, and Mary, see Patricia Rubin, "The Seductions of Antiquity," in *Manifestations of Venus: Art and Sexuality,* eds. Caroline Arscott and Katie Scott (Manchester and New York: Manchester University Press, 2000), 24–38.

[5] Seward, *The Symbolic Rose,* 23 n. 6.

[6] See Daniel Abrams, "When Was the Introduction to the *Zohar* Written?, with Variants in the Different Versions of the Introduction in the Mantua Print," *Asufot* 8 (1994): 211–226 (Hebrew).

[7] *Zohar* 1:1a; see 3:286b.

[8] *Zohar* 1:137a; see 3:37b–38a.

[9] See, for instance, *Zohar* 2:189b: "He began [to expound] and said, 'Like a rose amongst the thorns so is my beloved amidst the daughters' (Song 2:2). The Holy One, blessed, be He, wanted to make Israel in the pattern of what is above, to be the singular rose on earth as it is above. The rose that emits a fragrance and which is separated from all the other roses is naught but the one that is amongst the thorns." See also *Zohar* 3:38a, 107a, 233a.

[10] Tishby, *Wisdom of the Zohar,* 381.

[11] Ḥayyim D Chavel, ed., *Kitvei Ramban* (Jerusalem: Mosad ha-Rav Kook, 1964), 2:359.

[12] Tishby, *Wisdom of the Zohar,* 371.

[13] Gershom Scholem, *On the Kabbalah and Its Symbolism,* translated by Ralph Manheim (New York: Schocken, 1965), 107.

[14] For an analogous symbolic representation of the white and red rose in Christian hermeticism, see Carl G. Jung, *Mysterium Coniunctionis: An Inquiry Into the Separation and Synthesis of Psychic Opposites in Alchemy,* second edition, trans. R. F. C. Hall (Princeton: Princeton University Press, 1970), 305–307.

[15] On the dual nature of *Shekhinah* and her interchangeability with Lilith, see Scholem, *Kabbalah and Its Symbolism,* 105; idem, *On the Mystical Shape of the Godhead: Basic Concepts in the Kabbalah,* trans. Joachim Neugroschel, edited and revised by Jonathan Chipman (New York: Schocken Books, 1991), 190–192; Tishby, *Wisdom of the Zohar,* 376–379.

[16] *Zohar* 1:148a.

[17] For a more extensive discussion of this theme, see Elliot R. Wolfson, "Re/membering the Covenant: Memory, Forgetfulness, and the Construction of History in the Zohar," in *Jewish History and Jewish Memory: Essays in Honor of Yosef Hayim Yerushalmi,* edited by Elisheva Carlebach, John M. Efron, and David N. Myers (Hanover and London: Brandeis University Press, 1998), 214–246, esp. 216–222.

[18] See Elliot R. Wolfson, "Iconicity of the Text: Reification of Torah and the Idolatrous Impulse of Zoharic Kabbalah," *Jewish Studies Quarterly* 11 (2004): 215–242.

[19] *Zohar* 1:148b (*Sitrei Torah*).

[20] *Zohar* 3:107a

[21] Sharon F. Koren, "'The Woman from whom God Wanders:' The Menstruant in Medieval Jewish Mysticism," Ph.D. dissertation, Yale University, 1999, 150–208; idem, "Mystical Rationales for the Laws of *Niddah,*" in *Women and Water: Menstruation in Jewish Life and Law,* edited by Rahel R. Wasserfall (Hanover and London: University Press of New England, 1999), 101–21.

[22] *Zohar* 2:20a.

[23] The theme of walking has been central to my work on kabbalistic symbolism both in print and in teaching. For a fuller exposition, see the study done by my student, David Greenstein, "Aimless Pilgrimage: The Quotidian Utopian of the Zohar," Ph.D. thesis, New York University, 2003.

[24] *Zohar* 3:233a.

[25] The use of the eagle to symbolize divine providence is well attested in Jewish sources and finds its scriptural roots in Deuteronomy 32:11, a verse cited in the continuation of this zoharic homily.

[26] In a list of different designations of the *Shekhinah* found in *Zohar Ḥadash,* edited by Reuven Margaliot (Jerusalem: Mosad ha-Rav Kook, 1978), 31b, both "eagle" and "rose" are mentioned.

[27] *Zohar* 3:233a. For a slightly different explanation of the expression *shushan edut* in the latter strata of zoharic literature, see *Zohar* 3:223b (*Ra 'aya Meheimna*). In that context, the dual nature of *shoshan* signifies respectively the seventh and eighth emanations, *Hod* on the right and *Neṣaḥ* on the left, and the word *edut* is interpreted as a reference to the ninth emanation, *Yesod*.

[28] *Zohar* 3:238b (*Ra 'aya Meheimna*).

[29] *Zohar* 3:233a.

[30] *Zohar* 3:233a–b.

[31] The zoharic legend is drawing on a much older motif regarding Solomon's possession of a magic ring, which endowed him with the power to subdue demonic beings. See Pablo A. Torijano, *Solomon the Esoteric King: From King to Magus, Development of a Tradition* (Leiden: Brill, 2002), 76–86. See also Sarah Iles Johnston, "The *Testament of Solomon* From Late Antiquity to the Renaissance," in *The Metamorphosis of Magic from Late Antiquity to the Early Modern Period,* edited by Jan N. Bremmer and Jan R. Veenstra (Leuven: Peeters, 2002), 35–49, esp. 45–48.

[32] On the tradition about Solomon's acquisition of the occult arts of exorcisim, magic, and sorcery, see Torijano, *Solomon the Esoteric King,* 41–87, 192–224; Jan R. Veenstra, "The Holy Almandal: Angels and the Intellectual Aims of Magic, Appendix: *The Art Almadel of Solomon* (BL, ms. Sloane 2731)," in *Metamorphosis of Magic,* Bremmer and Veenstra, eds.,189–229.

[33] *Zohar* 3:233b.

[34] For fuller discussion, see Elliot R. Wolfson, "Light Through Darkness: The Ideal of Human Perfection in the Zohar," *Harvard Theological Review* 81 (1988): 73–95.

[35] *Zohar* 3:233b.

[36] Babylonian Talmud, Rosh ha-Shanah 17b. Rabbinic sages derived the thirteen attributes of mercy from Exodus 34:6–7.

[37] The intimate relationship of *Keter* (or *Atiqa Qaddisha*) and *Shekhinah,* depicted in the image of the rose, is also emphasized in *Zohar Ḥadash,* 51c.

[38] For discussion of this kabbalistic tradition, see Elliot R. Wolfson, "Beyond Good and Evil: Hypernomianism, Transmorality, and Kabbalistic Ethics," in *Crossing Boundaries: Essays on the Ethical Status of Mysticism,* edited by G. William Barnard and Jeffrey J. Kripal (New York and London: Seven Bridges Press, 2002), 119–120.

[39] The passage lends support to my argument in previous studies regarding the upper phallus lodged in the head, sometimes called in zoharic texts the *boṣina de-qardinuta,* which I render as the "hardened spark." See Elliot R. Wolfson, "Woman—The Feminine As Other in Theosophic Kabbalah: Some Philosophical Observations on the Divine Androgyne," in *The Other in Jewish Thought and History: Constructions of Jewish Culture and Identity,* edited by L. Silberstein and R. Cohn (New York: New York University Press, 1994), 179–182; idem, *Circle in the Square: Studies in the Use of Gender in Kabbalistic Symbolism* (Albany: State University of New York Press, 1995), 60–69.

[40] Babylonian Talmud, Rosh ha-Shanah 21b; Nedarim 38a.

[41] Palestinian Talmud, Berakhot 1:1; *Midrash Rabbah: Shir ha-Shirim,* edited by Shimshon Dunasky (Jerusalem and Tel-Aviv: Dvir Publishing, 1980), 6:14, 146; *Midrash Tehillim,* edited Solomon Buber (Vilna: Rom, 1891). 1:19, 9b.

[42] It is of interest to recall that Moses de León gave the title *Shushan Edut* to one of his Hebrew theosophic compositions. The explanation of the expression offered toward the beginning of the treatise is that *Shekhinah* attests to the unity of the divine, a unity that is construed explicitly in gender terms as a union of male and female enacted in the body of *Shekhinah.* See Gershom Scholem, "Two Treatise by Moses de León," *Qoveṣ al Yad* 8 (1976): 332 (Hebrew).

[43] In some zoharic passages, the sword is an androgynous symbol, and it is thus associated with *Yesod,* the phallic gradation, which comprises male and female. In other contexts, the image of the sword is associated more specifically with *Shekhinah* in the feminine guise of divine judgment, though even in this case the female is portrayed in decidedly masculine, even phallic, terms, the "sword that shall execute the vengeance of the covenant," *ḥerev noqemet neqam berit* (Lev. 26:25); alternatively expressed, the symbolic figuration of judgment is the feminine dimension of the male. See Tishby, *Wisdom of the Zohar,* 1365; Wolfson, *Circle in the Square,* 87 and 204 nn. 36–37.

[44] *Zohar* 1: 2216.

Botanical Progress, Horticultural Innovations and Cultural Changes

Botanical Progress, Horticultural Innovations and Cultural Changes

Botanical Progress, Horticultural Innovations and Cultural Changes

Botanical Progress, Horticultural Innovations and Cultural Changes

Links Between the Ottoman and the Western World on Floriculture and Gardening

Nurhan Atasoy

The first time the Ottomans encountered the Byzantine gardens, which were continuations of the earlier Roman garden traditions and whose development was manifestly influenced by Islamic gardens, was not in 1453 but, rather, earlier, when they first set foot in Rumelia and became the closest neighbors of the Byzantine world. Because the natural conditions remained the same, after the conquest Ottomans inherited the Byzantine garden tradition. They continued working on Byzantine land, gardens, farms, and vineyards by adding their own cultural knowledge, tastes, experience, and habits and by introducing innovations as circumstances dictated. A miniature from the 1600s shows an Ottoman vineyard and a figure trying to protect the grapes. Perhaps this vineyard was one of the former Byzantine vineyards that the Ottomans kept working on.[1]

Rather than adhering to a particular set of fixed rules, when creating their gardens, the Ottomans sought practical solutions that would suit topography, site, climate, and in general, the ambient conditions of the place where the garden would be. Instead of building watercourses, they created gardens where running water already existed. They embellished on what nature had already provided, planting trees and even putting in flowerbeds. Their additions and interventions, however, were not made according to some rigid plan.

Although Ottoman gardens lacked a strict formal organization, this does not mean that they were disorganized. They sought to preserve a setting that might have developed naturally. They would make the best possible use of the land when determining where a garden should be and the location and construction of the garden architecture. The location of terraces and embankments, the layout of watercourses, were never haphazard.

The Ottoman gardens did not conform to the plans and attitudes of Islamic gardens or gardens of other cultures. In order to understand why this was so, one only needs to look at the examples of the palace and privy gardens that they created in different places and how they conformed to their locations. The gardens of the Old Palace inside the Bursa citadel had rich water elements making use of the abundant supply of water. The gardens of the New Edirne Palace were built deliberately in between two rivers, whereas the gardens of the Topkapı Palace descended the mighty slopes of the Seraglio, surrounded by the sea on three sides. The Üsküdar Palace gardens had views looking from a modest height above the Marmara Sea; whereas the privy gardens like those at Küçüksu and Kağıthane running from the seashore entering the inlands and up the hillsides on either side of pleasant creeks on the Bosporus and the Golden Horn. When these are examined from the standpoint of their settings, one realizes that they were so utterly different from the other Islamic gardens in terms of their topography, prospect, and climate that one should not be surprised that the experiment with the Persian garden plan that was undertaken in the Karabali Garden on the European side of İstanbul does not really resemble any particular garden type at all.

Garden Pavilions

Many of the gardens were not designed as a setting for the impressive state ceremonies but, rather, as places where the sultans could spend a few enjoyable hours or even days in privacy. In the written sources, when the privy gardens in cities such as

İstanbul, Edirne, and Bursa are mentioned, it is usually in conjunction with the architectural elements that they contain. The most important of which were the pavilions, walls around the gardens, or, in some cases, supportive walls to create terrace gardens. The Ottoman garden pavilions exhibit an enormous amount of diversity in their architecture and they ranged in size from modest bowers to luxurious pavilions. An important element of a garden pavilion was the way it was located, so as to offer the best possible view of its surroundings. By far the most important feature of an Ottoman garden pavilion was that it should be open to its surroundings and blend into the garden as if it were a part of that nature.

The garden pavilion in Katibi Külliyatı,[2] a collection from the reign of Mehmed II (1451-1481) is just big enough for a single person, set in the middle of the scene and in an integral part of the garden. It stands on a stepped platform and its walls are covered with hexagonal wall tiles that are frequently seen during this period. Flanking the kiosk, there are walls decorated with plant motifs, which also forms a backdrop for the figures (Fig. 1). Also, in the Hamse of Nevai (1530),[3] there is a garden kiosk that has been constructed rather high so as to allow a better view of the garden. A small brook runs through this lovely garden, whose lawn is strewn with flowers in bloom (Fig. 2). In Şehinşehname (1592),[4] one of the

1. Raised garden kiosk from Mehmed II's reign (1451-81). Katibi Külliyati.

miniatures depicts a garden pavilion, this time in a Bosporus setting. The scene shows the sultan in his privy pavilion in the Kandilli Garden, receiving the news of his army's victory at the city of Revan (Yerevan). This small pavilion with a pyramidal roof, which is surrounded by cypress trees, is situated so as to provide an enjoyable view of the surroundings. The wall behind the pavilion encloses the garden property (Fig. 3).

Şehname-i Selim Han (completed in 1581), an illustrated account of the events of the reign of Selim II (1556-1574) and of the early years of the reign of his successor Murad III (1574-1595), shows Selim sitting in a garden pavilion at his palace in Silivri, where he has gone to take part in a great hunt (Fig. 4).[5] The archways of the kiosk, which we see with the Silivri citadel in the background, are open to the garden but there are red drapes that could be drawn for protection against the sun or bad weather. The kiosk has a portico and the sultan is shown seated on a stool, surrounded by members of his court. The fruit-filled dishes and a row of vases containing purple violets and white tulips and carnations with red centers are placed in front of him. The kiosk stands by the sea, from which it is separated by a red-painted latticework structure presumably designed so as to allow those inside to see out and restrain those outside from seeing through. Not all garden kiosks were elaborate or intended to be lived in for long periods of time. But even if it was just a simple bower that offered shelter only during daytime, every Ottoman garden was sure to have a structure whose purpose was to enhance the enjoyment of the garden (Fig. 5, a tray of flowers at the feet of the Grand Vizier, ca. 1784).

Cypresses and the Fountains

The two invariable elements of the Ottoman gardens were cypress trees and fountains consisting of a pool with one or more jets. In 1582, a festival lasting fifty-two days and fifty-two nights took place in Istanbul to celebrate the circumcision of Prince Mehmed, the son of Murad III. The festivities began with a parade proceeded by men transporting a large number of artificial trees (*nahil*) that were fashioned from iron and decorated with colored paper and objects made from wax.[6]

During the parade, one of the guildsmen exhibits a huge tulip made from paper.[7] Another was a "float" of the glassmakers' guild, which contained a working atelier. The float was surrounded by marching apprentices carrying examples of the craftsmen's work, the majority of which were flower vases.[8]

The members of the florists' guild marched

2. Garden kiosk elevated to allow a better view. Hamse of Nevai (1530).

carrying huge circular wooden trays with vases filled with different kinds of gorgeous flowers, turning the parade into a paradise. Some bore blossoming branches, whereas others had bowls filled with fruits like pomegranates and apples.[9]

The fruit-sellers then marched again, this time having been decorated with flowers themselves. An important thing about these miniatures is that they show us that, at this time, the members of this guild grew and sold both flowers and fruits. All of the fruit-sellers, for example, are depicted marching with vases full of flowers, or carrying individual flowers.[10]

The gardeners marched carrying their gardening tools set on their shoulders. They also brought a float, which consisted of a small rectangular garden on a wheeled cart. The "garden" is interesting because it is a good model of an idealized Ottoman garden at the time. On each corner there is a single cypress tree rising proudly, its summit slightly bent. The ground was planted with grass and was laid out with paths. In the very center was a pool with jets (Fig. 6).[11] Another group of gardeners showed up with a different model of a garden when it was their turn to march. Octagonal in shape, this one was enclosed by a red fence. Again, there were four cypresses and a pool with jets. Before the pool there were placed a row of vases filled with flowers.[12]

LINKS BETWEEN THE OTTOMAN AND THE WESTERN WORLD ON FLORICULTURE AND GARDENING

In another parade in 1720 that was held on the occasion of the festivities surrounding the circumcision of the sons of Ahmed III, marchers carried models of gardens that had been fashioned from sugar by the palace confectioners. These models reflect the garden landscaping practices of the day. The four models, which are huge and required about twenty burly arsenal-workers to carry each of them, had double-story garden kiosks that support balconies, shuttered windows, and a domed or pyramidal roof. One can also make out pools with jets and little boats, trees heavy with fruit with parterres of tulips and other flowers. Each garden was enclosed by a red wall (Fig. 7).[13] In the miniature showing the other two gardens, the "kiosk" of the garden in the upper part of the scene resembles a crenellated fortress and what looks like a panoramic balcony. Both gardens contain fruit-laden trees as well as cypresses, the latter of which are entwined by grapevines loaded with clusters of ripe grapes. The gardens are planted with tulips (Fig. 8).[14]

For millennia, the cypress tree (*Cupressus sempervirens*) has been much beloved throughout the Mediterranean region. Cypress was an important element of Byzantine gardens. Among the Ottomans, it was particularly popular. The example of the importance they attached to this tree can be seen in the original landscaping of the Arsenal grounds. After the conquest of İstanbul, Mehmed II ordered that a garden be created at the spot where he had made his first encampment during the siege. By his order, twelve thousand cypresses were planted here in a checkerboard pattern. The sultan himself planted seven of them. A

3. Süleyman the Magnificent sitting on a stool set on the dais in the garden of the Üsküdar Palace. Hünermane II.

miniature of this garden confirms the presence of cypresses.[15] In the miniatures depicting the spectacle on the Golden Horn in the Surname of Ahmed III, thick groves of these trees are to be seen in the Tersane gardens and along the shore.

The second courtyard of Topkapı Palace, which served as the setting for the court that housed splendid imperial ceremonies, parades, and gatherings, was practically a garden on its own with cypresses. According to Evliya Çelebi, some twenty thousand cypresses, planes, and other trees were planted around the various structures erected at the site palace between 1458 and 1467 during the reign of Mehmed II.[16] A document dated 1148 A.H. (1735), concerning trees to be planted in the palace gardens, orders the procurement of "vigorous and comely" saplings of plane, ash, linden, elm, mahaleb cherry, nettle, oak, bay, redbud, and wild pear and similar trees in quantities of four thousand from İzmit, Karamürsel, and Yalova.[17]

Flowers in the Gardens

Although flowers were cultivated by the Ottomans before the time of Süleyman I, the artistic renaissance that took place during his reign made itself felt in horticulture as well (Fig. 9). The flowers that they saw in gardens ranked high among the things that attracted the attention of foreigners traveling to the Ottoman empire, especially those visiting the city of İstanbul. Nearly all of them commented on the abundance of flowers in their writings and emphasized that this was a feature of the Ottoman gardens. Ottaviano Bon illustrates palace gardens planted with trees and flowers, as well as fruits, vegetables, and all kinds of different aromatic plants and herbs:

> The extensive palace gardens are indiscriminately planted with trees such as cypresses, boxwoods, bays, and myrtles that form paths of their own. These trees are never pruned. Along the paths one finds flowers together with scattered beds of cabbages, cucumbers, spinach, and melons as well as every sort of herb and vegetable in season.

4. *Köşk* in the Silivri Palace, Şehname-i Selim Han (1581).

Again, Ottaviano Bon describes the atmosphere and the necessary fundamental elements of a Turkish garden as such: "A few shady trees, a view, a rosebud, and the sound of a nightingale will transform any garden into a Turkish garden."[18]

Father Stephan Gerlach, a Lutheran chaplain accompanying the Habsburg delegation to the Sublime Porte describes the terraced Seraglio in about 1576:[19]

> On the first and second terraces of the Seraglio there are many woods, vineyards, and orchards of every kind of fruit tree. Also in this area are kitchen gardens in which there exists every kind of fruit tree and every sort of flower and herb.

After noting that there were sixty-one gardens within the palace in which five thousand gardeners were employed in 1600, of which 947 were palace officials, Petis de La Croix reports that this number increased to two thousand by 1677. He continues on narrating the decoration of the gardens during the spring festivals:[20]

The holiday celebrating the arrival of spring would begin when the tulips blossomed and at this time, the gardens would be decorated with lanterns and other finery.

According to the seventeenth-century historian and traveler Katip Çelebi, the tulips of Asia Minor were preferred for their color and specific form (Fig. 10):[21]

The tulips of Asia Minor are of a rare beauty in terms of their colors and petals. This is why gardeners come to Manisa in the spring and buy up the most beautiful ones to plant.

Ahmed III loved country life and he had an uncommon passion for flowers, for roses, carnations, lilacs, and jasmines, but most of all for tulips (Fig. 11). His fondness played a key role in the Ottoman capital as well as in his reign, which was for that reason called the "Tulip Period." The tulip was first introduced to Europe from Turkey in the sixteenth century. By the seventeenth century, the fad for tulips had become known as "tulipomania," which did not completely die out until the early nineteenth century. In the Ottoman empire, the interest was mainly in rare varieties, which could fetch astronomical prices. In Holland, by contrast, thousands of bulbs changed hands at inflated prices ramped up by speculators.

In the winter of 1554, Ogier Ghiselin de Busbecq, the Austrian imperial ambassador, reported seeing villagers transporting bunches of flower bulbs, such as jonquils, hyacinths, and tulips, while he was on his way from Edirne to İstanbul. On his return, he took some tulip bulbs with him as the flower was unknown in Central and Western Europe at the time. On each of his subsequent visits, Busbecq returned home with crates full of tulip bulbs as well as the seeds of rare plants. In 1562, a merchant from Amsterdam shipped the first recorded cargo of tulip bulbs to Europe from İstanbul.[22]

Another traveler, Jean-Claude Flachat,[23] who was in Turkey between 1740 and 1758, relates how tulips were planted with great care in the palace gardens and that no more than one variety was planted in a single bed. He also notes that rare bulbs were registered in a big inventory book, which also recorded any changes the flower might undergo and any variations. According to the author, the Turks had developed standards of beauty for tulips; that is, a perfect tulip should have a medium-sized calyx, a long stem, rather long and sharp petals, serrated leaves, and a magnificent color (Fig. 12). He adds that, at the time, it was the fashion in İstanbul to set a particularly rare example of such a tulip in a long-necked Venetian glass vase, to bestow poetic names on well-known varieties, and the poets were encouraged to compose lyrical poems exalting the beauties of the flower. Finally, Flachat expresses the opinion that the Turks regarded human life as less important than a thoroughbred horse or an elegant tulip.

According to another author, selling tulip bulbs anywhere but in Istanbul and exporting too many of them were offenses punishable by exile. He also relates the spring fetes during the reign of Ahmed III, which were held on two consecutive evenings during the month of April, preferably when there was a full moon. For these fetes, vases of tulips were arranged on shelves set up in the garden alternating with tiny lamps of colored glass and interspersed with glass globes filled with liquids of different colors to create a rainbow-like effect. Guests even dressed up in garments harmonizing with these colors. On the highest shelves, there were cages with canaries and rare songbirds. Colored candies were hidden among the flowers. Along with these events, the court's musicians played and dancers danced. At the end of the fete, the chief white eunuch distributed gifts, precious robes, jewels, and gold coins.[24]

Baron de Tott,[25] an adviser on military matters who accompanied

5. A tray of flowers at the feet of the grand vizier. (Choiseul-Gouffier, Fig. 98)

the French ambassador De Vergennes to İstanbul in 1755, says that tulips were very popular accompaniments to the evening parties held in the palace gardens and that a pleasurable atmosphere was created by illuminating them.

The astonishment expressed by foreigners at the abundance of flowers they saw both in the palaces around the city on their visits to İstanbul is evidence of how far European gardens lagged behind Ottoman ones considering the flowers at the time. The huge number of flower seedlings that were procured for the Ottoman palace gardens is also an indication of a great love for flowers.

As for the purchase of these flowers, a document from the reign of Süleyman I, dated 1526-1527, mentions purchase of "Kaffa tulips" as well as a large number of other unspecified flower species for the privy gardens.[26] Another example is a document dated 1564, which mentions great many expenditures, including "boxes of jasmines" and a "carnation courtyard."[27]

Another sixteenth-century register[28] lists the names of gardens that supplied roses for the palace confectionery among which figure Hasbahçe, Üsküdar, Göksu, Çubuklu, Beykoz, Paşabahçe, Haydarpaşa, and Fenerbahçe. The register lists the quantities of roses, while also providing us considerable information about the places where roses were being raised during Süleyman's reign. A document from the second half of the eighteenth century refers to the rehabilitation of one of the palace gardens,

6. Portable garden brought to the Meydan for the circumcision festival in 1582.

where 220 tubs of lindens, everblooming carnations, everblooming roses, "Frankish roses," yellow roses, "Frankish vines," jasmines, Virgin's bowers, and Double sweet Williams were purchased to be planted.[29]

Further evidence of the enormous quantities of flowers that were procured for Ottoman palace gardens is to be seen in a 1576 order for hyacinth bulbs from Aleppo and in a 1592 order for four hundred quintals (over twenty-two metric tons!) of red roses and another three hundred quintals of white roses from Edirne "as usual."[30] In the latter year an order was also placed for a hundred thousand hyacinth bulbs, fifty thousand white and fifty thousand blue, for use in the palace gardens. The order stipulates that bulbs are to be obtained from the "mountains and meadows" in Maraş.[31]

According to the documentary evidence, the gardens of the Topkapı Palace continued to be cultivated even after the sultan and his court relocated to more modern and comfortable accommodations on the Bosphorus in the early nineteenth century. From an expense register dated 1868,[32] the following plants were ordered for the palace grounds: azaleas, "cologne

7. Garden of candy, for the circumcision feast of the sun of Ahmed III in 1720 in the Surname-i Vehbi.

8. Garden of candy, for the circumcision feast of the sun of Ahmed III in 1720 in the Surname-i Vehbi.

flowers," and hyacinth bulbs; and pansies, rosebushes, carnation seedlings, and lindens "to be planted in pots." The order also included garden shears, flowerpots, and fertilizer.

In palace gardens, flowers were usually planted in beds. The bases of trees were often turned into flowerbeds as well. There are many examples of pictures showing this, although the practice seems to have become more prevalent in the eighteenth century. In one of the sunbursts on the lacquered covers of a copy of a history known as Tercüman-ı Destur (1728) that was illuminated by Abdullah Buhari (Fig. 13),[33] there is a garden scene with red flowers resembling tulips arranged in rectangular beds. The garden's walled enclosure is also noteworthy. There is a resemblance between this scene and the flowerbeds in the courtyard of Osman III in the harem quarter of the Topkapı Palace.

The Ottoman gardens were expected to be functional as well as beautiful. One of the most important features of the Ottoman gardens was that they were planted not only with flowers but also with fruits and vegetables.

9. From the pages of the anthology that Süleyman I wrote under the pseudonym "Muhibbi," illuminated by Karamemi.

Vegetables and fruits that were grown in the privy gardens and not needed for the palace were sold to the public. So, too, it appears that there were surplus flowers which were sold to the shops in town. A document from 1594 complains that there were more than a hundred shops engaged in this trade at the time and advises that this number should be reduced. The income secured from the sale of flowers, fruits, and vegetables raised in the privy gardens was directly lent to the privy funds, as documented in accounting registries.[34]

Within the eighteenth century, we begin to notice the European influence on the Ottoman garden landscapes. A miniature in the Atayi's Hamse (1721) shows two Muslim Ottomans taking part in a feast with a group of Christian nobles (Fig. 14). They are eating seated around a tray set on a platform resembling a fenced-in kiosk beneath a domed bower formed from sticks. In the background there is a pool and fountain between the bower and the garden fence. The upper parts of the fences surrounding the gardens through which a road passes further behind resemble a row of vases. Behind the garden fences are hills and a tree. Along the road is a row of cypress trees, a common feature of Ottoman gardens.[35]

10. Tulip. Abdullah Buhari. Şüküfename.

Ottoman Gardens and Europe

As briefly mentioned, the first wave of exotic plants to enter Europe in relatively

modern times corresponds to the years following the publication of Pierre Belon's account of his travels in the Ottoman Empire in 1546-1548. The French traveler observed the tulips, present in every Turkish garden, described them as a kind of lily, and noted that they were very different than those in his country.[36] A few years later, Ogier Ghiselin de Busbecq came across an illustrated copy of *Materia Medica*, a book that incorporated much ancient knowledge about medicinal plants, which he brought back with him to Vienna. With him, he also brought the first known specimens of tulip

11. Queenmother's bedroom in the Harem of the Topkapi Sarayi.

12. Tulip. Abdullah Buhari. Murakka (Album).

bulbs to Europe. It was not long before merchants who were encouraged by the demand for these flowers imported a flood tide of bulbs into Vienna, Antwerp, Paris, and London. Long before these bulbs made their appearance on European markets, however, they had arrived in the Middle East from China and elsewhere in Asia. In their new homes, the plants were quickly acclimatized and new varieties were created through grafting and hybridization. Horticulture and botany were quite advanced in the Islamic world at the time, not least because works on these subjects by ancient authors had survived by having been translated into Arabic and thus were available. Europeans' first major contact with this culture began in the eleventh century and the Crusades.

Georgina Masson relates the subject that William P. Stern had underlined in a conference held at the Royal Horticultural Society in 1965,[37] in her article on the gardens of flower collectors:[38]

13. Cover of Terüman-i Destur.

14. Meal in the Garden. Hamse-i Atayi. From Atasoy's *A Garden for the Sultan*.

The first substantial entry of flowers to western Europe was from the Turkish empire, which in that period encompassed the majority of Western Europe, in the 16th c. The division between western and Eastern Europe, due to the problems developed between the Roman and Greek churches in the 10th c., had occurred long before the Ottoman conquest of İstanbul in 1453. From then on the Turks expanded their territories into the Balkans, to the north and west; and even went as far west as to unsuccessfully siege Vienna twice. During the 16th and 17th centuries Western Europe was apprehensive of this expanding empire in the east.

Among Western Europeans who recognized the botanic riches of the Turkish Empire was Ghiselin de Busbecq, the ambassador of the Holy Roman Empire to İstanbul from 1554 until 1562. In awe of the variety of flowers he saw in Turkish gardens, Busbecq sent his friend in Vienna, Clusius (Charles de l'Ecluse), bulbs of tulips, hyacinths, anemones, and crown imperials. Clusius then moved to Leiden and brought along his new bulbs. He had already started sharing these bulbs with his large circle of colleagues in various parts of Europe.

Strangely, some of these plants actually had originated in Eastern Europe; yet it was not easy to determine this for early botanists. Western Europeans could begin to study these lands only after the decline of the Ottoman Empire in the nineteenth century.

15. Meryem *(cyclamen)*

16. Yellow tulip and roaming violet

17. Yellow narcissus *(daffodil)* and Algerian violet *(vinca)*

18. Single purple carnation

19. Double buttercup *(ranununculus asiaticus)*

20. Poppy. Murakka (Album)

Thus, after the 1560s, the flowers listed below were to be found in Europe as spread from the lands of the Ottoman Empire:

Crocus
Colchicum
Leucojum
Erythronium
Ornithogalum
Cyclamen (Fig. 15)
Allium
Hyacinth
Lily
Fritillary
Ranunculis
Tulips (Fig. 16)

It is possible to say that the roots of the development of a brand-new floral culture in Europe lie in the awe of traveling Europeans in face of the Ottoman love of flowers and Ottoman flower gardens.

Among the rather substantial numbers of flowers, trees and other plants that are definitely known[39] to have made their way into Europe either directly from Turkey or by way of Turkey before the 1600s are:

Crown imperial
Lilium candidum var.
Cernuum
Chalcedonicum
Muscari moschatum
Varieties of narcissus (Fig. 17)
Variegated anemones
Carnations (Fig. 18)
Iris pallida
I. susiana
Love-in-a-mist
Ranunculus asiaticus (Fig. 19)
Shrub cherry
Laurel
Syringa (Philadelphus coronarius)
Oleaster (Elaeagnus angustifolia)

Subsequent arrivals included the following:

Oriental plane tree
Black mulberry
Walnut (indirectly)
Hollyhock
White jasmine
Scarlet lychnis
"Female" peony
Opium poppy (Fig. 20)

They were later joined by:

21. Lacquerwork binding

Horse chestnut

Cloth-of-gold crocus

Galantus plicatus

Byzantine gladiolus

Day lily

Purple primrose

Sweet sultan

The cedar of Lebanon made its way into Europe soon after 1650. Sir George Wheler sent the first specimen of Hypericum calycinum in 1676. Weeping willow arrived in 1692.

During the reign of Ahmed III, Turkey became a horticultural center for the western world. Between 1703 and 1716, the great botanist William Sherard was the British counselor in İzmir, which is not far from Manisa (famous for its tulips), and he maintained a house and garden at Seydiköy, south of the city on the road to ancient Ephesus. That garden must have played a key role in enabling many species of plant to reach England and other English botanists through Sherard's brother. England became acquainted with the Turkish oak in 1735. In 1793, the plants listed here made their way to England:

Rhododendron luteum (Azalea pontica)

R. ponticum.

Much later arrivals included:

Galanthus elwesii in 1874

Chionodoxa luciliae in 1877

Crocus ancyrensis in 1879

Dwarf irises in 1887

I. histrio aintabensis in 1934

European gardens were deeply influenced by this influx of plants and especially by Ottoman gardens and their flowers. In a lecture delivered in 1965, the botanist William Stern said:[40]

In the period following the fall of Constantinople, Europe began taking a very close interest in the

Ottomans and both Catholic and Orthodox powers saw it as in their interests to maintain ambassadors and

representatives in Constantinople. These individuals were astonished by the wealth of West Asian flora that

they saw growing in Turkish gardens and they immediately set about introducing these plants into their own

countries. The result was a revolution in European gardens that began after the second half of the 16th c. (Fig. 21)

Working on the same subject several years later, Georgina Masson referred to Dr. Stern's paper, which was published in the journal of the conference's proceedings, while also revealing another important aspect of Ottoman gardens that went even beyond this influence:[41]

Dr. Stern pointed to the revolutionary effect upon western European gardens of the Turkish capture of

Constantinople and the fear of the Turkish menace that resulted in European powers sending ambassadors to

Constantinople. They saw Turkish gardens filled with flowers, many of which came from western Asia, and

Dr. Stern described the result as follows: "Never before or since has there been such a sudden astonishing

influx of colorful strange plants into Europeans' gardens, as when in the second half of the sixteenth century,

importations of unpromising onion-like bulbs and knobbly tubers from Constantinople brought forth tulips,

crown imperials, irises, hyacinths, anemones, turban ranunculi, narcissi, and lilies. They came at a time when

the painting of flowers and the publication of illustrated herbals in Germany and the Low Countries ensured

their portrayal.

Masson continues:

Significantly, Dr. Stern did not mention here the part played by Italy, and indeed her role seems to

have been sadly forgotten. But it lives, perhaps more vividly than anywhere else, in Ulisse Aldrovandi's observation books, still preserved in the Biblioteca comunale delli Archiginnasio in Bologna. The first tome in the series is undated, but the second was begun on the "Calends of April 1569," and the series continues up to Tome XXXI, begun on June 26, 1602. The last entry in it was made on the Botanical Garden at Bologna, where he died three years later. In these observation books Aldrovandi kept brief notes not only of the letters, lists of seeds and plants, and other objects of natural interest he himself received from his vast circle of correspondents, but also ones of outstanding interest received or sent by them. Thus in the first tome there is a note on the samples sent by Willem Quackelbeen to Pietro Andrea Mattioli. Quackelbeen was a doctor who accompanied the imperial ambassador, Ogier Ghiselin de Busbecq to Constantinople in 1554, where Quackelbeen died tragically of plague in 1561.

In Masson's book, there is yet another reference to the plants that Italian collectors obtained from Ottoman sources: one of the items in a list of flower bulbs for the Duke of Sermoneta, dated 1625, is "white narcissi from Constantinople."[42]

An interesting feature of the reign of Ahmed III is that it marked the beginning of a bizarre period, where the Ottomans, who had introduced the tulip to Europe in the first place, began importing the flowers from Holland and the European gardens, which once had been inspired by Turkey and had been developed with plants from there. Those gardens more and more served as models for the Ottoman gardens in the eighteenth century and continued to do so throughout the nineteenth century.

The influence of other world gardens on Ottoman gardening and horticulture developed through diplomatic relations; by economic exchange; because of the growing interest for exotic plants and animals; and because of the different nationalities of the gardeners who were working in the imperial and royal gardens, vineyards, and meadows of Istanbul.

In the earlier periods, those slaves appointed as gardeners working under imperial patronage—carrying their different garden traditions, horticultural expertise, gardening skills, and taste to the imperial capital—were usually from the Aegean Islands and the Balkans. In the later periods, gardeners invited from Europe influenced the development of Ottoman garden tradition in the nineteenth and early twentieth centuries. Archival documents lists the salaries of each gardener who was working in the imperial gardens, also mentioning their nationality.[43]

In the early eighteenth century, during the reign of Sultan Ahmed III, Ottoman ambassadors who traveled to France brought

22. Wall painting of a garden in the champer bo Mihrişah Sultan, Selim III's mother, in the harem of Topkapı Sarayı

back plans of the French gardens and palaces that they visited. These plans had been influential in the introduction of European garden tradition in the Ottoman world.[44] Ahmed's ambassador Yirmisekiz Mehmed Çelebi had written long accounts about the French gardens. During this period, tulip bulbs that had once been carried abroad to Holland were imported back to the Ottoman capital. Orange trees were sent from Europe as an imperial gift to the sultan as well.

The palaces and gardens that were built during the nineteenth century were influenced by European gardens (Fig. 22). These gardens not only introduced particular styles but also directly transferred new garden elements such as winter gardens. At the time, European gardeners also were commissioned for the design of the imperial or elite gardens. Documents dated 1861, 1877, and 1912 report that the imperial gardeners; Monsieur Sister, Monsieur Istafel, Hafiz Yakup Efendi, Anastas Nikolanin, Aziz b. Osman, and Osman b. Hüseyin, some of whom were foreigners, were given awards for their excellence in gardening.[45]

Another curiosity that can be observed during this period was the interest in exotic plants and animals. Breeding rare and exotic animals in the imperial gardens was an old Ottoman tradition. However, during the nineteenth and the early twentieth centuries, this tradition was promoted again on the envy for European gardens. Flowers, plants, and fruits were imported from Europe as a result of the fascination with the European style. During the eighteenth century, tulip bulbs were imported from Holland. Later in the nineteenth and early twentieth centuries, apple, almond, pear, and quince tress were imported from Europe.[46]

A document dated 1884 reports on caretaking of birds, pheasants, doves, goats, and other animals in the gardens of Ayazağa and Yıldız Palace.[47] Another document dated 1889 mentions doves imported from India.[48] An undated document gives information on several exotic animals raised in the garden of Yıldız Palace: gazelles and goats from Syria, birds from India. This same document also mentions coffee plants imported from Syria and another tree sapling blooming in violet flowers. A different document dated 1893 reports ginger being imported together with coffee plants.[50]

A document dated 1870 describes the sapling of a tree from Sudan, Africa; which grows so big that it is able to accommodate a hundred camels under its shade.[51]

In 1904, 76 fruit trees, more than 110 saplings of flowers, and a few packets of flower seeds were imported from Bombay, together with jars of marmalade and exotic birds.[52] Saplings from the different countries of the Far East and rare flowers from America were imported on the Sultan's command.[53]

NOTES

Abbreviations:

> BOA Başbakanlik Osmanli Arşivi (Ottoman Archives)
> IUK İstanbul Universitesi (İstanbul University Library)
> TSM Topkapı Sarayı Kütüphanesi (Topkapı Palace Library)

1. TSM, H1711, 9r.

2. TSM, R989, 93r.

3. TSM, H802, 134r.

4. TSM, B200, 98v.

5. TSM, A3595, 13r.

6. TSM, H1344, 1r.

7. TSM, H1344, 200r.

8. TSM, H1344, 32v.

9. TSM, H1344, 34v–35r.

10. TSM, H1344, 385v-386r.

11. TSM, H1344, 349r.

12. TSM, H1344, 196r.

13. TSM, A3593, 162v.

14. TSM, A3593, 161v.

15. İUK, T5461, 25v.

16. Evliya Çelebi, *Evliya Çelebi Seyahatnamesi I.,* Edited by Orhan Şaik Gökyay (İstanbul: Yapi Kredi Yayinları, 1996), 115; Orhan Şaik Gökyay, "Bahçeler," *Topkapı Sarayı Müzesi:Yıllık* 4 (1990), 12–13.

17. Refik Ahmet, *Onbirinci Asr-ı Hicri`de İstanbul Hayati (1000–200)* (İstanbul: Enderun, 1988), 133, n.161.

18. R. Withers, *Büyük Efendi`nin Sarayı.* Translated by Cahit Kaya (İstanbul: Pera Turizm e Ticaret A.S., 1996), 595–596.

19. Stephan Gerlach, *Stephan Gerlach des geltern Tage-Buch, der von Zween onvurdigsten romischen kaysern Maximiliano und Rudolpho* (Frankfurt: Johann David Zunner, 1674).

20. Frédéric Lacroix, *Guide du Voyageur a Constantinople et dans ses environs* (Paris, 1839); Barnette Miller, *Beyond the Sublime Port* (New Haven, Conn.: Yale University Press, 1931), 150–156.

21. Haji Qalfa, "Haji Qalfa or Hacy Halife Mustafa," *Cronologia Historia,* Translated by Rinaldo Carli (Venice: 1697).

22. Ogier Busbecq, *The Letters of Ogier Ghiselin de Busbecq, Imperial Ambassador at Constantinople, 1554–1562,* Translated by Seymour Forster (Oxford: Clarendon Press, 1968), 24–25.

23. Jean-Claude Flachat, *Observations sur le commerce et sur les arts de l'Europe, de l'Asie, de l'Afrique, et meme des Indes Orientales (1740–1758)* (Lyon: Jacquenaud père et Rusand, 1766).

24. Barnette Miller, *Beyond the Sublime Port* (New Haven, Conn.: Yale University Press, 1931), 122–126.

25. Baron François de Tott, *Mémoires du Baron de Tott sur les Turcs et les Tartares* (Amsterdam, 1784), 89.

26. BOA Kepeci 7097, 37.

27. TSM AD 5120.

28. TSM AD 10605.

29. TSM AE 105/10.

30. Ahmet Refik, *Onuncu Asr-ı Hicri`de İstanbul Hayati* (İstanbul: Enderun, 1988), 6 and 9.

31. Ahmet Refik, *Onbirinci Asr-ı Hicri`de İstanbul Hayati (1000–1100/1592–1688)* (İstanbul: Enderun, 1988), 3.

32. TSM AE 3268; TSM AD 9719.

33. TSM EH 1380.

34. Ahmet Refik, *Onbirinci Asr-ı Hicri`de İstanbul Hayati (1000–1100/1592–1688)* (İstanbul: Enderun, 1988), 18 and 19.

35. Baltimore Walters Art Gallery, W.666, 138r; Günsel Renda, "An Illustrated 18th Century Hamse in the Walters Art Gallery," *The Journal of the Walters Art Gallery* 39 (1981), 22-24.

36. Blunt 1950, 8.

37. William P. Stern, "Masters Memorial Lecture," *Journal of the Royal Horticultural Society* (August 1965), 326.

38. Georgina Masson, "Italian Flower Collectors' Gardens in the Seventeenth Century Italy," *The Italian Garden* (1972), 64–65.

39. John Harvey, "Turkey as a source of garden plants," *Garden History Society* IV/3 (1976), 21–22.

40. William P. Stern, "Masters Memorial Lecture," *Journal of the Royal Horticultural Society* (August 1965), 326.

41. Georgina Masson, "Italian Flower Collectors' Gardens in the Seventeenth Century Italy," *The Italian Garden* (1972), 37–61; 64–65. The page number for the reference in this article is mistakenly given as "322" rather than the correct "326."

42. Georgina Masson, "Italian Flower Collectors' Gardens in the Seventeenth Century Italy," *The Italian Garden.* (1972) 37-61; caption for ill. 6.

[43] Milli Sarayı Arşivi; Milli Saraylar Arşivi Defter 2949.

[44] Gül İrepoğlu, "Topkapı Sarayı Müzesi Kütüphanesindeki Batılı kaynaklar Üzerine Düşünceler" *Topkapı Sarayı Müzesi-Yıllık* 1 (1986.), 56–72; 174–197.

[45] BOA A. DVN. MHM 32/28; A.DVN.164–54;210-56;DH.MTV53/8

[46] BOA Y. PRK. HH 39/72

[47] BOA Y.PRK. TKM 7-32

[48] BOA Y.PRK.ESA 9/51

[49] BOA Y.PRK. ASK 259/71

[50] BOA Y.PRK.ZB 12/56; BOA Y.PRK. HH 27/23

[51] BOA Y.PRK.UM. 9–20

[52] BOA Y. PRK ESA 1/61

[53] BOA Y.PRK ESA 32/32; BOA Y.PRK ESA 9/17

BIBLIOGRAPHY

Refik, Ahmet. *Onuncu Asr-ı Hicri`de İstanbul Hayati*. İstanbul: Enderun, 1988.

Refik, Ahmet. *Onbirinci Asr-ı Hicri`de İstanbul Hayati (1000–1100/1592–1688)*. İstanbul: Enderun, 1988.

Refik, Ahmet. *Onbirinci Asr-ı Hicri`de İstanbul Hayati (1000–200)*. İstanbul: Enderun, 1988.

Arik, Ruchan. *Kubad Abad, Selçuklu Saray ve Çinileri*. İstanbul: Türkiye ıs Bankası, 2000

Atasoy, Nurhan. *A Garden for the Sultan, Gardens and Flowers in the Ottoman Culture*. İstanbul: Aygaz A.C., 2002.

Atasoy, Nurhan. "The Tower of Justice of the Topkapı Palace." *Art Turc-Turkish Art: 10th International Congress of Turkish Art, Geneva 17–23 September 1995*. Geneva: Fondation Mac van Berchem (1999): 93–101.

Busbecq, Ogier. *The Letters of Ogier Ghiselin de Busbecq, Imperial Ambassador at Constantinople, 1554–1562*. Trans. Seymour Forster. Oxford: Clarendon Press, 1968.

Frédéric, Lacroix. *Guide du Voyageur a Constantinople et dans ses environs*. Paris, 1839.

Çelebi, Evliya. *Evliya Çelebi Seyahatnamesi I*. Edited by Orhan Şaik Gökyay. İstanbul: Yapı Kredi Yayınları, 1996.

Flachat, Jean-Claude. *Observations sur le commerce et sur les arts de l'Europe, de l'Asie, de l'Afrique, et meme des Indes Orientales (1740–1758)*. Lyon, Jacquenaud père et Rusand, 1766.

Gerlach, Stephan. *Stephan Gerlach des geltern Tage-Buch, der von Zween oruurdigsten romischen kaysern Maximiliano und Rudolpho*. Frankfurt: Johann David Zunner, 1674.

Gökyay, Orhan Şaik. "Bahçeler." *Topkapı Sarayı Müzesi:Yıllık 4*. İstanbul, 1990.

Qalfa, Haji 1697. "Haji Qalfa or Hacy Halife Mustafa." *Cronologia Historia*. Translated by Rinaldo Carli. Venice: 1697.

Harvey, John. "Turkey as a source of garden plants." *Garden History Society* IV/3 1976: 21–42.

İrepoğlu, Gül. "Topkapı Sarayı Müzesi Kütüphanesindeki Batılı kaynaklar Üzerine Düşünceler" *Topkapı Sarayı Müzesi-Yıllık* 1. İstanbul: Topkapı Sarayı Derneği, 1986, 56–72; 174–197.

Masson, Georgina. "Italian Flower Collectors' Gardens in the Seventeenth Century Italy." *The Italian Garden*. Washington, D.C.: Dumbarton Oaks, 1972.

Miller, Barnette. *Beyond the Sublime Port*. New Haven, Conn.: Yale University Press, 1931.

Necipoğlu, Gülru. *Architecture, Ceremonial, and Power; The Topkapi Palace in the Fifteenth and Sixteenth Centuries*. Cambridge, Mass.: MIT Press, 1991.

Renda, Günsel. "An Illustrated 18th Century Hamse in the Walters Art Gallery." *The Journal of the Walters Art Gallery 39*. Baltimore: Walters Art Gallery, 1981, 15–32.

Stern, William P. "Masters Memorial Lecture." *Journal of the Royal Horticultural Society* (August 1965), 326.

de Tott, Baron François. *Mémoires du Baron de Tott sur les Turcs et les Tartares*. Amsterdam: 1784.

Withers, R. *Büyük Efendi`nin Sarayi*. Translated by Cahit Kaya. İstanbul: Pera Turizm e Ticaret A.S., 1996.

Botanical Progress, Horticultural Innovations and Cultural Changes

Botanical Progress, Horticultural Innovations and Cultural Changes

Botanical Progress, Horticultural Innovations and Cultural Changes

Botanical Progress, Horticultural Innovations and Cultural Changes

Precious Beauty: The Aesthetic and Economic Value of Aztec Gardens

Susan Toby Evans

The subject of the monumental parks of the Aztec empire may seem unfamiliar to most historians of landscape design—and to most archaeologists specializing in ancient Mexico—yet many people will have heard of Mexico City's Chapultepec Park, a great modern urban green space that was first established as an Aztec[1] dynastic pleasure park. Chapultepec's evolution as a monumental park was part of Mexican cultural history—it has been a valued, even revered piece of real estate since before the time of the Aztecs, and this importance continues into the present.[2] It was probably not the first elaborate garden in ancient Mesoamerica, but it initiated a tradition of monumental park building on a scale that almost certainly had no precedent there. We have good documentation of this tradition[3] and it provides landscape history scholarship with a unique view of how monumental gardens evolved along with the trajectory of general cultural development in one of the great empires of the ancient world, that of the Aztecs.

In this chapter, I interpret the coevolution of the Aztec empire and its monumental parks from the theoretical perspectives of cultural evolution and cultural ecology. These explanatory frameworks help us to understand how the design and elaboration of great gardens are themselves a major diagnostic of civilization; garden design takes its place along with such well-known other features of mature state-level societies as cities, palaces, and monumental civic and ceremonial architecture.

This study begins with a brief review of the concepts of cultural evolution and cultural ecology and the engine that drives the evolutionary process, demographic growth. Then we will examine conditions leading to the rise of the world's six great earliest civilizations, with emphasis on Mesoamerica. With this background, we can consider, in detail, the rise of the Aztec empire and the history of Aztec monumental parks. This will demonstrate that as the population of the Aztec core area rose dramatically in the centuries before the Spanish conquest in A.D. 1521, the ensuing competition among ruling dynasties for resources spurred the expansion of the Aztec empire. Furthermore, a series of environmental crises precipitated episodes of empire expansion and thus led to the development of new parks, and the wealth generated by this process permitted the growth of the system of monumental parks designed to glorify the empire's rulers.

Cultural Evolution and Cultural Ecology

The changes in Aztec society took place over the few centuries before the European intrusion into the New World, which culminated, in Mexico, with the Spanish conquest of the Aztec empire in 1521. The historical processes aptly demonstrate the interplay of several important principles: cultural evolution and cultural ecology.

The concept of cultural evolution conjoins *culture*, the human adaptive strategy, with *evolution*, meaning descent with modification, powered by selective pressures on a population that is consistently pressing—or exceeding—the limits of its resource base. The principles of biological evolution[4] are so generally familiar that, of late, the "Darwinian" vogue has even hit such remote academic byways as literary criticism. However, a crucial point distinguishes humans from other organisms as to our evolutionary role: when faced with changing conditions and selective pressures, we respond with cultural solutions, innovations, rather than having to wait for a random genetic mutation to give rise to a better-adapted organism. Cultural evolution is the dynamic outcome of the operation of these principles with *cultural ecology*, which is the commonsense notion that any culture and its environmental setting are interactive—a change in one may provoke a change in the other, which may cause a responsive reaction, and so on.[5]

When anthropologists look at how cultures adapt to their environments over many centuries, we see clearly that there are some broad patterns that are played out repeatedly in human history, revealing general societal similarities. Twelve thousand years ago, at the end of the Pleistocene Ice Ages, all of our ancestors were mobile hunter-foragers. Over the succeeding millennia, in various regions of the world, food production was developed and sedentary farming villages were established. In time, population growth led to settlements of larger size and density, and to competition over resources. As populations grew in size and density, wealth differentials—differences in access to key resources—increased steadily.

For example, consider those six key regions of the world where great civilizations first arose: Mesopotamia (southwest Asia), the Nile Valley (northeast Africa), the Indus Valley (Indian subcontinent), the river valleys of northern China (northeast Asia), northwestern South America, and Middle America (Fig. 1). In the larger regions of these cradles of civilization, occupation always began with mobile hunter-foragers living on wild resources. Some bands of hunter-foragers would begin to occupy certain campsites all year long, and to produce their own food. Centuries later, some permanently occupied farming villages would have grown larger than others, and would have become central places for their regions. This was a characteristic pattern of chiefdoms like those known from accounts of travelers and ethnographers in the contact era, for example, of the ancient Hawaiians or Native Americans of the Northwest coast.

There is an enormous difference, in cultural evolution terms, between the chiefdom, which is essentially a big family (even

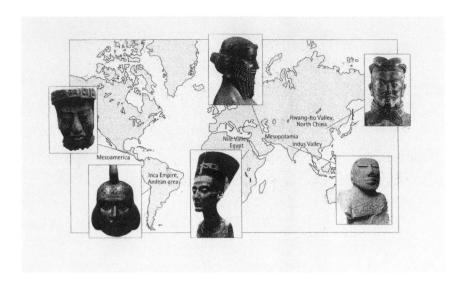

Figure 1. Locations of the world's earliest civilizations and examples of their respective Great Traditions of art (illustration, Evans 2004: 22)

if some of the kin are very distant poor relatives), and the socially stratified state. A chief can call for labor for a community granary or community ritual building, but no chief can command the labor to build and maintain a palace (which would represent personal aggrandizement), much less a private monumental park. In contrast, the head of a stratified state—a king, for example—can demand labor and materials, and severely punish those who won't cooperate. As a rule, in more complex cultures access to important resources is concentrated in the hands of fewer people, and those people maintain their power over resources through coercive force.

States are familiar to all us, because we all live under their authority, and we know the features of states—big populations, complex economy, taxation, writing, organized religion, social stratification, large cities with monumental architecture, including palaces for the rulers—and elaborate designed landscapes for the enjoyment of elites. From the six cradles of civilization in which the first states arose, the cultural format of the stratified state spread, over the centuries, to encompass the whole globe. Today there is no place that is not claimed by some nation-state, and few people who are not taxed by a state-level government.

Cultural Evolution and Monumental Gardens

These trends tell a cultural evolutionary story, and monumental gardens are a part of this story, because they only emerge in the most complex cultures. In fact, they serve as a diagnostic of complex culture. Archaeologists regard the palace as a true sign of highly complex societies,[6] but elaborate designed landscapes certainly provide the same kind of evidence. The great monumental gardens of the world—Chapultepec, Versailles, Babylon, for example—are all also diagnostics of the state, for several reasons.[7] In the first place, they represent the ruler's ability to devote the state's resources to his own private pleasure, and on a grand scale.

A second reason why a great garden expresses a high degree of societal complexity is that monumental gardens are one aspect of the development of a distinctive and mature artistic tradition—what some anthropologists have termed a "Great Tradition" of masterful works of intellectual achievement.[8] This is another diagnostic of this hypertrophied cultural evolutionary situation. Chiefdoms produce some vigorous expressions of representation—think of carved totem poles from the Pacific Northwest coast, or Maori carvings from New Zealand. However, it is state-level societies that achieve such artistic mastery and style that when you see an example of that style, you not only can readily identify the culture from which it came, but you also intuitively understand it—not as "folk art"—but as a highly refined object or painting, or piece of music, or culinary achievement. Or, a great garden.

And, in fact, monumental garden style is a distinct expression of a particular Great Tradition, and it can also embody an idealization of the particular culture-ecological relationship that characterizes the society's adaptation to its environment.[9] Mimesis in garden design is an extremely common means of honoring the important resources of a landscape in a representational form,[10] and one could perceive such mimetic effects as encapsulating essential components of cultural ecology.

Mesoamerican Culture History and Cultural Ecology

Mesoamerica, one of the world's major areas of ancient civilization, is geographically defined by the limits of growth, within a contiguous area, of its most important crop, maize (corn). The first people established themselves in this region over ten

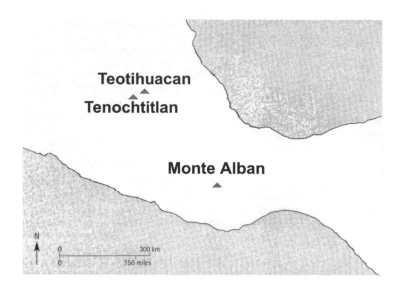

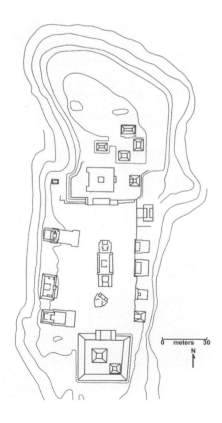

Figure 2. Map of Middle America, showing the locations of the Early Classic period (A.D. 250–600) cities of Teotihuacan and Monte Albán and the Late Postclassic city of Tenochtitlan (founded ca. A.D. 1325).

Figure 3. Map of the Valley of Oaxaca, showing the location of Monte Albán.

Figure 4. Air view of Monte Albán, from the east, showing the civic-ceremonial complex, with arrow pointing to the South Pyramid (photo, courtesy of Dean Snow).

Figure 5. Plan of Monte Albán, showing distribution of monumental architecture.

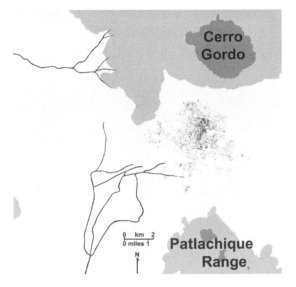

Figure 6. The Teotihuacan Valley, showing the location of the city of Teotihuacan and major geographical features.

Figure 7. Teotihuacan, looking north up the Street of the Dead (photo by S. Evans).

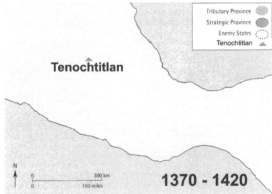

Figure 9. Map of Mesoamerica showing the location of Tenochtitlan.

Figure 8. Teotihuacan, looking south down the Street of the Dead.

thousand years ago, but the culture historical story really doesn't begin, for those of us interested in the designed landscape, until about four thousand years ago, when the first traditions of maize-based village agriculture became widespread.

Mesoamericans had a perspective on the relationship of humans to their environment that both contrasts radically with that of modern Westerners and also shares some modern Western attitudes. Like us, they saw the landscape as something to exploit for their own needs, but rather than having the capitalistic attitude that the earth is essentially a passive repository of potential wealth, ancient Mesoamericans saw the biophysical world around them as vibrantly alive. This animistic perspective encompassed not only flora and fauna but also mountains and caves, rivers and springs, thunder and lightning. This view is understandable, given the volatile nature of the Mesoamerican environment, with its earthquakes, active volcanoes, hurricanes and torrential storms, and its closely juxtaposed contrasts of snow-covered mountains and tropical jungles.

Mesoamericans revered these geographical features as forces and embodiments of supernatural power, and saw in their designed landscapes a means of bringing order into the world and revering its animism. The farm field—*milpa* to the Aztecs—was their sustenance and it played a central role in honoring the gods, because they assumed that the earth was the *milpa* of the gods. Their settlements were designed to pay homage to the living environment around them, and we see this in Mesoamerica's first great cities, which arose around two thousand years ago (Fig. 2). Two of these cities, Teotihuacan and Monte Albán, clearly display the principles of appropriating landscape features—honoring the living landscape—in their design. In a sense, this is monumental landscape design on a much larger scale than the imperial gardens of the Aztecs—which also incorporated elements of the surrounding countryside into their design—but these cases are instructive in terms of conditioning our perceptions about Mesoamerican spiritual and aesthetic attitudes.

Monte Albán

This ancient site sits atop a geographical prominence—an island—in the center of the Valley of Oaxaca (Fig. 3). It is surrounded by mountains, and was founded by an ethnic group who called themselves the "Zapotecs," the "cloud people," which is appropriate because their capital is positioned at the level of the clouds resting on the heights of the valley's perimeter of mountains.

The whole promontory of Monte Albán indeed appears to be a mountain, and it is crowned by a miniature mountain, a tall pyramid on the south end of the site (Fig. 4). The site's plan (Fig. 5) is a plaza ringed by a set of platformed ceremonial buildings and in the center of the plaza is a central "ridge" of more ceremonial buildings. Thus, Monte Alban's layout echoes the promontory's position in the valley—the ring of platforms represents the surrounding mountains, the plaza is the valley, and the site's central structures are the promontory itself.

Teotihuacan

Another ancient site, Teotihuacan, is located about thirty miles (ca. fifty km) northeast of modern Mexico City. Here, the setting is quite different (Fig. 6). Teotihuacan is situated on a broad plain, and, although, in a technical sense, mountains surround it, many of them are very distant. Two, however, are quite close, and frame the site; they are Cerro Gordo on the north, and the Patlachique Range to the south.

Cerro Gordo looms over the city's two great pyramids; Figure 7 shows the pyramid of the Moon, in the distance in front of Cerro Gordo, and the Sun pyramid, at right. The two pyramids are nearly identical in proportion, and when modern visitors to the site learn that the Moon Pyramid is substantially smaller than the Sun Pyramid (although the summits of the two are at the same level) they are perplexed, because this seems counterintuitive: shouldn't the larger pyramid be at the end of the ceremonial causeway, as its correct cognitive conclusion? It would seem that this positioning of the two pyramids is quite deliberate, a specific effect designed by Teotihuacan's planners. The placement of the smaller but identically proportioned Moon Pyramid at the end of the causeway is a little like the effect evoked by garden designers who grant depth to a garden by placing at a distance those plants with foliage that are smaller but similarly shaped to plants in the foreground—it increases the sense of distance. But here the effect also causes a perceptual disjuncture, because rather than fooling the eye with a background of infinite space, the background is a substantial mountain, similar in form to the two effigy mountains, the pyramids. The relative smallness of the Pyramid of the Moon pulls the mountain forward, making it look even larger, impressing the viewer with its importance to Teotihuacan.[11]

Another optical effect that links the city's architecture to its surroundings is seen in the opposite direction, the view south down the causeway, from the top of the Moon Pyramid (Fig. 8). The Sun Pyramid is on the left, and in the distance is the Patlachique Range, its shape replicated in the slopes and terraces of the pyramid. Once again, the ancient city has mimicked the attributes of the surrounding environment, the Moon and Sun Pyramids serving as smaller versions of the mountains in the background.

From these examples, it is clear that ancient Mesoamericans would have understood cultural ecology, although their spiritually based principles of environmental motivation[12] would not have permitted our modern, scientifically based view of causality. For example, the volcano Popocatépetl, now actively smoking, forms the southeastern part of the rim of mountains that circumscribe the Basin of Mexico.[13] In the past several millennia, Popocatépetl has erupted and these environmental crises have precipitated important cultural changes. Teotihuacan's huge size—about 125,000 people by A.D. 300—was thought to have resulted, in large part, from influxes of refugees from an early eruption of Popocatépetl and other volcanoes in the southern part of the Basin.

At other times, humans themselves cause the culture-environmental dynamic to begin. The Teotihuacan Valley was effectively deforested by building Teotihuacan and supplying it with lumber and firewood. This led to erosion that stripped the valley's slopes down to hardpan, and clogged the essential drained field cultivation system southwest of the city, requiring major investments in labor to keep this system functioning. Food production declined even more in the sixth century, with the eruption of proto-Krakatoa in Indonesia, which brought on a kind of "nuclear winter" even in parts of the Americas.[14]

Fall and Rise

Mesoamerican history, like that of other parts of the world, is thick with rise and fall stories. Teotihuacan and Monte Albán fell, and other cities and regions rose in importance. After Teotihuacan, the power vacuum in the Basin of Mexico was gradually filled by a set of much smaller city-states which slowly grew, both from intrinsic growth and migrations. Among the most important of the migrants were the Aztecs, a diverse set of ethnic groups who settled in the Central Highlands around A.D. 1200.[15] The Aztecs were Nahuatl speakers who claimed origins in the semimythical homeland, "Aztlán" (meaning "place of the white heron" or "place of whiteness"), an island home that they had abandoned decades before. And among the least impressive of these migrating Aztecs were the Mexica,[16] who would build Tenochtitlan, a capital as large as Teotihuacan, and become the leaders of the Aztec empire and of the monumental park building trend. Tracing their trajectory as an example of how cultural evolution reflects the interplay of human populations and their environments, we will see that as the disasters struck the Aztecs, their empire expanded, as outlined in Table 1.

Antecedents and Beginnings

Before the Mexica Aztecs had their gardens, their empire, and their great capital, Tenochtitlan, they were an uncouth and aggressive ethnic group who migrated to the Basin of Mexico in the late twelfth or early thirteenth century. At the time they arrived in the Basin of Mexico, it was well settled, its population, totaling fewer than two hundred thousand people, living in a set of city-states distributed over the alluvial plain that forms a ring around the central lake system. Each city-state was governed from an urbanized town serving as the capital for a noble dynasty, which drew tributes from the population of farmer-artisans living in villages in the surrounding countryside.

The sophisticated city-dwelling rulers of the Basin's city-states viewed the rough and quarrelsome Mexica Aztecs with trepidation.[17] The Mexica thought they had found a refuge on the summit of Chapultepec, but their position was betrayed by an estranged member of their group. They were evicted; they sacrificed the traitor and threw his heart into the boggy swamplands of Lake Texcoco, just northeast of Chapultepec. On this spot they would later come upon a vision of an all-white world in the mist.[18] Somewhat later they returned to this site and found the sign that they could establish their city: a cactus plant growing from a rock, and resting upon it was an eagle with a snake in its mouth.[19] This they took to be confirmation that they had reached their destination, their new Aztlán. The rulers of the Tepanec domain in the Basin of Mexico, who controlled that area, permitted them to settle on these boggy islands in return for their service as mercenary soldiers, to help the Tepanecs expand their modest city-state into a confederation of tribute payers. The traditionally accepted date of the establishment of Tenochtitlan is 1325, and the population of the Basin of Mexico had grown to about 365,000.

By 1375, the Mexica had become respected as diligent and belligerent soldiers, and were rewarded by being permitted to establish their own ruling dynasty. Thus, this marks the beginning of their history as a legitimate political entity in Mesoamerica, and this history would end about 150 years later, when Tenochtitlan was destroyed by the joint forces of the Spaniards and tens of thousands of their native Mesoamerican allies, many of them former tributaries of the Mexica-run Aztec empire who were seeking liberation and retribution.

Beginning with dynastic inception, Mexica history can be divided into four important phases, each with critical ecological events, political strategies, and attendant development of monumental gardens, set against a background of population growth. From 1375 to 1430, the Mexica were part of the Tepanec confederation and established a pleasure park at Chapultepec. In about 1430, the Mexica and their allies took over the Tepanec confederation and consolidated it within the Basin of Mexico, a process that lasted until about 1450 and was accompanied by the development of other pleasure parks, including botanical gardens.[20] From about 1450 to the 1470s, the Aztec tribute empire expanded beyond the Basin of Mexico, including the establishment of the important Mexica park at Huaxtépec, in a tropical climate outside the Basin. The last phase began in the 1470s. The period to 1500 saw the further development of the empire and the establishment of urban pleasure parks in Tenochtitlan and its allied capital, Texcoco. From 1500 to 1519 (the beginning of the Spanish intrusion) the empire expanded further and the system of monumental gardens and pleasure parks was fully mature.

1375–1430: The Mexica Dynasty and the Chapultepec Pleasure Palace

From A.D. 1375 to 1430, the population of the Basin of Mexico increased from about 450,000 to nearly 600,000, and all of the city-states known from Colonial era documentary sources had been established. Tenochtitlan had grown from a fishing village into a substantial town, and its need for fresh water quickly outstripped its few local springs. In 1420, the Mexica built a pleasure palace at Chapultepec, their one-time refuge. Mexica use of it depended on the generosity of their overlords, the Tepanecs.

Chapultepec's springs had become essential for Tenochtitlan's survival—an ecological crisis calling for a cultural evolutionary solution. The general cultural context was clearly that of state-level political organization with a high degree of social stratification and wealth differentials. Tenochtitlan was a small cog in this system, with no empire (Fig. 9). However, such confederations operate dendritically, and tribute-paying towns would themselves have been able to mobilize labor and materials from their own tributary populations for their own projects such as monumental buildings and landscape development, a factor that encouraged such city-states to try to remain independent, or to subjugate others.

Identifying Key Features of the Aztec Monumental Garden: Chapultepec

The development of a pleasure palace at Chapultepec in about 1420 was one such project, perhaps as a way of further securing Tenochtitlan's use rights to Chapultepec and its resources. The palace's designer was Nezahualcoyotl, a cousin of the rulers, the son of the ruler of Texcoco, the most important capital in the eastern Basin of Mexico.[21] Designing and overseeing construction of the first documented pleasure palace in Mesoamerica, he also oversaw the building of the first aqueduct that brought water from Chapultepec's springs to Tenochtitlan.[22]

Nearly a century later, in 1519, when the Spaniards first saw it, Chapultepec was the dynastic pleasure park of the most powerful rulers in an empire of five to six million people, and as such it was the focus of lavish development.[23] The promontory's heights offered wonderful views, which not only pleased the eye but also were important for communication in the pre-modern era. However, Chapultepec's most important asset was its set of freshwater springs. Around them were built bathing pools, palaces, and shrines. And from them came Tenochtitlan's water supply. Long before 1519, when the city's population was over one hundred thousand, the city required Chapultepec's springs in order to survive, and access to this environmental resource played a major role in Aztec history. In ancient Mesoamerica, water is revered, and the most precious substance was not gold but jade, in part because of its resemblance to water. The drop of water itself was a symbol of

Figure 10. Motecuzoma II sits for his sculpted portrait at Chapultepec (Durán *Atlas*).

preciousness. This ideological expression of value highlights our appreciation of Tenochtitlan's appropriation of this necessary resource, and the rulers' prescience in claiming the magnificent setting of the springs for their own pleasure park.

Chapultepec shows another typical feature of Aztec monumental gardens: the lavish use of sculpture, particularly bas reliefs. Many reliefs were portraits of the rulers; Figure 10, a sixteenth-century drawing, shows a ruler's portrait being carved into the cliff face at Chapultepec, and this custom, maintained by successive rulers, established a visual message of the dynastic family that controlled the Aztec empire.[24] Other depictions showed plants that grew in the empire, but could not be cultivated in this chilly high-altitude environment. Professional gardeners and landscapers worked hard to grow as wide a range of plants as possible at an altitude of about seventy-five hundred feet (ca. twenty-three hundred meters) in a region with a frosty winter, and then the sculptors added representations of those that were too fragile to survive the climate. Thus, the garden mimicked the empire's wealth of resources with its combination of actual examples and artistic depictions, all governed by living rulers and portraits of dead ones. In its representation of political domain, this garden format constituted a kind of green encyclopedia—a botanical garden in the modern sense of the word.

Processional paths were marked by lines of trees of a particular kind, the Montezuma cypress,[25] a fast-growing Taxodium species that could achieve sixty meters in height—thus on a par with coast redwoods. So thoroughly does the ecological ideal merge with political power in this one tree that the tree's Aztec name, *ahuehuetl*, was a metonym for king.[26]

In spite of there only remaining a few shreds of the Aztec landscape designer's concept, we can perceive some essential features of the Aztec garden, and get a sense that the Aztec designer, working with his wealthy patron, would have had some common concerns with André Le Nôtre or Lancelot "Capability" Brown: idealize the domain and express it in the monumental garden, use plants that suggest nobility, and include luxurious features that puff up the patron's ego and coddle the patron's desire for comfort.

Now let us return to Aztec history and examine how Chapultepec and other great Aztec parks functioned as lavish expressions of the cultural evolutionary position of the mature Aztec empire, and the culture-ecological relationship of the Aztecs to their physical world. Finally, we look at how features of the Aztec monumental garden tradition were adopted by Europeans, and spread into global gardening practices, often without any sense of their exotic origins.

1430–1450s: The Birth of Empire and Development of Dynastic Parks

Control over the springs at Chapultepec was an important precipitant of conflict between the Tepanec overlords and their tributaries. In the 1430s, the Tepanec Wars were fought, resulting in the takeover of the Tepanec confederation by the Mexica and their allies. From 1430 to the early 1450s, the population of the Basin of Mexico continued to increase, and most people paid tribute to the Mexica and their allies. In addition, the Aztec alliance expanded to the west and to the south, where they brought into their nascent empire regions at much lower altitude than the Basin, regions with a tropical climate (Fig. 11).

The most important allied city-state was Texcoco, now under control of Nezahualcoyotl, who reestablished his family's dynasty in this, their ancient capital. He also began work on his second dynastic park design project, this time for himself, at Texcotzingo.

Texcotzingo

Nezahualcoyotl's family had long used—and revered—a hill called Texcotzingo ("Little Texcoco"), about three miles (five kilometers) northeast of Texcoco. In fact, Texcotzingo would become another redesigned mountain mimicking a political domain. Nezahualcoyotl must have planned Texcotzingo while he was working on Chapultepec.

From the heights of Chapultepec, he could look in a direct line and see Tenochtitlan, five kilometers away, and across the lake, his dynastic capital, Texcoco, and, five kilometers behind it, his family's retreat, Texcotzingo (Fig. 12; Fig. 13). This geographical symmetry would have tremendous appeal for Mesoamericans, who held sacred the principle of duality. Texcotzingo was even more ambitious than Chapultepec.[27] Its design incorporated a total system of rock-cut platforms and shrines, rooms and baths, sculpture and fountains (Fig. 14), which were fed by an aqueduct five miles long and in places two hundred feet high that brought water from higher mountains and then sent it splashing down in channels and waterfalls, eventually feeding the fields of the farming villages below. Imagine the labor involved in carving those baths into the solid rock—with stone tools—and consider also the constant need for maintenance, provided by rotating crews of tribute-paying villagers.

Chapultepec and Texcotzingo were dynastic parks for these two related Aztec families—over the subsequent ninety years of empire building, ending with the Spanish conquest, these dynasties intermarried repeatedly—and their shared passion for garden development amounted to a status rivalry contest on a massively expensive scale. In terms of cultural evolution and monumental gardens as a marker of complex society, the establishment of these dynastic parks was the cornerstone of a much larger program of park development that included, at this time, two other types: horticultural nurseries and game reserves, dotted around the Basin of Mexico.

Horticultural nurseries

Gardens established for the purpose of growing plants actually served several functions. First, they supplied the landscaping needs of the dynastic parks, and of the palaces and public spaces of the new capital cities, Tenochtitlan and Texcoco. Second, these nurseries were, themselves, pleasure parks. For example, Acatetelco, north of Texcoco, featured allées of ahuehuetl trees surrounding a huge square pond filled with water from two rivers whose channels had been radically rerouted for this purpose. The pleasure palace built there was called Ahuehuetitlan, "in the place of the ahuehuetl trees." In layout, Acatetelco may have resembled the Menara gardens of Marrakesh, Morocco, a pleasure park that was also an orchard around water features such as a square pond and canals.[28]

In such nurseries, tree saplings and flowering perennials would achieve sufficient size for planting out, and permanent cutting gardens provided flowers and greenery for palaces and temples. As the major cities expanded, so did the palace and temple complexes, and having plants ready for landscaping would have been essential to royal plans for impressive displays of status. Kings vied with each other to acquire rare plants, nurturing them under optimum conditions.

Game reserves

The third kind of park was the game reserve. Aztec kings, like their contemporaneous counterparts in Europe, enjoyed the hunt, and early on established areas where they could go with their teams of beaters and bearers and bring down a few deer or

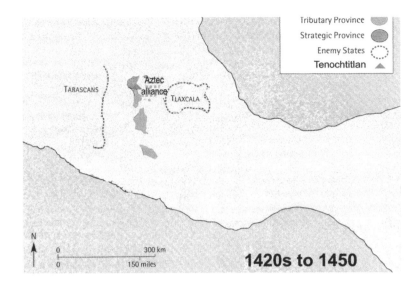

Figure 11. Map of Mesoamerica showing the Aztec empire, 1430–1450.

Figure 12. View from Chapultepec toward the northeast, adapted from a late-nineteenth-century painting by José María Velasco.

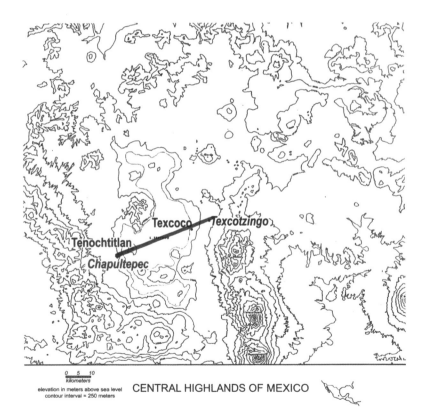

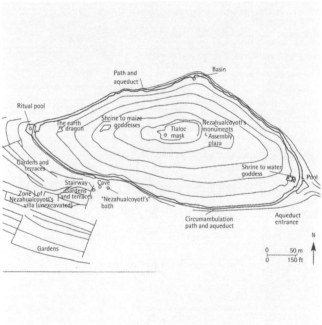

Figure 14. Plan of Texcotzingo (Evans 2004: 476).

Figure 13. Map of the Basin of Mexico, showing the direct sightline between Chapultepec and Texcotzingo, framing the cities of Tenochtitlan and Texcoco.

Figure 15. Map of Mesoamerica showing the Aztec empire, 1450s to circa 1470.

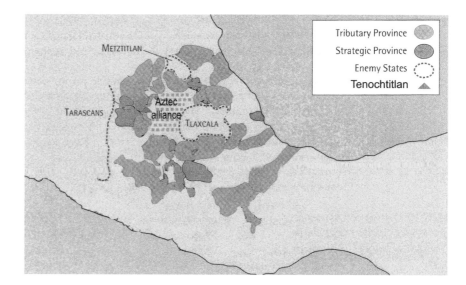

Figure 16. Map of Mesoamerica showing the Aztec empire, 1470s to 1500.

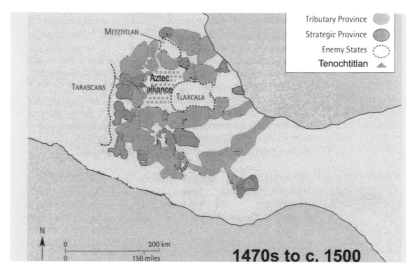

Figure 17. Map of Mesoamerica showing the Aztec empire, 1519.

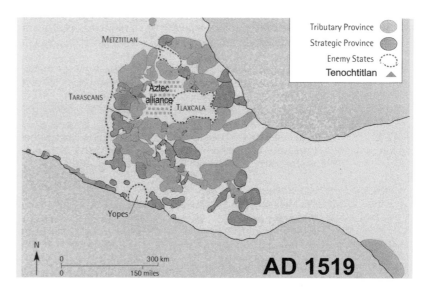

other game. Two of the favorite hunting reserves for Aztec kings at the time of the Spanish Conquest were islands in Lake Texcoco, and both islands featured palaces for the comfort of their visitors. These spots also came to be coveted by the Spaniards—Cortés, for example, claimed Tenochtitlan's island game reserve for himself and imported llamas from Peru to raise there. Like the horticultural nurseries, game reserves also were pleasure parks in the sense that luxurious accommodations were provided. Little is known about landscaping at the game reserves, but some excavations have revealed evidence of palace architecture, indicating a concern to insure a setting worthy of the kings.

Empire and Garden Expansion from about 1450 to the 1470s

Twenty years after Tenochtitlan's kings began building their own empire, a set of crises befell them. The 1450s were a period of trials: amid constant population growth (the Basin's population would have been about 660,000 in 1450) were crop failures that caused famines so severe that commoners sold themselves into slavery down on the Gulf coast to avoid starvation. For the kings, achieving security for their families required a more drastic solution. As soon as the crises abated, the Aztec kings expanded their empire into the hot lands to secure access for themselves to regions they called "The Land of Food" (Fig. 15). Given the lack of beasts of burden in ancient Mesoamerica, it was not practical to import food into the Basin of Mexico from more than about 150 kilometers (about ninety miles) away, because the porters bearing it would need to eat as much as they could carry in order to survive the trip. Thus, empire expansion was not motivated by the need for a food supply for the Basin's population.

But there were other reasons to expand the empire besides providing food, or a safe haven for royal families should hunger return. The distant hot lands supplied valuable things that kings needed: jade, gold, cacao (chocolate was the beverage of choice for royalty), vanilla, quetzal feathers for their royal headdresses, jaguar skins for royal costumes and accoutrements, woven cotton fabric (only the nobles could wear cotton), and raw cotton to sell to the commoners for them to weave into fabric and give to the nobles as tribute.

Thus it was after 1450 that the system of tributaries expanded out of the regions around Basin of Mexico and began extending over much of modern Mexico. These tributes—luxury goods from afar, and more utilitarian goods and services from the core region—provided an immense income for the Aztec emperors, and their capital cities were expanded, with new palaces and new gardens. This urban renewal project in Tenochtitlan came on the heels of a devastating flood, and provided public works projects for the commoners in need of food. Some of the population therefore received needed sustenance while the city was expanded and beautified, and the aqueduct from Chapultepec was rebuilt.

It was also during this period that Tenochtitlan's kings established a dynastic pleasure park and horticultural garden at Huaxtépec, in the tropical Valley of Morelos, about one hundred kilometers southeast of Tenochtitlan.[29] According to the Spaniards, it was the most beautiful garden they had ever seen. The tropical locale of this extensive park permitted a much wider range of plantings than was possible in the Basin of Mexico. Huaxtépec combined the facilities of the pleasure park—lakes and baths, palaces and shrines—with the practical value of the horticultural nursery. Tenochtitlan began demanding rare tropical plants in tribute from their new vassal states in the Gulf lowlands. The plants were first delivered to Tenochtitlan "in great quantities, with the earth still about the roots, wrapped in fine cloth" and from there they were "taken to Huaxtépec and planted around the springs."[30] Professional gardeners accompanied the delivery of these plants, to insure that they were properly tended, including carrying out the blood sacrifices at the time of planting.

The Basin of Mexico continued to grow, and so did the empire. By 1470, the Basin's population would have reached about 730,000. Tenochtitlan was beset by a new crisis, a severe earthquake that leveled many of Tenochtitlan's buildings, and prompted a new round of public works projects and urban beautification. Expansion of the empire continued toward the east and south, with more regions in the tropical Gulf lowlands and in Guerrero and Oaxaca coming under Aztec control (Fig. 16).

At this time, a new type of monumental garden came into being, the urban amusement park. These were found in the two imperial capital cities, and, like the other pleasure gardens, were developed by and for the ruling families. In both capitals, zoological gardens were established. These extended the horticultural themes of gathering together examples of the plants grown in the empire, and gathered animals as well.

In fact, in Tenochtitlan there developed particularly meaningful zoological parks. There were several different facilities, in at least two locations. One was located east of the Great Temple, where there were "kept separate cages of lions and tigers [jaguars], ounces [lynx], wolves, and foxes. In other courtyards, in a different type of cage, he kept many kinds of falcon and hawks and all manner of birds of prey. . . . Also in large earthenware vessels there were many snakes and vipers, and all of this was merely a form of grandeur. In this house of beasts he kept men and women monsters, some crippled and others dwarfed or hunchbacked."[31]

To the European sensibilities of Andrés de Tapia, phenotypically non-normative humans were "monsters."[32] However, to the Aztecs such individuals enjoyed a special spiritual status, reflected in the belief that the patron god of dwarves, Xolotl, was the twin brother of the great god Quetzalcoatl. This duality established a privileged relationship, and the distinctive otherness of dwarves provided a psychological refuge for lords, in that they could not aspire to any normal role in life. In fact, some parents would deliberately deform their children in order that they might have a future at court.[33] In addition to being sought after as exhibits in the royal zoo of wild beasts, hunchbacks could be advantaged, becoming the special attendants of the emperor and other lords[34] and in this capacity were in the lords' confidence; they were their valets, messengers, jesters, curers, and seers. They would have been visible members of royal and noble entourages, and may have accompanied their lords to visit less fortunate dwarves, caged in the zoo of the wild beasts. However, the two groups of dwarves would have shared one of the same roles in the pleasure parks, serving as a psychological pathway to otherness, at the same time that they represented the economic power of the empire to collect and display such individuals.

When their empire began to encompass a huge area and millions of tribute-payers, the Tenochca could command in payment (and suggest as appropriate gifts to themselves) all manner of living oddity, and in addition to hunchbacks, albinos were sent to the capital from throughout the empire. These individuals were gathered into another special park in Tenochtitlan, a facility that was several city blocks west of the royal palace. They shared this facility with an array of waterfowl, and also with storehouses for precious goods collected in tribute.[35] The zoo of wild beasts also included tribute storehouses, but these were for more common items like woven lengths of cloth. At the waterfowl-albino facility, conquistador Andrés de Tapia saw "a hall and two other chambers full of gold and silver, and green stones."[36] Thus, these urban pleasure parks juxtaposed portable, negotiable wealth and the far more subtle displays of affluence provided by landscaped expanses of expensive city land, filled with living rarities, costly to acquire and maintain.

Of the two facilities, the wild beast-hunchback display was probably the more ominously unsettling because of the obvious presence of life-threatening animals and the proximity of the Great Temple, where human life was regularly sacrificed. Aztec nobles were thought to have had a closer relationship to the gods than did the commoners, and took seriously their role of setting an example in autosacrifice. The atmosphere of the wild beast zoo would have combined the charnel house ("and it was a sight to see the amount of meat fed to all these birds and beasts" wrote de Tapia[37]) and the madhouse, with caged humans alert for opportunities to call attention to themselves and thus upgrade their position from zoo to palace.

The waterfowl-albino park, on the other hand, would have been more soothing and refreshing to the spirit. Cortés himself, who found that the residences there were "only a little less magnificent" than Motecuzoma's royal palace, is worth quoting at some length, from his description of:

> . . . a very beautiful garden with balconies over it; and the facings and flagstones were all of jasper and very well made. In this house there were rooms enough for two great princes with all their household. There were also ten pools in which were kept all the many and varied kinds of water bird found in these parts, all of them domesticated. For the sea-birds there were pools of salt water, and for river fowl of fresh water, which was emptied from time to time for cleaning and filled again from the aqueducts. . . . Above the pools were corridors and balconies, all very finely made, where Mutezuma came to amuse himself by watching them.[38]

The waterfowl were exhibited in displays that mimicked their natural habitats, much like the best modern zoos, and had specially trained keepers and veterinarians who worked fulltime to care for them: "Without fail, more than six hundred men were kept occupied in the care of these fowl. There was, besides, a place where the sick birds could be cured."[39] In such details, recounted by a member of Cortés's company who was noted for his lack of exaggeration, we have a strong insight into the substantial resources of the Aztec empire that could be devoted to a royal folly.[40]

Recall that when the Mexica founded Tenochtitlan, they claimed that it was because of a vision of whiteness they saw emanating from their future island home, an echo of their ancestral island home, whose name implied whiteness.[41] The vision may have been the effect of swamp gas, but this zoo with its special albino exhibit, a focus of comment by the conquistadores, was no doubt meant to honor their ancient beginnings. This is a highly refined example of the combined forces of economics and ideology being used by the rulers to display the essence of their heritage, at the same time that they were producing a cultural genre, a living display that educated onlookers about the range of Aztec wealth and power. The onlookers were, of course, nobles themselves, but the Tenochca often entertained as guests the rulers of enemy states, as well as allied kings. In both cases, the imperial message would be conveyed.

1500–1520: Final Years of the Aztec Empire

The final period before the arrival of the Spaniards in 1519 saw the Aztec empire at its most extensive. New territories were conquered in the tropical coastal lands, and historical records indicate that the Aztecs were reaching further east into the Maya domain (Fig. 17). By 1500, the Basin of Mexico's population had reached nearly 850,000. Again, an ecological crisis prompted urban redevelopment, with another devastating flood.[42] The Great Temple was rebuilt for the sixth time, assuming the size (height, 30.7 meters or 101 feet) that the Spaniards saw. The new emperor, Motecuzoma II, built himself a new palace, now underlying modern Mexico's Palacio Nacional. All of the city's denizens, rich and poor, were charged with the tasks of rebuilding their houses and planting trees and gardens.

By 1519, the empire had expanded down to the Pacific coast, with outposts hundreds of kilometers from the Basin of Mexico. Millions of people were sending goods and labor service up to Tenochtitlan and its imperial partners, a huge funneling of wealth into the coffers of a very few families. And by 1519, the system of royal pleasure parks and gardens in the region right around the capitals included about twenty different sites, representing the four basic imperial garden types: the great imperial dynastic retreats (Chapultepec and its counterpart, Texcotzingo), horticultural nurseries, urban amusement parks, and game reserves.[43]

Thus, in the short century between the establishment of Chapultepec in 1420 and the arrival of the Spaniards in 1519, the Aztec rulers had used monumental gardens as one of the most prestigious means of displaying royal wealth. This trend would have ramified throughout the culture. Of course, the highest royals would have had the largest and most elaborate gardens and the greatest variety of plants and decorative motifs on display, but, all down the social ladder, there would have been a keen awareness of the value of gardens. The sumptuary laws that reserved certain rights of residential decoration for those who had earned them in service to the king also established such displays as markers of status and, inevitably, engendered a sensitivity to these trends in artistic taste.

Even the commoners were not immune from this passion for gardening. The native informants of the sixteenth-century chronicler Sahagún told him that garden design was among the chief pleasures of kings,[44] and other sources indicate that certain kings encouraged all residents of Tenochtitlan to cultivate beautiful surroundings. And people from all over the Basin of Mexico would have seen royal monumental gardens, in spite of their exclusivity, because of the labor service for the kings, and typical assignments for villagers in the region around the capitals were in the palaces and gardens. From each village in tribute-paying areas, dozens of farmer-artisans would have spent a few weeks as a part of the "palace people" as they were known in the Aztec language. They would have been cleaners and porters, basically assigned to manual labor. But still, their tasks would have made them aware of the most expensive and fashionable styles in interior design, cuisine, clothing, and, of course, gardens. Their ability to imitate such patterns was of course limited by both the circumstance of poverty and the laws that forbade even the nobles from living beyond their station. This situation of peasant access to royal and noble lifestyles was not uncommon in archaic agrarian societies, and would have created a more uniform aesthetic than we might imagine, considering the wealth differences between the peasants and the palace.

Furthermore, the Spaniards seem to have shared this appreciation of the style of Aztec gardens, at least those that bore a categorical resemblance to the gardens of Europe. They lavished praise on Huaxtepec[45] and many other gardens, and claimed for themselves those that fit into European standards of pleasure parks. By contrast, they had little interest in Texcotzingo, Texcoco's great dynastic pleasure park, because it too boldly expressed the culture-specific mimetic principle, and also was too far from Mexico City, the dazzling Colonial capital. The Spaniards didn't want Texcotzingo—the only Spaniard who cared about it was the evangelical archbishop Zumárraga, who was determined to destroy every devil-worshipping image on it.

Of the other parks, transformation from Aztec aesthetic values to those of Spaniards was a relatively straightforward process. Tenochtitlan was largely destroyed in the siege that conquered the Aztec empire, and thus few of the urban amusement parks were left. The grounds of the albino "Place of Whiteness" zoo were, appropriately, assigned to the Franciscan order for the establishment of their convent. Chapultepec, we have seen, continued in service as a royal pleasure park, and Cortés claimed Huaxtepec for himself. New gardens established in and around mansions and convents used many plants and artistic prototypes of the Aztec culture.[46]

Aftermath of the Aztec Empire

Mexico in the sixteenth and seventeenth centuries was almost as unknown by the outside world as it had been in the pre-contact fifteenth century, so protective were the Spaniards of their New World colonies. Few non-Spanish visitors penetrated New Spain, and the aim of the Spanish government was to transfer as much wealth as possible as rapidly as possible from the New World to the Old. Gold was the only real cure for the Spanish disease, Cortés told one of the native rulers along the Gulf coast, but there were other treasures that came to have real value. Culinary, medicinal and decorative plants all were disseminated from Mexico and Central America to Europe, and eventually to the rest of the world. When we enjoy flowers like marigolds, cosmos, and dahlias, we should imagine their ancestral forms gracing Aztec royal gardens. Mexican trees and shrubs were similarly spread.[47]

The legacy of Aztec monumental parks may be perceived as well in one of the great garden developments of Europe in the Age of Discovery: the botanical garden.[48] Although Europe had a long tradition of horticultural gardens of various kinds,[49] the "green encyclopedia"—the living compendium of known plants—emerged in Italy in the 1540s (at Pisa in 1543, at Padua in 1545).[50] In fairness to the creative and systematizing intellectual climate surging through Renaissance Europe, I would not argue that the idea of the botanical garden was simply a copy of the Aztec prototype. But European botanists and garden designers—and their royal patrons—were as eager as Aztec kings for new plant material and new ideas about gardens,[51] in part to express, mimetically, their own scope of knowledge about and economic interest in the wider world. The Spaniards had described the gardens they had seen, both in print and in exhaustive debriefings, and of course the plants themselves were exported from the Americas. Given the timing of these events, and the prevailing European spirit of cataloguing the world's curiosities and valuables, it is not unlikely that Aztec gardens provided a source of inspiration for what has become an essential component of the world's corpus of monumental gardens.

Aztec monumental gardens will never be as well known to us as the great gardens in the European or Asian or Islamic traditions, but important aspects can be identified, and still perceived. Like their Old World counterparts, Aztec gardens represent a cultural evolutionary development in terms of artistic mastery and expression of societal complexity. They reveal the refined aesthetic sensibilities of one of the world's most memorable civilizations as well as refined skill in horticulture, and the gardens are still somewhat visible, if we know where to look and what to look for, in some of modern Mexico's great parks. Furthermore, their legacy lives on in the range of plants we use, all over the world,[52] and in all botanical gardens. For these contributions, we should be grateful to the Aztec rulers and the empires that made their gardens possible.

NOTES

1. "Aztec" refers to the Nahuatl-speaking ethnic groups dominant in the Central Highlands of Mexico during the Late Postclassic period (ca. A.D. 1430–1521). The "Aztec Empire" was the most extensive political domain in the culture history of pre-Columbian Middle America (a geographical region extending from the modern U.S.–Mexican border down to the juncture of South America with Panama). Within Middle America lies the culture area, Mesoamerica, consisting of much of Mexico, plus Guatemala, Belize, and the western portions of Honduras and El Salvador. For a recent overview of pre-Columbian Mesoamerican culture history and archaeology, with extensive discussion of the Aztecs and their political and economic history, see Susan Toby Evans, *Ancient Mexico and Central America: Archaeology and Culture History* (London and New York: Thames and Hudson, 2004).

2. Luciano Cedillo Álvarez, "Chapultepec: Recurso Para el Siglo XXI," *Arqueología Mexicana* 10, no. 57 (2002): 62–65; Miguel Angel Fernández, "El Jardín de Limantour," *Arqueología Mexicana* 10, no. 57 (2002): 54–55; Amparo Gómez Tepexicuapan, "Los Jardines de Chapultepec en el Siglo XIX," Arqueología Mexicana 10, no. 57 (2002): 48–53; Victor Manuel Ruiz Naufal, "Los Jardines de Chapultepec y Sus Reflejos Novohispanos," *Arqueología Mexicana* 10, no. 57 (2002): 42–47; Mario de la Torre, *Chapultepec, Historia y Presencia*, (Mexico City: Smurfit Cartón y Papel de México, SA de CV, 1988); Lorenza Tovar de Teresa and Saúl Alcántara Onofre, "Los Jardines en el Siglo XX: El Viejo Bosque de Chapultepec," *Arqueología Mexicana* 10, no. 57 (2002): 56–61.

3. Susan Toby Evans, "Aztec Royal Pleasure Parks: Conspicuous Consumption and Elite Status Rivalry," *Studies in the History of Gardens and Designed Landscapes* 20 (2000): 206–228; Alain Musset, "Les Jardins Préhispaniques," Trace no. 10 (1986): 59–73; Zelia Nuttall, "The Gardens of Ancient Mexico," *Annual Report of the Board of Regents of the Smithsonian Institution* (Washington, D.C.: Government Printing Office, 1923), 453–464.

4. The essential principles of biological evolution are: (1) all species have the potential to produce more offspring than the environment can support (the Law of Biotic Potential); (2) offspring may vary, such that those bearing traits rendering them better-adapted to a particular environment will survive, and others will not (survival of the fittest); (3) insofar as such traits are the result of genetic mutations, they may be passed to the succeeding generation, and over time the species may change in the direction of greater adaptation: evolution will have occurred. These principles are readily adaptable to the situation of cultural evolution by substituting "innovation" for "genetic mutation." Humans have not *physically* evolved for the last 100,000 years; since the emergence of fully modern humans at that time, we have used culture as our means of adaptation. We innovate and then share what we know with others. The information we share can extend our adaptive patterns as quickly as we express them.

5. Robert McC. Netting, *Cultural Ecology* (Menlo Park: Benjamin/Cummings Publishing Co., 1977); Julian H. Steward, *Theory of Culture Change* (Urbana: University of Illinois Press, 1955).

6. William T. Sanders, "Chiefdom to State: Political Evolution at Kaminaljuyu, Guatemala," in *Reconstructing Complex Societies, An Archaeological Colloquium*, ed. Charlotte B. Moore, Supplement to the Bulletin of the American Schools of Oriental Research no. 20 (1974): 97–113; Kent V. Flannery, "The Ground Plans of Archaic States," *Archaic States* (Santa Fe: School of American Research, 1998), 15–57.

7. Unfortunately for archaeologists dealing with very ancient societies, such monumental gardens are far more ephemeral than the elite buildings they once surrounded.

8. David L. Webster, Susan Toby Evans and William T. Sanders, *Out of the Past: An Introduction to Archaeology* (Mountain View: Mayfield Publishing Co., 1993), 167–169.

9. Scholars have documented this abundantly for the Old World; two examples (of many) are Stephen Daniels and Denis Cosgrove, "Introduction: Iconography and Landscape," in *The Iconography of Landcape: Essays on the Symbolic Representation, Design and Use of Past Environments*, ed. D. Cosgrove and S. Daniels (Cambridge: Cambridge University Press, 1988): 1–10; and Vincent Scully, *The Earth, The Temple, and the Gods: Greek Sacred Architecture* (New Haven, Conn.: Yale University Press, 1979).

10. John Dixon Hunt, *Greater Perfections* (Philadelphia: University of Pennsylvania Press, 2000): 98.

11. Susan T. Evans and Janet Catherine Berlo, "Teotihuacan: an introduction," in *Art, Ideology, and the City of Teotihuacan*, ed. J.C. Berlo (Washington, D.C.: Dumbarton Oaks, 1992): 1–26. As the map of the Teotihuacan Valley (see Fig. 9) shows, the site was built on the lower edges of the southern slope of Cerro Gordo, an ancient volcanic cone. This slope is a basalt shelf, and from under it seep the springs that permitted a large and densely settled city to be established in this semi-arid area (William T. Sanders, "Ecological Adaptation in the Basin of Mexico: 23,000 b.c. to the Present," in *Supplement to the Handbook of Middle American Indians*, 1, *Archaeology*, ed. J.A. Sabloff [Austin: University of Texas Press, 1981], 147–197. Teotihuacanos seem to have recognized the vital importance of Cerro Gordo to their very existence; their name for the mountain may have been "Mother of Stone" (Stephen Tobriner, "The Fertile Mountain: an Investigation of Cerro Gordo's Importance to the Town and Iconography of Teotihuacan," *Teotihuacan: Onceava Mesa Redonda*, II [Mexico City: Sociedad Mexicana de Antropología, 1972], 103–114).

12. For a discussion of the Aztec deities governing the natural world, see Ana María L. Velasco Lozano, "Dioses y naturaleza," *Arqueología Mexicana* 10, no. 57 (2002): 34–35.

13. An enclosed hydrological basin until the early 1600s, when Spanish engineers drained it by building an extensive canal system, the pre-Columbian "Basin of Mexico" is called the "Valley of Mexico" when referring to the period after A.D. 1600.

14. Richardson Benedict Gill, *The Great Maya Droughts: Water, Life, and Death* (Albuquerque: University of New Mexico Press, 2000), 293. An even more dramatic and widespread culture-ecological disaster in ancient Mesoamerica involved the fall of Maya civilization. By the eighth and ninth centuries A.D., the Maya population had grown so large that agricultural intensification led to environmental degradation, resulting in a demographic collapse so severe that the heartland of this vital civilization remained virtually uninhabited for hundreds of years (David L. Webster, *The Fall of the Ancient Maya* [London and New York: Thames and Hudson, 2002]).

15. Michael E. Smith, "The Aztlan Migrations of the Nahuatl Chronicles: Myth or History?" *Ethnohistory* 31 (1984): 153–186.

16. The Mexica [pron. *maySHEEkah*] were further immortalized when their name was applied to the name of the modern nation, Mexico, and its capital, Mexico City, built on the ruins of the Mexica Aztec capital, Tenochtitlan.

17. One ruler gave them his daughter as the bride of their leader; the Mexica sacrificed her to their gods and when her father arrived for the wedding they proudly displayed her flayed skin.

18. Fray Diego Durán, *The History of the Indies of New Spain* (Norman: University of Oklahoma Press, 1994 [1581]), 40.

19. This was adopted as the central motif of the modern Mexican flag.

20. Edelmira Linares, "Los Jardines Botánicos de México, Su Historia, Situación Actual y Retos Futuros," *Revista Chapingo, Serie Horticultura* 2 (1994): 29–42; Doris Heyden, "Jardines Botánicos Prehispánicos," *Arqueología Mexicana* 10, no. 57 (2002): 18–23.

21. In 1418 the Tepanecs had had the Texcocan king killed, and Nezahualcoyotl (pron. *netsahwahlCOYoht,* meaning "Fasting Coyote") went into exile, eventually finding asylum in Tenochtitlan. Nezahualcoyotl would live to retake his throne, make his Texcocan kingdom, the Acolhua domain, an important partner to Tenochtitlan in the Aztec empire, become famed as a poet and a civil engineer, and a political survivor of great skill.

22. Mexico City's great modern boulevard, Paseo de la Reforma, traces the ancient course of the aqueduct.

23. Beatriz Braniff Torres and María Antonieta Cervantes, "Excavaciones en el Antiguo Acueducto de Chapultepec," *Tlalocan* V. 5 (1966):161–168, 265–266; Susan Toby Evans, "Chapultepec Park," in *Chicago Botanic Garden Encyclopedia of Gardens, History and Design*, ed. C.A. Shoemaker, 1 (Chicago and London: Fitzroy Dearborn Publishers, 2001), 261–263; Miguel León-Portilla, "Chapultepec en la Literatura Nahuatl," *Revista de la Universidad de México* 24 no. 11 (1970): 1–10; María de la Luz Moreno and Manuel Alberto Torres, "El Origen del Jardín Mexica de Chapultepec," *Arqueología Mexicana* 10, no. 57 (2002): 41; Felipe Roberto Solís Olguin, "Chapultepec, Espacio Ritual y Secular de los *Tlatoani* Aztecas," *Arqueología Mexicana* 10, no. 57 (2002): 36–40; Torre, *Chapultepec, Historia y Presencia.*

24. Henry B. Nicholson, "The Chapultepec Cliff Sculpture of Motecuhzoma Xocoyotzin," *El México Antiguo* 9 (1961): 379–443.

25. Now the national tree of Mexico, see Aurora Montúfar López, "Ahuehuete: Símbolo Nacional," *Arqueología Mexicana* 10, no. 57 (2002): 66–69.

26. Fray Bernardino de Sahagún, *Rhetoric and Moral Philosophy*. Book 6 of the Florentine Codex. (Santa Fe: The School of American Research and The University of Utah, 1969 [1569]), 252.

27. Miguel Medina, *Arte y Estética de el Tetzcotzinco: Arquitectura de Paisaje en la Época de Netzahualcóyotl* (Mexico City: Universidad Nacional Autónoma de México, 1997); Miguel Othon de Mendizábal, "El Jardín de Netzahualcoyotl en el Cerro de Tetzcotzinco," *Obras Completas*, 2 (Mexico City: Imprenta del Museo Nacional de Arqueología, Historia y Ethnografía, 1946), 443–451; Richard Fraser Townsend, "The Hill of Texcotzingo Mapping Project," *National Geographic Society Research Reports* 20 (1979): 755–760.

28. A third function of horticultural nurseries is documented by several other gardens of this type, which supplied medicinal herbs. The medical knowledge of the Aztecs with regard to plants was truly impressive: Xavier Lozoya, "Arqueología de la Tradición Herbolaria," *Arqueología Mexicana* 3, no. 14 (1995): 3–9; Xavier Lozoya and Mariana Loyoza, *Flora Medicinal de México* (Mexico City: Instituto Mexicano de Seguro Social, 1982); Bernard Ortiz de Montellano, "Aztec Medicinal Herbs: Evaluation of Therapeutic Effectiveness," in *Plants in Indigenous Medicine and Diet,* ed. N. Etkin (Bedford Hills, New York: Redgrave Publishing Co., 1986), 113–127; Bernard Ortiz de Montellano, *Aztec Medicine, Health, and Nutrition* (New Brunswick, N.J.: Rutgers University Press, 1990). Modern tests of the value of Mesoamerican herbal remedies have determined that a high percentage were effective in treating the conditions for which they were prescribed. In 1570 the Spanish government sent the great natural historian Francisco Hernández to New Spain, to visit the Aztec gardens and catalogue the medicinal herbs (Francisco Hernández, *Cuatro Libros de la Naturaleza y Virtudes Medicinales de las Plantas y Animales de la Nueva España*, [Morelia: Escuela de Artes, 1888 {1571}]).

29. Enrique Juan Palacios, *Huaxtepec y sus Reliquias Arqueológicas* (Mexico City: Publicaciones de la Secretaría de Educación Pública, 1930); Druzo Maldonado Jiménez, *Cuauhnáhuac y Huaxtepec* (Cuernavaca: Universidad Autónoma de México Centro Regional de Investigaciones Multidisciplinarias, 1990), Octavio Rocha Herrera and Susan Toby Evans, "Huaxtépec," in *The Archaeology of Ancient Mexico and Central America: An Encyclopedia*, ed. S. T. Evans and D. L. Webster (New York: Garland Publishing Co., 2001), 349–350; Susan Toby Evans, "Huaxtépec," in *Chicago Botanic Garden Encyclopedia of Gardens, History and Design*, ed. C. A. Shoemaker, 2 (Chicago and London: Fitzroy Dearborn Publishers, 2001), 609–611.

30. Durán, The History of the Indies of New Spain, 244–245.

31. Andrés de Tapia, "The chronicle of Andrés de Tapia," in *The Conquistadores*, ed. P. de Fuentes (New York: The Orion Press, 1963 [ca. 1534]), 40.

32. Of course, dwarves were also a part of European court life, as shown in Diego Velasquez's painting "Las Meninas" (1656).

33. Motolinía (Fray Toribio de Benavente), *History of the Indians of New Spain* (Washington D.C.: Publications of the Academy of American Franciscan History, 1951 [1541]), 269.

34. Sahagún, *Kings and Lords,* Book 8 of the Florentine Codex (Santa Fe: The School of American Research and The University of Utah, 1979 [1569]), 30.

35. Tapia, "The chronicle of Andrés de Tapia," 40–41.

36. Tapia, "The chronicle of Andrés de Tapia," 40.

37. Tapia, "The chronicle of Andrés de Tapia," 40.

38. Hernan Cortés, *Letters from Mexico* (New Haven, Conn.: Yale University Press, 1986 [1519–1526]), 109–110.

39. Tapia, "The chronicle of Andrés de Tapia," 40.

40. In the English landscaper's sense of the word, pertaining to a pleasure-garden feature.

41. Durán, *The History of the Indies of New Spain*, 40.

42. The sixteenth-century chronicler, Durán, cites the beautification program for Tenochtitlan that followed a disastrous flood in about 1499 (Durán, *The History of the Indies of New Spain*, 373).

43. Evans, "Aztec Royal Pleasure Parks," 210.

44. Sahagún, *Kings and Lords,* 30, also Ana María L. Velasco Lozano, "El jardín de Itztapalapa," *Arqueología Mexicana* 10, no. 57 (2002): 26–33.

45. Huaxtépec "where is the garden which I have said is the best that I have ever seen in all my life, and so said . . . our Cortés" (Bernal Díaz del Castillo, *The Discovery and Conquest of Mexico* [New York: Farrar, Straus, and Cudahy, 1956 {1560s}], 375). Cortés mentioned it in his letters to the king of Spain (Cortés, *Letters from Mexico*, 196).

46. Jeanette Favrot Peterson, *The Paradise Garden Murals of Malinalco: Utopia and Empire in Sixteenth-Century Mexico* (Austin: University of Texas Press, 1993).

47. Maguey, the century plant (*Agave* spp.), is so widely grown that it hardly seems exotic when found all around the Mediterranean.

48. "It is impossible to say positively whether, as Conde Carli first suggested in 1777, the first botanic gardens established in Italy in the Sixteenth Century were in fact based on much earlier Aztec models." (Frank J. Lipp, "A Heritage Destroyed: The Lost Gardens of Ancient Mexico," *Garden Journal* 26, no. 6 [1976]: 188). Gian Rinaldo Carli published "Lettere Americane" (1777) covering a wide variety of topics (Ivan Markovic, "Gian Rinaldo Carli, Istriani Illustri" [http://www.istrians.com/istria/illustri/carli/bio.htm {2000}]). See also Lucile Brockway, *Science and Colonial Expansion: The Role of the British Royal Botanic Gardens* (New York: Academic Press, 1979), 72.

49. Elizabeth Barlow Rogers, *Landscape Design: A Cultural and Architectural History* (New York: Harry N. Abrams, Publishers, 2001), 118–124.

50. F. Nigel Hepper, "Botanic Garden," in *The Oxford Companion to Gardens*, ed. G. Jellicoe, S. Jellicoe, P. Goode, and M. Lancaster (Oxford and New York: Oxford University Press, 1991), 67–68.

51. Phillip II, King of Spain, who had sent the great naturalist Francisco Hernández to New Spain to document the flora, himself had a strong interest in natural history and gardens (Aurora Rabanal Yus, "Felipe II y los Jardines," *Felipe II y el Arte de Su Tiempo* [Madrid: Fundación Argentaria, UAM Ediciones, and Visor, 1998], 401–424).

52. Xavier Lozoya, "El Oro Verde de América," *Arqueologia Mexicana* 1, no. 6 (1994): 6–11.

Botanical Progress, Horticultural Innovations and Cultural Changes

Botanical Progress, Horticultural Innovations and Cultural Changes

Botanical Progress, Horticultural Innovations and Cultural Changes

Botanical Progress, Horticultural Innovations and Cultural Changes

Perfume and Power from the Ancient Near East to Late Antiquities

Yizhar Hirschfeld

Perfumes and Propaganda

In 10 B.C., Augustus ordered two obelisks to be brought from Egypt to Rome; one was erected in the Circus Maximus and the other in the Campus Martius. The obelisk set up in the Circus Maximus is depicted on numerous perfume bottles manufactured throughout the Empire in the days of Augustus. One of the finest of these is a glass bottle with a narrow neck and a pear-shaped body, currently in the Getty Museum in Malibu.[1] The bottle, 7.5 cm high, has a relief decoration of an obelisk, cult symbols and altars in white on the cobalt blue background (Fig. 1). In Egypt, the obelisk symbolized the power of the sun god, who in Greek mythology is depicted as riding in a chariot; hence Augustus' order to place it in the Circus Maximus. The monumental obelisks and the perfume bottles were part of Augustus' imperial propaganda.

Why did perfume bottles play a role in political propaganda? Perfumes were believed to have the power of seducing, beguiling and evoking strong religious feelings. Different qualities of fragrances—fresh and light, sweet and heavy, exotic and aphrodisiac—could be combined in endless ways, an art that demanded knowledge, imagination and taste. In antiquity the secrets of the art were closely guarded and passed down from father to son. Perfumes and incense, as well as spices, were luxury commodities and their weight was measured in silver and gold.[2]

Figure 1. Perfume bottle decorated with an obelisk and cult symbols, first century B.C. (after Harden 1987: No. 36).

Incense and fragrances played a major role in the worship of the gods in the temples of ancient Mesopotamia and Egypt. Alexander the Great used the power of perfumes in his propaganda. Plutarch tells how as a child Alexander was sacrificing in the company of his tutor Leonidas and scattered frankincense liberally on the altar with both hands. Leonidas told him: "When you have conquered the perfume-producing land then you can offer lavishly".[3] When later he conquered Gaza, "the principal city of Syria," Alexander sent 500 talents of frankincense (λιβανωτόν) and 100 talents of myrrh (σμύρνης) to his tutor in Macedonia and instructed him to use it lavishly. At that time, Gaza was the port for the Arabian trade in frankincense and myrrh, which were components of incense. The quantities sent by Alexander were considerable: a talent weighs about twenty-four kilograms and consequently the delivery consisted of twelve thousand kilograms of frankincense and twenty-four hundred kilograms of myrrh. Alexander's action was emulated by later rulers. Incense played a part in the cult of rulers throughout the Hellenistic-Roman world. Thus, for example, in the great inscription of Antiochus I of Commagene in the first century B.C., the king orders the burning of incense to himself and his divine ancestors.[4] The image of Alexander was close to the heart of Augustus too; he sent an expeditionary force to conquer the source of incense—the kingdom of Sheba in South Arabia. Although the campaign failed because of a lack of water, from that time on the Arabian peoples became "friends of the Roman people".[5]

The transformation of perfume into a symbol of the cult of rulers finds expression in the New Testament story of the three Wise Men who brought gold, frankincense, and myrrh to the infant Jesus (Matt. 2:13). From a contemporary point of view, this was a prophecy of Jesus's future role as king and Messiah. Ironically, the birth of Christianity in the first century A.D. heralded the beginning of the end of the intensive use of perfume that characterized the Hellenistic-Roman world. The association with the old pagan religions and the cult of rulers was too strong, and the emperors of the fourth century prohibited the use of fragrances for any use other than the burning of incense in churches, which continues to the present day.

The Demand for Perfumes and Incense

There was an immense demand in the Greek and Roman worlds for perfumes and incense as components of cosmetics, for use in temples (such as the Temple of Jerusalem) and as accompaniments to the burial of the dead. Bottles of perfume, intended to purify the air, have been found in many burial caves. Josephus's description of the funeral of Herod the Great of Judea mentions "five hundred of Herod's servant and freedmen carrying spices".[6]

The Roman Empire, the "Empire of Pleasure" as it was termed in Dalby's book *Empire of Pleasure: Luxury and Indulgence in the Roman World* (2002), generated a huge demand for perfumes. The Romans perfumed their bodies, sprinkled perfume over the guests at symposia, and perfumed the walls of bathhouses. Pliny noted disapprovingly that the tough legionary soldiers anointed their hair with perfumed oil. The Roman custom of cremation increased the demand for perfumes. Thus, for instance, we hear that during the cremation of Nero's beloved wife Poppaea, the streets of Rome were made fragrant with a quantity of perfume that equaled the entire annual Arabian trade.

In antiquity, perfumes were made entirely from natural ingredients, some imported from distant lands and some produced in such small quantities that their price was high. Among the most precious perfumes of the ancient world was balsam, which came from Judea. According to Pliny, the price of one *sextarius* of balsam (about half a liter) ranged between 300 and 1,000 *denarii*.[7]

Because of their high price, cosmetics were marketed in small quantities. Consequently, there was a flourishing industry of beautifully designed small containers, made with luxurious materials like alabaster that apart from their aesthetic appeal insulated the contents and preserved their fragrance. When glass-blowing was introduced in the first century B.C., the perfume industry adopted the prestigious material and began to manufacture perfume bottles. Among the finest of these products are blown glass bottles, about ten centimeters high, in the shape of a dove (Fig. 2), manufactured in Italy in the first century A.D.[8] After the insertion of the perfume the bottles were sealed by melting the glass, making it necessary to break the tail of the dove a little to extract the contents.

Frankincense and Myrrh: The South Arabian Incense Trade

Frankincense and myrrh were and still are the main ingredients of incense. Tiny drops of resin that are tapped from the trees harden into crystals after contact with air. When burned, these crystals emit clouds of fragrant white smoke.

Figure 2. Perfume bottle in the shape of a dove, first century A.D. (after Harden 1987: No. 37)

The demand for these products created a flourishing trade centered on South Arabia, the starting point of marine and land routes that reached the far corners of the ancient world. Because of the problems of supply and transport, the kings of Egypt attempted to import the plants from which perfumes and incense were produced. In the fifteenth century B.C., Queen Hatshepsut sent an expedition to Punt (today's Somalia) to bring frankincense and myrrh trees to Egypt.[9] The expedition is documented in the reliefs of the temple at Deir el-Bahri. Assyrian sources mention frankincense and myrrh among the tribute and taxes sent from Arabia. When the Queen of Sheba in South Arabia arrived in Jerusalem, she was accompanied by camels bearing spices, gold, and precious stones (I Kings 10:2).

Frankincense also was used in the domestic cult. A small altar of the fifth century B.C. inscribed with the word "frankincense" in Hebrew was found in the excavations of Lachish. Among the items discovered in the caves of the Judean Desert was a bronze shovel for burning incense in the home, dating from the second century A.D. On each side of this object was a kind of small candlestick in which frankincense and myrrh could be burned.[10]

Myrrh was used in its natural crystal form in incense or in the form of a perfumed oil. This was the main ingredient of "holy anointing oil," mentioned frequently in the Old Testament as the oil for anointing priests (for instance, Exodus 30:23–25) and later kings. Myrrh was also used for medicinal purposes. Before the crucifixion of Jesus, he was offered "wine mingled with myrrh," perhaps to relieve his pain (Mark 15:23).

The production, transport, and marketing of frankincense and myrrh were traditionally in the hands of the Arab tribesmen of the Hejaz. The Hellenistic period saw an expansion of this trade, in which Arab tribes, particularly the Nabateans, played a leading role. Pliny describes the Arabs as "the richest people in the world." In his words, Rome spent an annual sum equivalent to 100 million *sesterces*, or about 25 million *denarii*, on goods from the East—pearls and perfumes supplied by the Arabs.[11]

Trade was conducted over both land and marine routes. Land transport was principally by camels, whose domestication in the twelfth and thirteenth centuries B.C. gave a boost to Arab trade. Excavations carried out at sites along the Perfume Road

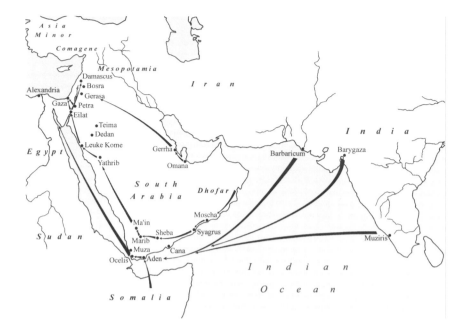

Figure 3. Map of routes in the spice and perfume trade (after Dayagi-Mendels 1993: 119)

Figure 4. *Boswellia sacra* growing on Mt. Dhofar in Oman (photograph: James Aronson)

in the Negev have uncovered camel bones in much larger frequency (72 percent) than those of horses and donkeys (28 percent).[12]

The land route led north along the western side of the Arabian peninsula, parallel to the Red Sea. This route, which set out from Sheba (today's Oman), was in use in the tenth century B.C. and brought the Queen of Sheba to Jerusalem. At its northern end the road branched off in several directions, to Egypt via Leuke Kome, to Gaza via Petra, and to Damascus via Gerasa and Bosra in the Hauran (Fig. 3).

In botanical studies, frankincense is generally assigned to various species of *Boswellia* of the *Burseraceae* family.[13] The main species yielding frankincense resin is *Boswellia sacra*. This is a tall evergreen tree that grows on the rocky mountain slopes of Dhofar in Oman, about eight days' walk from Sheba, the capital of the Hadhramaut (Fig. 4). The tree's thin papery bark emits a strong aromatic fragrance. The trunk is covered by four layers of whitish bark and the interior is soft and crumbly. A deep cut in the trunk releases the flow of a whitish transparent resin,[14] which hardens and forms crystals. Cuttings from the tree were transplanted to British botanical gardens in Aden and Bombay and took root, demonstrating that the tree can thrive in artificial conditions.

Myrrh is usually identified with *Commiphora myrrha*, also *Burseraceae*, which grows on the mountain slopes of desert regions of Sudan, Somalia, southern Arabia, Iran and northern India. This is a thorny shrub, two meters high on average.[15] Its branches grow densely in all directions and bear small trefoil leaves and small white flowers (Fig. 5). Its trunk, covered with thin papery bark, is greenish in color and contains the reddish aromatic resin, which is tapped during the summer. The resin is stored before its sale in baskets, well wrapped to preserve its aromatic qualities, in dry caves with well-swept floors.[16]

Boswellia and *Commiphora* are similar in that they both grow in arid regions and have a papery bark. However, *Boswellia*

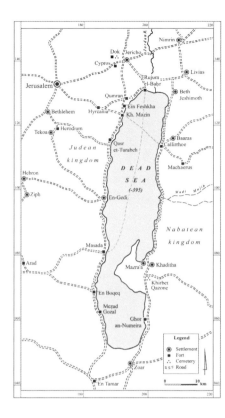

Figure 5. *Commiphora myrrha* growing in Somalia (photograph: James Aronson)

Figure 6. Map of the Dead Sea Valley during the early Roman period

differs from *Commiphora* in lacking thorns and being evergreen. The resin of the two species differs in color; *Boswellia* is whitish and *Commiphora* is reddish.

The Balsam Enigma

The cultivation of balsam and its sale as aromatic oils and medicines were a very valuable source of income to the kingdom of Judea in the Hellenistic and Roman periods. Josephus noted that balsam was the finest of perfumes, whereas Pliny[17] described it as the best of the fragrances. According to the sources, balsam grew nowhere in the world but in the Dead Sea region. It was cultivated, together with other aromatic plants, in the desert oases of Jericho, En-Gedi, and Qumran (Fig. 6). In the words of Josephus: "Here [in Jericho] too grow the juicy balsam, the most precious of all the local products, the cypress and the myrobalanus".[18] "Cypress" is the henna mentioned in the Song of Songs (1:14): "My beloved is unto me as a cluster of henna-flowers in the vineyards of En-Gedi." Henna is a reddish aromatic substance that is used today to color the hair, the nails and the palms of the hands.

According to Pliny, the resin, or "juice of the balsam," was extracted by a special knife not made of iron. This is called *opobalsamum*, "which is extremely sweet in taste".[19] Pliny also mentions *xylobalsamum*, "wood of the balsam," consisting of the branches, leaves, seeds, and bark of the shrub. These were boiled with olive oil to produce an aromatic oil of high price, although not as high as that of *opobalsamum*.

Because of its remarkable qualities and great value, balsam is mentioned frequently in the ancient sources. Diodorus of Sicily, in the first century B.C., noted that it grew only in the Dead Sea region and was used in the preparation of medicines.[20] The sources describe it as a tree with evergreen foliage; Theophrastus, in the late fourth century B.C., described it as a tall tree

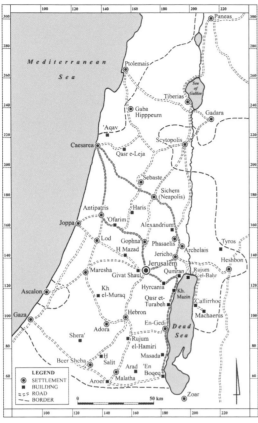

Figure 7. Map of routes in the balsam trade from the Dead Sea Valley to Caesarea (in gray)

Figure 8. Herodian juglet containing balsam oil from a cave near Qumran (photograph: J. Patrich)

as high as a pomegranate tree, that is, three to four meters.[21] However, in later sources, such as Pliny of the first century A.D., balsam is defined as a low shrub, only one meter high. The discrepancies between the descriptions are usually explained as being the result of improvements introduced into the plant over time by its cultivators.[22] The Madaba mosaic map of the early seventh century A.D. shows dark green shrubs growing in the Dead Sea region. The creator of the mosaic may have been attempting to depict the balsam shrubs that, together with palm trees, were part of the typical vegetation of the area.[23]

Balsam was cultivated in the Dead Sea region in artificial plantations. In the fourth century B.C., Theophrastus mentions two balsam plantations, one larger than the other. The larger one was apparently in Jericho and the smaller one in En-Gedi. These two plantations are also described by Pliny, whereas Strabo records the existence of a "balsam park" next to the royal palace in Jericho.[24] The Greek word he uses, *paradeisos*, derives from the Persian *pardessa*, meaning a walled garden. In later times, this word was adopted by European languages as a synonym for the Garden of Eden.

The origin of the balsam was apparently the kingdom of Sheba in South Arabia. According to Josephus,[25] it was the Queen of Sheba who brought balsam plants as a gift to King Solomon. The dry and frost-free climate of the Dead Sea region, as well as its rocky soils, were apparently particularly suited to the cultivation of the aromatic plant. However, archaeological finds demonstrate that Josephus's testimony is anachronistic. Excavations conducted at En-Gedi have shown that at the time of King Solomon in the tenth century B.C. the oasis was uninhabited[26] and that settlement began there only later, during the reign of

Figure 9. Installation for perfume production found at 'Ein Feshkha, looking northwest

Figure 10. Installation for perfume production found at En-Gedi, looking east

King Josiah of Judah (640–609 B.C.). It seems likely that balsam was indeed introduced from South Arabia, but in the reign of Josiah in the second half of the seventh century rather than in the time of King Solomon. The Song of Songs, attributed to the period of Josiah, mentions the "beds of spices" (*arugat ha-bosem* in Hebrew) of En-Gedi (5:13). The Hebrew word *bosem* was originally the specific word for balsam and was only later applied generally to perfumes in general. A later Talmudic tradition testifies that it was Josiah who substituted balsam oil for the myrrh oil previously used to anoint the heads of the kings of Judah.[27] Mazar's excavations of Tel Goren demonstrated that the oasis of En-Gedi flourished in the late seventh and early sixth centuries B.C., and settlement continued there after the Babylonian conquest. The balsam plantations are attested from that time on through the Babylonian, Persian, Hellenistic and Roman periods.

After the battle of Actium in 31 B.C., Augustus transferred control of most of the Dead Sea area to Herod the Great. From this time onward, the region enjoyed increased prosperity. The area devoted to cultivation of balsam was extended to small oases such as 'Ein Feshkha near Qumran, 'En Boqeq near Masada and Callirrhoe ('Ain ez-Zara) in the eastern shore of the Dead Sea.

Herod made balsam one of the kingdom's main sources of income. He established an international harbor at Caesarea and built two roads connecting Caesarea with the Dead Sea (Fig. 7). One of them passed through Jerusalem, itself a major consumer of perfumes, and the other through Sebastia in Samaria. In this way, Herod controlled production, transport, and

Figure 11. Pottery vessel used in perfume production found at En-Gedi, fifth–sixth centuries A.D.

Figure 12. Mezad 'Arugot: a large installation for perfume production in the En-Gedi oasis, fourth–sixth centuries A.D.

marketing to the West and could compete successfully with the Nabatean trade in perfumes from the Far East and South Arabia. The revenue that flowed into the royal coffers enabled Herod to carry out his grandiose building plans.

After the annexation of Judea to the Roman Empire in 6 A.D., the balsam plantations in Jericho and En-Gedi passed to the Roman *fiscus*. During this period, changes were made in the cultivation of the plant. Pliny testifies that the balsam plantations spread up the mountain slopes. It appears that the Roman cultivators, with their extensive agrarian expertise, were able to propagate balsam by cuttings and to grow it on trellises like a vine. During the Great Revolt, the Jewish rebels attempted to destroy the balsam plantations; according to Pliny, the rebels and the Roman legionaires fought over each plant.[28] In the victory parade held by Vespasian and Titus after the revolt, balsam plants were proudly displayed in the streets of Rome.

The cultivation of balsam in the Dead Sea region continued in the Late Roman and Byzantine periods. Galen, the celebrated physician of the second century A.D., commended the balsam known as "En-Gedi" after its place of cultivation. The Church Fathers Eusebius in the fourth century and Jerome in the fifth century mentioned the balsam grown in En-Gedi and Zoar at the southern end of the Dead Sea. The mosaic floor of the sixth-century synagogue of En-Gedi contains an inscription that mentions the "secret of the village." In one interpretation, this is none other than the secret of the cultivation and production of the precious perfume.[29] The excavations that I have carried out at En-Gedi show that the village was destroyed in a fierce fire in the late sixth century and was abandoned by its inhabitants. From this time on we hear no more of balsam; the villagers of En-Gedi apparently took the secret with them.

In recent years, sites around the Dead Sea have yielded finds that illustrate the scale of the local perfume industry. A jar discovered at Masada was apparently inscribed with the Aramaic words *miz balsana*, probably meaning "balsam juice." If so, this is an exact translation of the Greek word *opobalsamon*, meaning the same thing.[30] A small jug of the Herodian period discovered in a cave near Qumran still contained remains of a dark viscous liquid.[31] It was found, wrapped in palm fibers, between some rocks (Fig. 8). Chemical analysis showed that the contents were a vegetable oil of a kind that is unknown today, making it quite possible that the oil was balsam oil.

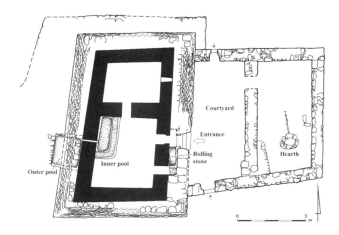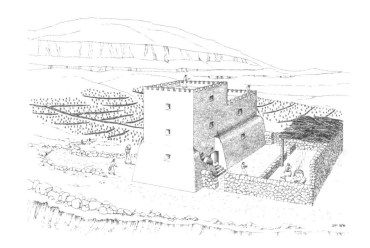

Figure 13. Plan and proposed reconstruction of Mezad 'Arugot.

At 'En Boqeq excavations revealed the remains of a workshop for the production of perfume.[32] A similar estate house built around an enclosed courtyard was uncovered at 'Ein Feshkha south of Qumran.[33] In the estate complex at 'Ein Feshkha and in the royal estate of the Herodian winter palace at Jericho there were installations that were probably connected with the production of perfume essences. The installations contain a soaking pool, a holding vat, and heavy cylindrical stones for crushing branches pruned from the balsam trees (Fig. 9). After the branches were crushed, they were soaked in the pool and the resulting perfume essence drained into the holding vat was put into jars.

Similar, although smaller, installations were found in domestic buildings of the Late Roman and Byzantine periods at En-Gedi. These installations consist of a stone-built basin and a jar sunk in the floor, with a plastered channel connecting the two (Fig. 10). The basin was used for soaking and the sunken jar received the perfume essence. Close to one of these installations we discovered a unique pottery vessel that was also used in the local perfume industry (Fig. 11). This is a medium-sized jar with a wide mouth. Around the shoulder of the vessel was a double wall, its base pierced by nine small drilled holes. The purpose of these holes was to drain back into the jar any of the precious liquid that was spilled. In the ancient world, perfume essence was produced by wringing cloth above a container; the jar from En-Gedi was apparently used in this way.

Among the ancient agricultural terraces of the oasis of En-Gedi are the well-preserved remains of Mezad 'Arugot. In front of the entrance is a stone-walled courtyard (Fig. 12). Excavation of the courtyard uncovered a hearth surrounded by a thick layer of ash. The structure, which dates from the fourth to sixth centuries A.D., has thick walls and its entrance could be sealed by a rolling stone. Inside was a plastered pool like a bathtub, with a larger rectangular pool next to it (Fig. 13). The two pools are connected by a channel built within the wall. This unique arrangement was probably not a regular water storage system but was more likely an installation for the production of perfume essence. In the first stage, crushed balsam branches were soaked in the larger pool. After this, the oily liquid was transferred by flotation via the channel to the smaller plastered pool. The perfume essence was then heated with olive oil in the courtyard. Because of its large size, it seems likely that Mezad 'Arugot was an industrial structure in common ownership that served the entire village of En-Gedi, which reached its peak of prosperity in the fourth to sixth centuries.

A question that remains unsolved is that of the botanical identification of balsam. Most modern botanists identify it with *Commiphora balsamum*.[34] However, this species does not match the description of balsam given in the sources, since it is not evergreen and is distinctly thorny. By contrast, *Boswellia sacra*, from which frankincense is produced, is an evergreen tree found almost exclusively in South Arabia. As we have seen, it has been transplanted successfully, demonstrating that the story of the Queen of Sheba bringing it as a gift to King Solomon is not totally implausible. Was *Boswellia sacra* the archetype from which the balsam that grew in the Dead Sea region was developed?

In the recent excavations carried out at En-Gedi, we excavated dozens of ancient agricultural terraces in an attempt to find physical remains of balsam shrubs. However, despite our efforts, this attempt has so far been unsuccessful; all the botanical remains discovered were identified as modern species. If we do discover remains of a plant that is unknown to botanists, it would be worthwhile to attempt a DNA analysis, which could be compared to *Boswellia* and *Commiphora*. However, the botanical identification of balsam must remain a mystery for the time being.

In conclusion, a chain of direct and indirect evidence connects the Dead Sea region with the cultivation and production of balsam perfume over almost a millennium. From the evidence at our disposal, we may conclude that:

1. Balsam was introduced from South Arabia

2. It was an aromatic plant

3. It was cultivated over the years in artificial plantations

4. It was unique to the Dead Sea region

5. It was used in the production of costly perfumes and medicines

6. It became extinct in the late sixth or early seventh century

Will we ever be able to reconstruct the balsam and reintroduce it to the Dead Sea region? To do this, we will have to continue excavating the terraces of En-Gedi in our search for physical remains of the plant and transplant saplings of *Boswellia* and *Commiphora* from Oman and Somalia to plantations in En-Gedi, Jericho, and Zoar. Each of these is in a different country today, En-Gedi in Israel, Jericho in the Palestinian Authority, and Zoar in Jordan. It seems to me that this would be a worthwhile international project, creating three attractive green havens and perhaps renewing one of the most important sources of income of ancient times.

Notes

[1] Donald B. Harden and others, *Glass of the Caesars* (Milan: Olivetti, 1987), 83-84.

[2] Mikhal Dayagi-Mendels, *Perfumes and Cosmetics in the Ancient World* (Jerusalem: Israel Museum, 1993), 9-10.

[3] *Plutarch's Lives,* VII, *Alexander.* Bernadotte Perrin, trans. (London and New York: Loeb Classical Library, 1929), XXV, 3-5.

[4] Stephen Mitchell, *Anatolia I: Land, Men and Gods in Asia Minor* (Oxford: Clarendon Press, 1993), 177.

[5] Glen Warren Bowersock, *Roman Arabia* (Cambridge, Mass.: Harvard University Press, 1983), 46-47; ibid, "Perfumes and Power," *Profumi d'Arabia* ed. A. Avanzini (Roma: Atti del convegno 1997), 543–556.

[6] Josephus, *The Jewish War.* Henry St. John Thackeray, trans. (Cambridge, Mass.: Harvard University Press, 1927) I, 673.

[7] Pliny, *Natural History.* H. Rackham, trans. (Cambridge, Mass.:Harvard University Press, 1945), XII, 54.

[8] Harden, *Glass of the Caesars,* 93.

[9] F. N. Hepper, "An Ancient Expedition to Transplant Living Trees," *Journal of Royal Horticultural Society* 92 (1967): 35–38.

[10] Yigael Yadin, *The Finds from the Bar Kokhba Period in the Cave of Letters* (Jerusalem: Israel Exploration Society, 1963), 48-49.

[11] Pliny, *Natural History,* XII, 81.

[12] Zohar Amar, "The Ancient Trade in Incense, Perfumes and Spices." *The Nabateans in the Negev* ed. R. Rosenthal-Heginbottom) (Haifa: Hecht Museum (in Hebrew with English summary), 2003), 61–66, see 62.

[13] F. N. Hepper, "Arabian and African Frankincense Trade." *Journal of Egyptian Archaeology* 55 (1969): 66–72; Shanina A. Ghazanfar, 1998. "Plants of Economic Importance." *Vegetation of the Arabian Peninsula.* S. A. Ghazanfar and Martin Fisher, eds. Dordrecht: Kluwer Academic Publishers, (1998): 241–264.

[14] J. P. Mandeville, "Frankincense in Dhofar." *Journal of Oman Studies* 2 (1980): 87–89.

[15] F. N. Hepper, "Trees and Shrubs Yielding Gums and Resins in the Ancient Near East." *Bulletin of Sumerian Agriculture* 3 (1987): 107–114, see 110.

[16] Mandeville, "Frankincense in Dhofar," 87-88.

[17] Pliny, *Natural History*, XII, 111.

[18] Josephus, *The Jewish* War, 469.

[19] Pliny, *Natural History*, XII, 117.

[20] *Diodorus of Sicily.* Charles Henrri Oldfather, trans. (Cambridge, Mass.: Harvard University Press, 1935), II, 48.9.

[21] Theophrastus, *Enquiry into Plants.* A. Hort, trans. (Cambridge, Mass: Harvard University Press, 1916), IX, 6.

[22] Joseph Patrich and Benny Arubas, "A Juglet Containing Balsam Oil (?) from a Cave near Qumran." *Israel Exploration Journal* 39 (1989): 43–59; Hannah M. Cotton, and Werner Eck, "Ein Staatsmonopol und seine folgen Plinius, Naturalis Historia 12, 123 und der Preis für Balsam," *Rheinisches Museum für Philologie* 140 (1997): 153–161.

[23] F. N Hepper and J. E. Taylor, "Date Palms and Opobalsam in the Madaba Mosaic Map," *Palestine Exploration Quarterly* 136 (2004): 35–44.

[24] Strabo, *Geography.* H. L. Jones, trans. (Cambridge, Mass.: Harvard University Press, 1930), XVI, 41.

[25] Flavius Josephus, *Jewish Antiquities.* Henry St. John Thackeray, trans. (Cambridge, Mass.: Harvard University Press, 1930), VIII, 174.

[26] Benjamin Mazar, "En-Gedi." *The New Encyclopedia of Archaeological Excavations in the Holy Land.* E. Stern, ed. (Jerusalem: Israel Exploration Society, 1993), 399–405.

[27] Babylonian Talmud, *Krituth* 5,2.

[28] Pliny, *Natural History*, XII, 118.

[29] Dan Barag, "En-Gedi: The Synagogue," *The New Encyclopedia of Archaeological Excavations in the Holy Land.* E. Stern, ed. Jerusalem: Israel Exploration Society, 1993, 405–409.

[30] Yigael Yadin and J. Naveh, *Masada I: The Aramaic and Hebrew Ostraca and Jar Inscriptions* (Jerusalem: Israel Exploration Society, 1989), 47.

[31] Joseph Patrich and Benny Arubas, "A Juglet Containing Balsam Oil (?) from a Cave near Qumran," *Israel Exploration Journal* 39 (1989): 43–59.

[32] Moshe Fischer, Mordechai Gichon and Oren Tal, *En Boqeq, Excavations in an Oasis on the Dead Sea, II: The Officina,* (Mainz am Rhein: Von Zabern, 2000).

[33] Yizhar Hirschfeld, "Excavations at 'Ein Feshkha, 2001," *Israel Exploration Journal* 54 (2004): 37–74, see 55–65.

[34] F. N. Hepper, "Trees and Shrubs Yielding Gums and Resins…" 111.

Botanical Progress,
Horticultural Innovations
and Cultural Changes

Botanical Progress,
Horticultural Innovations
and Cultural Changes

Botanical Progress,
Horticultural Innovations
and Cultural Changes

Botanical Progress,
Horticultural Innovations
and Cultural Changes

Horticultural Changes and Political Upheavals in Middle-Age Andalusia

Mohammed El Faïz

During recent years my own research on the medieval Islamic economy uncovered evidence—behind the massive transformation of agricultural activities—for a true "agricultural revolution."[1] Even though this revolution was initiated in the Muslim Middle East between the eighth and tenth centuries, it only came to fruition in Andalusia during the eleventh and twelfth centuries. These two centuries constitute the "Andalusian period" within the general progress of Arabic horticulture. Sevilla, after Cordoba and Toledo became an agricultural capital and the Mekka of agronomers.

How do horticultural changes in the wake of the "agricultural revolution" relate to political and social upheavals? We have selected the historic and geographic frame of Muslim Andalusia (eighth to fifteenth centuries) to study these relationships and to unravel their interactions with science and technology.

However, the analysis of the complex phenomenon of Arabic medieval horticulture demands preliminary attention to religious, political, social, and cultural influences. As we know, such factors, in the past as in the present, are tantamount to the success or failure of agricultural innovations. Islam, Arabic language, state policy, and political decentralization constitute some of the issues that will be shown to be conducive to horticultural changes.

Once this frame has been introduced, we will turn to the scientific and technical evolution of horticulture. Botany, agronomy, and hydraulics will be shown to be emerging sciences that lay at the basis of garden art and of the new horticulture. We shall pay special attention to the first "tropicalization" of the Mediterranean Andalusia, before the discovery of America favored a second larger "tropicalization" beginning in the sixteenth century.

The Role of Agronomy: The Texts and Terminology of the Agronomists

If the number of agronomic publications contributes to agricultural expansion, the rural history of Moslem Spain from the ninth to the fourteenth centuries appears especially fecund. Indeed, during this period, six agronomic treatises were published, giving rise to an agronomic movement that extended to several cities (Toledo, Sevilla, Granada, and Almeria). Let us concentrate on the most interesting titles: *The convincing book about agronomy*, by Ahmed Ibn Hajjâj in Sevilla in 1074; *The Book of the Goal and Proof*, by Mohammed Ibn Bassâl (1074-1085?); *The book of Agriculture*, by Abu L-Khayr in Sevilla (eleventh century?); *The Book of the Garden Flowers and of Spiritual Walk*, in Granada by Mohammed Al-Tighnari (1107-1110?); *The Book of Agriculture in Sevilla* by Yahya Ibn Al- 'Awwam (end of the twelfth century); *The book of the Parade of Beauty and Achievements of Fruitfulness*, about the foundation of the art of agronomy, by Sa'd Ibn Luyûn (1282-1349) in Almeria in the fourteenth century. The examination of the probable dates of these agronomic publications reveals outstanding moments during a long period of continuous scientific creativity. Indeed, in a relatively short period of time (from 1074-1110), Hispano-Arabic agronomy reached its apex. In less than a half-century, four agricultural treatises came to light (Ibn Bassâl, Ibn Hajjâj, Abû L-Khayr, Al-Tighnarî). Nearly one century elapsed before Ibn Al-'Awwâm wrote his encyclopedia of rural economy. Even a longer time went by before Ibn Luyûn brought to conclusion the publication of his agricultural poem, the glittering array of Andalusian agronomists. This goes a long way to show how the ebb and tide of agronomic reflection was bound to change with the economic situation. The agronomic writers under study comprised a community adhering to the same scientific paradigm that earned its name of "Agronomic School," by which it is generally known. However, the term Andalusian agronomists, which I use here for brevity, calls for some clarification. Indeed, they are agronomists in the restricted sense of specialists in technical aspects of husbandry. However important the technical contents of their publications, it was not the sole aspect of importance. They also took into account legal and economic issues, mingling sometimes the magico-astrological tips with considerations of a botanical pharmacological or medicinal nature.[2]

The word "agronomy" comes from two Greek roots meaning "field" and "law." It indicates in its current acceptation a part or totality of the sciences applied to agriculture. The Greeks also used the word *Geoponica* to indicate all that related to land tillage. As for the Romans, they chose the narrower terms of *re rustica* and *de agricultura* in order to indicate the vast field of rural economy.[3]

One can distinguish among the Arab agronomy treatises either a narrow or a broader understanding of agriculture. The first group restricts itself to the technical activities of agriculture, thus indulging in studies of the agricultural factors of production (water, ground, manure), the description of food, aromatic plants and tree cultivation. Only on the rare occasion can we find a digression relating to the ecosystem or to legal matters. On the contrary, other works supply information about poultry, apiculture and cattle. For instance, Ibn Al-'Awwam, in the twelfth century, subscribes to this broad understanding of rural activity. The Al-Andalus world owes to this writer its first large "Maison Rustique" or the country farm. This monumental work addresses agriculture and such related activities as breeding, veterinary medicine, poultry breeding, and bee-keeping.[4] Beyond the quest for a definition of agriculture, Arab agronomists debated whether agronomy was an art (*fann*) or a science (*'ilm*). However, Ibn Bassal did not wonder whether agriculture could be based on scientific research as he adopted a normative perspective. We are especially indebted to Ibn Hajjâj for considering agricultural activity both as a trade and a science focused on the analysis of cultivated lands.

Abû L-Khayr requires that the farmer be intelligent, sharp-witted, capable of availing himself of basic knowledge relative to agriculture and determining the causes and intricacies of this activity. Ibn Luyûn adds that agricultural trade demands careful study of its four basic factors, namely: grounds, water, manures and tillage. Following along the same vocabulary, Ibn Al-'Awwâm introduced the word "art" to designate agricultural trade. The Arab agronomists were practically unanimous in seeing agriculture as a professional practice resting upon the accumulation of a specialized knowledge. Yet only Al-Tighnarî demonstrated, more so than the previous agronomists, keen attention to definitions and agrarian vernacular. With him, agronomy reaches scientific accuracy and encompasses all of the rules and laws that apply to the improvement of any agricultural land. He doesn't forget, however, to mention how pleasurable for the eyes it should also be. Thus, when discussing agriculture, Hispano-Arabic agronomists defined and specified the contents of a comprehensive agriculture, integrating various aspects of rural economy. Such discussions do not reach for pure speculation, but rather seek to locate clearly agriculture among human activities. Following in the agronomists' footsteps, Ibn Khaldûn defines agricultural science as a branch of physics concerned with the production of food and grain. Abû L-Khayr Tashkoprozada (d. 1554), a turkish scientist, chooses the same classification, but gives agronomy a definition and a specific responsibility: it must not only ensure the growth of cereals, fruit trees, or other plants, but also must constitute a vital contribution to human life. It explains why its name *Al-filaha* derives from the root of the word *falah*, meaning perenniality. It also comprises, concludes this author, curious activities like the production of out-of-season fruits and of new varieties resulting from grafting.[5]

This conclusion provides the link to horticulture, an important topic in the teaching of Arab agronomists. Ibn Al-'Awwâm devotes a large part of its treatise of agriculture to garden art, fruit trees, cultivation, herb and kitchen gardens, aromatic and sweet-smelling plants, naturalizing and grafting. However, we have to wait until the publication in 1943 of the dictionary of agricultural terms by Moustapha Chehabi to find a scientific definition of horticulture considered as *bastana*, or garden culture, with all its developments (floriculture, kitchen gardening, fruit-growing, arboricultural, and ornamental trees).[6]

Social and Cultural Sources of Expansion Within the Arab World

Let us first review briefly some development factors that apply to the whole of the Arab world: first, the role of Islam that made a single religion and literary and scientific language into a source of unity for huge countries until then either insulated or fractured by internal war; second, the role of Arabic language spoken by all people from Baghdad to Cordoba that allowed the circulation of knowledge; third, the role of the Islamic State in disseminating the use of Arabic, translating important texts into Arabic, financing of academies, research teams and libraries, and supporting efforts at economic development.[7] In this context of cultural opening and voluntarist intervention, agriculture engaged in changes characterized by the improvement of farming methods and the development of horticulture. The latter appears more and more dependent on urban growth and demand issuing from cities that became, in the Middle Ages, windows for commerce and opulence. On top of these common development factors, the Al-Andalus region went through a rather original political situation, marked by a succession of centralization and decentralization of power. Although it is still difficult to determine the exact role of these political changes, the period of domination of the Tayfa kingdoms succeeded, it seems, in promoting competition between the larger Andalusian cities and stimulating horticultural progress in their garden-suburbs. Al-Andalus thinkers also singled themselves out, providing an interpretation of Islam favorable to agricultural and horticultural development. Agronomists, living in a context of economic growth, had no difficulty extracting from the corpus of Muslim legislation the recommendations necessary to the development for agricultural productive forces. In an effort to establish an ethnic construction conducive to the progress of rural economy, they called as well on tradition coming from the Prophet as Arabic pieces of advice and proverbs. The whole work offers a methodical presentation of recommendations and advice concerning various aspects of agricultural progress. In this study, we will only point to pieces of advice geared to encouraging tilling or to ensuring better farm-management. Al-Tighnarî and Ibn Al-'Awwâm gather a number of hadiths (sayings attributed to the Prophet) that set agricultural activities in a favorable light, establishing a system of values for agricultural activity. "Allah loves the Believer who practices a trade;" "the production of whoever plants trees or sows seeds later used to feed a man, a bird or an animal will be regarded as alms;" "Allah will reward whoever plants a tree in proportion to the tree's production." The reward is promised to the plowman and the horticulturist even if the product was intended for family needs.[8]

The fulfillment of this exalted trade required the observation of the rules of Islamic morals that demand virtuous behavior in the pursuit of trade activities. Al-Tighnarî turns to the sociological concept of *niyya* (good faith, pure intention) as the basic principle for human action that must help farmers carry out the ideal of Muslim solidarity because all men are not able to devote themselves to agriculture, especially the sick ones and the children. Followers of this principle would behave in harmony with the hadith that says, "Allah helps its admirers as long as those help their kin and kith." Thus, we observe that the main goal of agricultural activity is social stability rather than productivity. In all likelihood, the awareness of the significance of legal and theological factors to gain access to the means of production encouraged the Andalusian agronomists to integrate in their treatises all the precepts of Islam related to land-tillage and horticulture.[9]

The history of mentalities also suggests other factors that explain horticultural dynamics. The revision of relationships between professional and manual work and the valorization of the mechanical arts constitute two major contributions of the Arab scientists of the Middle Ages. A cultural turn took place in the ninth and tenth centuries allowing the emergence of mechanical thinking to seek links between speculative and practical geometry, and ways of using mathematics to solve engineering problems. The advent of the Islamic civilization freed science from the rigid frames of thought imposed by slavery on Hellenistic scientists who could not, in spite of their theoretical discoveries of economic production, move beyond Aristotelian approaches issued from two distant but enterprising cultural capitals. Baghdad, first, where scientists opened scientific applications to the new and promising domains of craftsmanship and public works; then Seville, where agronomists succeeded in implementing changes in and improvements of rural practices.[10] Once mental blocks to technological progress were partly lifted, Arab scientists and engineers were set free to contemplate applications of the mechanical arts to horticulture crafts and hydraulics, in a playful pursuit of usefulness.

Evolution of Horticulture in Al-Andalus Hydraulics

It is most interesting to note the emergence of horticulture as a domain of knowledge and innovation in Al-Andalus. A whole body of knowledge was gradually evolved from explorations in the different fields of hydraulics, botany, and agronomy, the auxiliary sciences that stimulated horticultural progress.

Arabo-Muslim hydraulics went through a "golden age" between the ninth and eleventh centuries. Not only did the most important figures of the Arab School of hydraulics live during this period, but practical experience and progress in the mastery of underground water increased tremendously.[11] We propose to approach hydraulics in the Iberian peninsula as part of the broader development of Arab hydraulics. However specific and privileged it is, we can only understand and interpret it correctly with respect to the formation of the Arab School of hydraulics and the constitution of the Almoravid empires. It is likely that hydraulics originated in Al-Andalus from the corps of Civil Engineers traveling with the Arab armies. This assumption does not exclude, however, possible contributions from Middle Eastern civilians (Syrians, Iraqi, Iranian, Yemenis) and Iberic and Berber populations. After a first phase, lasting until the end of the Caliphate Omeyyade (VIII 2nd to XI 2nd), signs pointing to the constitution of a well-structured hydraulic administration accumulate. Starting in the late eleventh century, one can even speak of the formation of an Andalusian school of hydraulics, not only able to continue the development of the hydraulic infrastructure built during the former centuries, but also to send experts to other areas of the empire. There is at present a large number of studies devoted to hydraulics in Moslem Spain demonstrating both surface and underground water control. Arabs, as soon as they settled in the Iberian peninsula, succeeded in marshaling out river waters for the benefit of irrigation mills and drinking water for the new cities. Dams became, in this context, a major regional development. After the study by Norman Smith, the contribution of the Moslem world and Al-Andalus to the history of dams became relatively well known.[12] In this domain, Moslem engineers could easily tap into the knowledge accumulated in the regions at the heart of innovation and hydraulic engineering since high antiquity, thanks to the vehicular role of the Arab language. The ideas eked out on the Tiger and Euphrates banks could thus quickly reach the most western part of the Moslem world (Sicily, North Africa, Andalusia).

The oldest dams are found on the Guadalquivir River. Arab engineers developed both the technique of the derivation dams and that of weirs (gravity dams). Smith quotes the multiple functions of the Cordoba dams setting mills into motion, protecting the city against floods, providing it with drinking water, and so on. Less spectacular dams were built as well in Valencia, Murcia, and Granada. Archaeology also documented some weirs in the provinces of Jaen and Almeria. All of these dams, linked to a wide and complex network of seguias (canals), provided the main transport and distribution system of water for irrigation. These irrigation channels fitted with more or less sophisticated dispatching devices also provided for water mill operations. We also observe, besides these devices, the use of aqueducts and siphons for overcoming terrain difficulties. Beyond accumulation, transport, and distribution of water, the implementation of mechanical devices set Al-Andalus hydraulics apart from the rest of the Arab world. Textual and archaeological evidence demonstrate widespread use of waterwheels (*Norias*), the scoop-wheel (*saquiya*), or chain of pots, and the other hydraulic systems, such as *shadufs*. But the exploitation of underground water only reached a climax with the adoption of the Persian technology of *Qanat* (underground drainage galleries), introduced in Madrid at its foundation in the ninth century, which gave the impression of a city built on a fresh water sea. The study of *Qanat* is sufficiently advanced today. Mr. Barcelo and his team recently emphasized the great achievements in Majorca, where the network of collecting galleries reached its highest density. The upheavals brought by new water technologies reached beyond the plains and *huertas* (garden-suburbs), mountain zones, where the Arabo-Berber settlers introduced and developed terrace cultivation.[13] Such was, in outline, the course of progress of Andalusian hydraulics at the origin of new horticultural advances.

Diffusion of Plants and the First "Tropicalization": Botany and Garden Art in the Service of Horticulture

The Caliphate of Cordoba (929-1031) stands out as a period of medical, pharmacological, and botanical progress. Historical-biographical sources elicit the presence of botany and horticulture among the new branches of knowledge. Scientists' biographies often refer to the occupations of botanists *(nabâtî)* and horticulturists *(shajjâr)*. The development of botanical knowledge in Andalusia was stimulated at its beginnings by travels to the East, reading of the *Treatise of Plants*, by Al-Dinawari (d. 895), and the translation from Greek to Arabic of the *De Materia Medica* by Dioscorides. A team of Moslem, Jewish and Christian scientists collaborated on this translation which was published in 948 and followed by a large number of revisions and commentaries.[14] Far from resting satisfied with this heritage, the Andalusian scientists succeeded in improving it and augmenting it with their personal discoveries and observations. The list of written works grew richer from the contributions of successive generations of botanists who lived during the times of rulership by the Tayfas kings (1035-1088), reunification by the Almoravides and the Almohades (1090-1229), and the beginning of the Nasride dynasty in Granada (1237-1492). One remembers the names of Abu Ja' far Al-Ghafiqi (d. 1165), Abu 1-'Abbas ibn Al-Rumiyya (d. 1240), and Ibn Al-Baytar (d. 1248), who was one of the most prominent botanists of Islam. His book of Herbs comprises 2,330 paragraphs and refers to 1,400 medicinal drugs, about a third of which are new medicinal drugs introduced into the Arab pharmacopoeia. The progress of botanical studies was closely related to the birth and development of garden art. Thus, people in Al-Andalus lavished specific attention on gardens. Historical sources mention, as early as the eighth century, the first botanical garden of Al-Andalus that Abd Al-Rahmân I (756-788) created in his Palace Al-Rusâfa to remind himself of Syria, where the Omeyyad Caliph sent two emissaries, Yazîd and Safar (whence the name "safari") to fetch the seeds and rare seedlings for naturalization and later dissemination into the whole of Moslem Andalusia.

Safar Ben 'Ubayd, already famous for creating a variety of pomegranate, also initiated the propagation of lemon trees. One can also add to the credit of the Omeyyad dynasty the construction of the Madinat Al-Zahrâ, and its gardens by 'Abd Al-Rahmân III (921-961). These two palatial centers, with their splendid architecture, hydraulic network and array of plants, were used at the same time—as experimental gardens, production orchards and recreation and pleasure grounds—until their destruction in 1010. Far from disappearing with the fall of the Omeyyad dynasty, botanical gardens continued to play an integral role in the diffusion of new plants, especially during the kingdoms of Tayfa,. In Toledo, institutionalization began to such an extent that for Mamun DI-I-Nun (1043-1075) felt the need to employ for his royal garden a medical doctor and botanist (Ibn Wafid) and an agronomist (Ibn Bassâl). Similar developments took place in Sevilla with the construction of the garden of Ibn 'Abbad (1069-1090) called Ha'it Al-Sultan (The Sultan Enclosure); in Almeria where Ibn Sumadih (1085) built a large garden named Al-Sumadihiyya; in Saragossa, Valenica, Tortosa, Granada, and all localities where patrons could support the fashion for gardens. The testimony of Ibn Bassâl, as reported by Al-Tighnari, testifies to the intensity of rare seeds and citrus fruits seedlings being circulated between these cities. The Andalusians did not spare any means to enrich the range of plants cultivated in Spain. And one will always remember the behavior of the cordouan Al-Ghazal, who sent in embassy to Byzantium (ninth century), managed to smuggle out of the country a variety of fig tree prohibited to export.[15] The Almohad rulership (1130-1269) brought about the invention and diffusion, in several cities of the Western Muslim world, the garden model of the *Agdal* or *Bouhayra*. G. Deverdum defines the *Bouhayra* as "a huge enclosed orchard, supplied with a large basin to bring bountiful waters for irrigating the fruit trees." Thus, it designated at first a reservoir and then the meaning was extended to cover the garden and its outbuilding. As for the Berber word Agdal, its etymological meaning designated a private, enclosed park with a water-basin. In fact, *Agdal* and *Bouhayra* refer to the same thing: royal gardens near the Palaces of the Sultans. They are usually very large, divided into several enclosures and surrounded with fortified walls.

This model of gardens with huge water basins achieved its highest geographical extension during the Almohad empire, especially during the reigns of the Caliphs 'Abd Al-Mumin (1133-1163) and Abû Yaqûb Yûsuf (1163-1184). From Marrakech to Seville, including Rabat, Ceuta, and Gibraltar, the Almohads established these decorative gardens supplied with abundant hydraulic resources. The *Agdal* of Marrakech was founded in 1157 by 'Abd Al-Mumin. (Fig. 1) It spreads on almost five hundred hectares of orchards. The *Bouhayra* of Sevilla was built in 1171 by Abû Yaqûb Yûsuf. The size of this garden-making activity is striking. Hydraulic mastery allowed growing and rearing of tens of thousands of fruit and decorative trees.[16] Alongside the royal gardens the countryside gardens contributed to scientific activity. Allow me to pinpoint here that the variety of pomegranates, imported from Syria, was initially acclimatized in an agricultural village in the district of Rayyah (near Malaga) before being sent to the garden of the Al-Rusâfa Palace. Its name perpetuated the memory of Safar who was at the origin of its acclimatization and appears as a pioneer with respect to horticultural experimentation. The

Fig. 1. Grand basin at the *Agdal* in Marrakech, also know as the "Bouhayra" or "Little Sea".

Aljarafe was one among a whole series of experimental stations that spread throughout Andalusia and brought about an agronomic revival and botany between the eleventh and twelfth centuries. Thanks to its natural assets and the vicinity of a large urban metropolis (Sevilla), this region became a center for horticultural research.[17] The plants that migrated from the Eastern parts of the Islamic world toward its most Western parts were numerous and varied (rice, sugar cane, cotton, banana tree, saffron, eggplant, spinach, watermelon, etc.). The sugar cane and the banana tree, for instance, were acclimatized along the Mediterranean coast of Andalusia, from Almeria to Gibraltar. In these warm and mild regions of Al-Andalus the Arabs could, thanks to their hydraulic mastery, carry out the first "tropicalization" in the history of Southern Europe, before the discovery of America would facilitate a second more diverse tropicalization.[18] Historical and geobotanical sources enable us today to better grasp this process, seemingly starting in the tenth century, that led to the creation of a tropical "environment" on an area ranging from the Mediterranean coasts to the valley of the Guadalquivir River.

Classification of Trees and Citrus as "Sour Fruits"

Hispano-Arabic agronomists and botanists did not limit themselves to the description of the species of citrus. They also sought to integrate citrus fruits in a methodical classification. Ibn Bassâl succeeded at the beginning of the eleventh century, in classifying trees into four groups called the "mother of genres" (*ummahat Al-ajnas*): oleaginous trees (olive tree, common bay-tree, the Ben, lentish pistachio tree); gum trees (peach, apricot, almost trees); milky trees (fig, oleander . . .); and aqueous trees (apple, pear, quince, pomegranate, vine . . .). The author distinguishes, in addition to these capital classes, a fifth one, comprising evergreen trees with aqueous sap. Some of these trees, such as the pine and cypress, share gum tree features, others, such as the lemon tree, share aqueous tree features; and still others like the lime tree share oleaginous tree features; finally others like the rose laurel share milky tree features. Abû L-Khayr, who adopted the same system of classification, added to oleaginous

Fig. 2. "Citron" from the ancient citrus fruit collection in Marrakech.

trees the oleander, the terebinth and the walnut tree; and the ficus sycamore, the caprifig tree and the mulberry tree to milky trees. Ibn Al-'Awwâm, building on the expertise of contemporary horticulturists, further refined the taxonomy worked out by his predecessors. He divided the aqueous trees into two categories: trees with heavy water (olive tree, common bay tree, oak, myrtle) and trees with light water that lose their leaves in winter (apple and quince trees, vine, pomegranate . . .). These texts thus outline a rational classification of trees (and more particularly citrus fruits) based on sap properties. (Fig. 2) Following this principle, Andalusian agronomists not only succeeded in integrating the various species of citrus in the kind of aqueous trees, but distinguished between those with purely liquid and those with oily sap.[19] It is worth nothing that G. Gallesio could, in the early nineteenth century, recognize in the writings of medieval Arab scientists the designation of citrus fruits by the common term of "acidic fruits." Contemporary information corroborates this fact. The agronomic treatise, by Ibn Luyun, composed in 1348, makes explicit reference to the species of citrus trees (lime, south orange tree, lemon tree, zanbu) as "sour fruit" trees (*dawat Al-khulul*). This discovery authorizes us to establish a link with the generic term "*agrumi*" (from the arabic radical for "sour," khal) that appeared in Italian in the sixteenth century. However, it is only in 1922 that Gallesio's wish came true and the word "*agrumes*" was adopted in French. Today, citrus fruits are still called in Arabic by their common name of "sour fruits" (*hawamid*), harking back to the middle ages. [20]

Citrus: Transplantation, Environmental Accommodations, Nurseries, Grafting and the Various Uses of Oranges and Lemons in the "Citrus Mania"

In the Middle Ages, generations of Arab agronomists and horticulturists met the challenge of transporting and transplanting various species of citrus from the Asian monsoon climate to the less lenient Mediterranean one in moderately dry and semiarid areas. How did this transfer take place? And, especially, on what ecological knowledge did the men engaged in the citrus fruits ventures rely on, not only to ensure transportation of unknown species, in the Western Muslim world, but to achieve an even more difficult acclimatization? Arabic geographers and historians account for various attempts (ninth and tenth centuries) at acclimatizing in the East new varieties either of citron tree (round citron) or of Seville orange. These attempts ended in half-failures in localities probably insufficiently prepared to receive them. Further inquiry into the causes of these half-failures always reveals a lack of attention to ecological factors. Given the major difference between the climate to the West and East of the Mediterranean sea—less rain fall that came mostly during the months of vegetative rest, long summer drought, and dangerous

Fig. 3. "Tank of Dar al-Hana" (or Resting Pavilion) in the *Agdal* garden in Marrakech.

cold periods (especially in Andalusia)—numerous transplantation failures could be expected. Such would have been the case if the Andalusian agronomists had not elaborated from the very beginning an adequate system of information on citrus fruits ecology. Ibn Bassâl, an agronomist well informed of the environmental circumstances of Eastern agronomy, established the basic premises for citrus fruit ecology. Taking stock of the seven climate areas that the various species of citrus can thrive in, he identified the third climate (India, China, Persia, Iraq, Syria, and North Africa), and noted that their culture is practically excluded in the fourth climate (France, Northern Spain, Roman provinces). Abû L-Khayr recommended avoiding mountainous areas for the lime tree and Seville orange cultivation because "if one plants them in these regions," he says, "they will not be very productive and will fail more rapidly." Elsewhere he adds that "The citrus owes its life to water." This sentence expresses in a pithy way an awareness of new ecological realities, in which citrus fruit trees demand an artificial cultivation to which men must provide as much water as possible. (Fig. 3)

To protect citrus fruits against freezing and the cold, enemies no less dangerous than drought, Andalusian agronomists recommended sheltering them with plaits and covering their feet with leaves and ashes of courge. To succeed in cultivating the new species of citrus and other water-demanding plants, the Arabs of the Western world launched into a huge endeavor for the mastery of river and underground water. But this leap would not have been thinkable in the absence of some initial understanding and capacity for control of the new plants' ecology. Arabic historical sources generally mention transportation of citrus seeds from one place or one country to another. This is probably the privileged form, although not the only one, through which citrus fruits traveled from East to West in the Muslim world. Hispano-Arabic agronomists contribute to the

Fig. 4. "Citron Tree" from the ancient citrus fruit collection in Marrakesch.

improvement of the methods of selection and multiplication of the various aspects. Their teaching embraced all phases of citrus fruit tree cultivation from transportation up to tree-fructification and product transformation. Ibn Hajjâj must be credited with clarifying the terminology of nurseries and their function. The word for nursery "*Al-tarmidanat*," is borrowed from Yunius, who claims it is derived from a Greek word. It applied to places where subjects are first planted before they are taken away and replanted elsewhere. There cannot be a better definition for nurseries that remains: a place where young plants are grown before being transplanted. From an agronomist's point of view, the nursery is twofold: the seed bed and the nurturing bed for cuttings and young seedlings. Pips are first raised in newly made clay pots (*qasari*) or large pots with a hole in the bottom. The transportation of the citrus pips is forwarded directly to these pots or large clay vases. But, before engaging in plant propagation, one should find a site and prepare the nursery. The nursery for citrus fruits must be established on a piece of ground uncultivated for at least two years. It must, moreover, be very well protected from winds. The use of special tools, smaller and better suited to the care of young seedlings often laid out in tight rows is also recommended. The multiplication of citrus fruits by sowing of seeds is certainly "good and efficient" as explains Ibn Bassâl, but it takes too much time, hence the use of cuttings. Several ways of transplanting from nursery to orchard were experimented with; when the plant could not be removed with all its roots, horticulturists used cuttings by bed (*takbis*) and of cuttings out of pot or funnel (*istilaf*) for their transplantation. Grafting was employed both to multiply and to improve the various citrus varieties. Often graftings were transported from one country to another or covered great distances between two regions. For the voyage, the grafting shoots were stored "in a vase with a narrow opening that had never contained oil, but only salt water before this use." The graftings were placed dry in these vases and locked up to prevent being harmed by the wind.

It is likely because of these precautions that citrus graftings could make increasingly long voyages and spread more quickly to new grounds. Besides farming technique progress, Arab citrus cultivation developed in the Middle Ages a practical sense of nurture. The agronomists' teachings provide for each variety of citrus a kind of technical guideline containing all of the useful elements relative to the choice of soil, tilling method, manuring, irrigation, and shearing size of citrus fruits.[21] Thanks to the progress of horticultural techniques, Muslim Spain during the eleventh and twelfth centuries went through a kind of citrus mania, similar in intensity to the "tulipomania" that seized Dutchmen a few centuries later (seventeenth century).[22] Arabs at this time engaged in horticultural luxury, cultivating citrus fruits not only for using their fruits, barks, and pips, but also for decorating and embellishing their gardens. (Fig. 5) The number of known uses of the different citrus varieties defies the imagination. Nabatean agriculture reckoned with twenty-one uses of the citron tree: thirteen for its fruit, six for its leaves, and two for its branches and roots. The Sevilla orange tree was appreciated for the beauty of its foliage, the quality of its juices and syrups, the fragrance of its flowers used to manufacture perfumed essences, and for the use of its bark in medicine. An anonymous botanist of Sevilla (eleventh century) wrote the first description of lemonade and showed the value of lemon essence to treat "cold illnesses," especially those striking fishermen and navigators. Yet the most complete description of the various uses of lemons is to be found in

Fig. 5. Ancient citrus fruit collection, Marrakech.

the *Treatise of Lemon* by Ibn Jami, translated as of 1602 into Latin. This fruit described six varieties of drinks, ranging from sweetened lemonade to lemon liquors and vinegars.[23] Uses of citrus fruits spread and diversified so much that all means to multiply them and improve their production were sought. This fever that we call the "citrusmania" could only be alleviated by creating gardens used not only for acclimatization places, multiplication, and amelioration of citrus fruits, but also as places where the creative imagination of Arab agronomists and gardeners could indulge in the quest for the strangest grafted trees, for fruits with extraordinary forms and for ornamental uses of the different citrus species and varieties. Botanical gardens, beginning in the eighth century, sponsored by huge fortunes and stimulated by a deep desire to propagate the new plants, provided the initial impulse. These gardens demonstrated possibilities that spread like fire to the social elite first and later to an increasingly broad population that could not resist the temptation to propagate citrus fruits in their gardens, and even later the courtyards of modest dwellings.

Conclusion

Reaching the end of this study, we can gauge from the topics covered how much the history of horticulture would gain from being reexamined. Can we think today of the Mediterranean world without its citrus landscapes, sugar cane, and rice plantations? Can we imagine it without the innumerable plants that the Arabs introduced in the south of Europe (Andalusia, Sicily) in the Middle Ages? Yet these horticultural changes did not prevail without conflicts and resistance. These innovations grew to such an extent that they did not fail to stir opposition not only from traditionalists, but also from some scientists. Al-Tighnari, the most brilliant agronomist of Granada, fell prey to the folk superstition that attributed all the misfortunes of Andalusia in the eleventh century to the propagation of the Sevilla orange tree. According to him, whenever it grows it brings only ruin and desolation to its owners. Yet he was the author who wrote with the greatest competence about citrus cultivation. Superstitions about the Sevilla orange tree in Al Andalus in the eleventh century call to mind the misfortune of the potato in sixteenth-century Europe. The plant of American origin will be made responsible for all evils. We may come to believe that all of these resistances of a psychological and mental nature are the tribute that horticultural innovations must pay, throughout history, to succeed in changing food practices and our cultural landscapes.

NOTES

1. Mohammed El Faïz, "La révolution agricole dans l'Espagne musulmane est-elle mesurable?", in *Histoire et Mesure*, vol. XIII, 1998, édition CNRS, Paris; A. M. Watson, *Agricultural Revolution in the early Islamic World* (Cambridge University, London–New-York, 1983).

2. Mohammed El Faïz, "L'apport des traités agronomiques hispano-arabes à l'histoire économique d'Al-Andalus," in *Ciencias de la naturaleza en Al-Andalus. Textos y estudios* III, ed. Expiración Garcia Sánchez (Granada: Consejo Superior de Investigaciones Cientificas [CSIC], 1994), 406.

3. Victor-Donatien de Musset Pathay, *Bibliographie agronomique* (Paris: édition Colas, 1810), 97.

4. El Faïz, "L'apport des traités agronomiques hispano-arabes à l'histoire économique d'Al-Andalus," op. cit. 412.

5. Ibn al-Awwâm, *Le Livre de l'Agriculture, trans. J.J.* Clément-Mullet, revised edition with an introduction by El Faïz (Arles: Edition Actes-Sud, 2000), 15.

6. L'Emir Moustapha Chehabi, *Dictionnaire des termes agricoles arabe-français* (Le Caire: Imprimerie Misr, 1957).

7. Ahmad Y. al-Hassan and Donald Hill, *Sciences et techniques en Islam, trans.* Hachem el Husseine (Paris: Edifra, UNESCO, 1991), 7-9.

8. El Faïz, "L'apport des traités agronomiques hispano-arabes," op. cit., 416-17.

9. Ibid., 419.

10. El Faïz, *Histoire de l'hydraulique arabe* (Paris: Actes Sud,, 2005), 317.

11. El Faïz, "L'apport des mécaniciens arabes à l'hydraulique médiévale," in *La Houille Blanche* (Revue internationale de l'eau), no. 4/5 (2002): 89-90.

12. Norman Smith, *A History of Dams* (London: P. Davies, 1971).

13. El Faïz, *Histoire de l'hydraulique arabe*, op. cit., 225-26.

14. Max. Meyerhof, "Esquisse d'histoire pharmacologique et botanique chez les Musulmans d'Espagne," in Al-Andalus III (Madrid-Granada: CSIC, 1935).

15. El Faïz, "The Aljarafe of Sevilla: an experimental garden for the agronomists of Muslim Spain," in *The Authentic Garden: A symposium on gardens,* eds. Leslie Tzon Sie and Erik de Jong (Leiden: Clusius Foundation, 1991); Expiración Garcia Sánchez and A. Lopez y Lopez, "The Botanic Gardens in Muslim Spain," in *The Authentic Garden*, 79.

16. El Faïz, *Les jardins historiques: de Marrakech: mémoire écologique d'une ville impériale* (Florence: EDIFIR, 1996); *Les jardins de Marrakech* (Arles: Actes-Sud, 2000).

17. El Faïz, "The Aljarafe of Sevilla . . . ," op. cit., 139-40.

18. Jean Sermet, "Acclimatation: les jardins botaniques espagnols au XVIIè siècle et la tropicalisation de l'Andalousie," in *Mélanges offerts à Fernand Braudel* (Toulouse: Privat, 1973).

19. El Faïz, "Les agrumes dans les jardins et vergers de l'Occident Musulman (VIII-XIVè siècle)," in Il Giardino Delle Esperidi, eds. Alessandro Tagliolini and Marghertia Azzi-Visentini (Florence: EDIFIR, 1996); Françoise Aubaile-Sallenave "La greffe chez les agronomes andalous," in *Ciencias de la Naturaleza en Al-Andalus* III, ed. Expiración Garcia Sánchez (Granada: CSIC, 1994).

20. Ibid., 116-17.

21. *Il Giardino Delle Esperidi*, op. cit.118-21.

22. D. Onno Wijnands "Commercium Botanicum: the diffusion of plants in the 16th century," in *The Authentic Garden: A symposium on gardens*, op. cit.,79.

23. Andream Bellunensen, *De Llimonibus tractatus Embitar Arabis* (Paris: édition P. Chevalier, 1602).

*Botanical Progress,
Horticultural Innovations
and Cultural Changes*

*Botanical Progress,
Horticultural Innovations
and Cultural Changes*

*Botanical Progress,
Horticultural Innovations
and Cultural Changes*

*Botanical Progress,
Horticultural Innovations
and Cultural Changes*

Cultural Values and Political Change: Cherry Gardening in Ancient Japan

Wybe Kuitert

Japan has a young and volcanic geology, also the erosion of slopes is always high, due to significant precipitation. This generates a dynamic and intricate topography that formed a setting for primary forests where differing patterns of vegetation evolved in isolated areas. One and the same cherry species could develop separate varieties in regions that were isolated, but actually not so far removed from each other. In its isolation, any cherry variety remained quite variable, and ready to hybridize whenever it was brought together again with another variety. With human occupation, the cherry became a follower of civilization. Specific cultural values became attached to flowering cherries as a desirable plant, and man started to transport them in ancient Japan. For political reasons, hitherto isolated varieties of *Prunus serrulata* Lindley were brought together, finally inducing hybridizations leading to cherished garden forms. In the early seventeenth century, collections of such cherries were assembled, forming the start of a second phase of cherry gardening on an even wider scale. Some ancient, singular garden forms circulate at present as narrowly defined clonal cultivars. This later history is not treated here.

Records are not generous with facts when it comes to such details as gardening with cherry hybrids. Portraying circumstances this case study nevertheless intends to show how political change and cultural values provided foundations for hybridization and gardening with cherries in ancient Japan.

Varieties of Flowering Cherries

Three varieties of *Prunus serrulata* are introduced here as leading characters of the following pages.[1]

The Japanese Mountain Cherry (*P. serrulata* var. *spontanea* (Maximoxicz) Wilson), is found in the southern half of Japan. It is a tree of the more open forests of the foothills where it comes up with a rather ascending tree shape. It is quite variable: peculiar specimens of the Japanese Mountain Cherry can be found among its seedlings in the wild. Above all, the deep red coloring of young sprouts and all parts of the flower, except the white or pinkish petals, can be spectacular in the blossom season. Flowers are not, or just slightly, fragrant; double-flowered forms have been found (Fig. 1).

The Ōshima Cherry (*P. serrulata* var. *speciosa* (Koidzumi) Koehne) is found on the island Ōshima and neighboring islets and coasts of eastern Japan. Adapted to the young and volcanic geology of the region, it matches an easy germination of the seeds to an even greater variability than the variety above. It has a broad and spreading tree shape, adapting to the island forests. Typical are the bristled leaves and large, white flowers. In the wild, one may find double flowering forms, even with a pink shade. Flowers of the average Ōshima Cherry are fragrant.

1. Distribution of *Prunus sargentii* and *P. serrulata* var. *spontanea*.
o= *P. sargentii*. ●= *P. serrulata* var. *spontanea*. Adapted from the
Flower Association of Japan, *Manual of Japanese Flowering
Cherries* (Tokyo: Nihon Hana no Kai, 1982).

A third cherry dealt with here is the Korean Mountain Cherry (*P. serrulata* var. *pubescens* (Nakai) Wilson). Westerners, knowing this cherry only from Korea, have given it this confusing English name. It is also native to a wide region of Japan. It typically has hairs on details like leaf and flower stalk; fall colors can be striking. The Korean Mountain Cherry shows variability in its native habitat; indeed, double-flowered forms may be found; flowers are not, or hardly, fragrant (Fig. 2).

In untouched nature, flowering cherries, whatever species or variety, are trees from the open forest edge or from clearings. A clearing appears when an old, big tree falls, or when man comes in and starts felling trees. Japan's colonization by a fifth-century wave of immigrants from the continent went together with large-scale clearing of the shady primary forests, giving new opportunities for cherries. They increased in number inevitably as a follower of civilization.

Cherry Appreciation in the Nara Period (710–784)

In the early eighth century, a stable political system led to the flourishing of the capital Heijōkyō, the modern city of Nara that gave its name to the Nara period (710-784) (Fig. 3). An urban society came to flourish around the imperial court that was set up after Chinese models. Court culture was steeped in the Chinese example, so that we find a continental vision of the cherry as well. Cherries, when combined with the willow, served to assure the coming of spring, seen in association with the firm ruling of an emperor. In fact, the spring show of pinkish cherries and fresh green willows was one of the pointers demonstrating the Chinese emperor's mandate of heaven to rule the empire.

But poetry of the Nara period, as can be seen in the Man'yōshū (a compilation of poetry done in ca. 770), gives a far more complex image of the cherry. The complexity arises from the free and whole-hearted approach of the Man'yōshū. It compiled basically everything from previous centuries that was considered poetry, taking folk songs alongside poems that are instant scribbles or serious compositions. In any case, the cherry is associated with brightness and cheerful beauty. Some poems describe the cherry in the landscape where it is seen from a distance in or at the hills, simply describing the landscape.[2] But in many a case such a serene landscape evokes memories of some beloved or beautiful person. Then the cherry comes to stand as a parallel, acting almost as a representative of the one beloved. An example is the poem by Harima no Otome. She is awaiting her lover, the minister of Ishikawa who after many years will return to the capital now. On the occasion she presented her poem; it is spring in the year 719 and cherries are in bloom on mount Tayuraki:

> *Tayuraki no yamano onohe no sakurabana*
> *sakamu haruhe ha kimishi shinohamu*
> Cherries in bloom on mount Tayuraki,
> Every spring they blossom, I think of you dearly in love.

2. Distribution of *P. serrulata* var. *pubescens* and *P. serrulata* var. *speciosa*. ● = *P. serrulata* var. *pubescens*. ▲ = *P. serrulata* var. *speciosa*. Adapted from the Flower Association of Japan, *Manual of Japanese Flowering Cherries* (Tokyo: Nihon Hana no Kai, 1982).

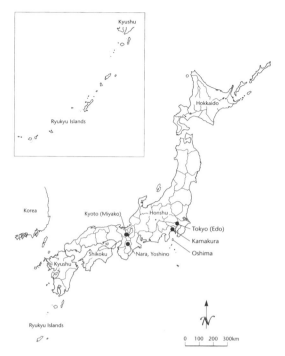

3. Map of Japan with selected cities and islands important in cherry ecology and history.

Other poems of the Man'yōshū follow similar imagery, matching the sight of cherries on a hillside to a memory of something beautiful, like a beloved.[3] In the landscape, a secondary forest had replaced the primeval forest, and the cherry was increasingly seen on the mountains surrounding a valley where humans lived. Together with the cry of the deer in autumn or other seasonal details it assured that the world was simply beautiful and right.[4]

We come across the cherry not only in poetry but also in plain records. There is clear evidence that a double-flowered cherry tree was presented several times to the temple Kōfuku-ji in Nara city. This simple fact is repeated in abundance in later history, where fact and fiction often interplay to form a most exciting legend. It tells of the historic emperor Shōmu (in power 724-749) who left the city in spring for an outing to nearby Mount Mikasa. In a valley at the side of a path, he was struck by the splendid beauty of a peculiar double-flowered cherry just in bloom. Having returned home he told his wife empress Kōmyō about it.[5] The emperor's account excited her and she let it be known that she wished to have a branch of the tree, so that she also could enjoy this unusual beauty. Servants were sent out to fetch a branch. But finding the tree, they dug it up, root and all, and brought it back, planted it in the palace garden to be a joy in spring for every year to come. But not for long. In the reign of Kōken (in power 749-758), daughter of Shōmu and Kōmyō, Nara culture was flourishing and the big temples, such as Kōfuku-ji, were indeed at the zenith of their power. Kōfuku-ji was the family temple of the Fujiwara clan, always entangled in intricate intrigues of power and political marriages with the imperial court. The priests of Kōfuku-ji, knowing that the court had taken a cherry from mount Mikasa, were not amused. They simply took it back and planted it in front of one of their temple halls, to become a famous cherry and the pride of the priests for years to come. Whether the story is true up to the last detail is not so important at this point. It is at least the first time in history that cherry beauty was the object of jealousy and greed in circles of the highest power. We will return in detail to the double-flowered cherry from Nara later.

Cherries for a Courtly Capital

After the Nara period had come to an end, the court was moved to a new city, about seventy kilometers to the north in 794 A.D. The new capital was known as Heiankyō, the modern city Kyoto, a name we will use here. Kyoto functioned as a capital for the almost four centuries of the Heian Period (794-1185). A stable political structure with tributary agricultural provinces again ensured a flourishing of urban culture.

4. The cherry woods at Yoshino were kept up by continuous planting of cherry saplings by pilgrims. *Wazakura Yoshinoyama shōkeizu,* 1713. (Collection of Kyoto University).

The Nara Chinese models were slowly molded into highly refined, truly Japanese expressions; also the status of the cherry was elevated from straight poetic to an even stronger symbolic meaning. Somewhere between 834 and 848 A.D. the plum tree (*Prunus mume* Sieb. & Zucc.), traditionally planted together with a citrus tree in front of the main hall of the imperial palace, was replaced by a cherry. When the palace buildings were destroyed by fire in 960, the tree perished. At the rebuilding a new cherry was planted, this time clearly documented as a tree brought from the Yoshino mountains close to Nara.[6] Yoshino has a seminatural forest with dominant presence of the Japanese Mountain Cherry. Ever since the seventh century the trees were considered holy therefore forbidden to be cut. Apart from a few large-scale planting actions, pilgrims used to plant cherry saplings that were sold year in and year out by villagers. It made for a small-scale but steady management in keeping the cherry woods up (Fig. 4). There are no reasons to believe that the botanic identity of the present cherries differs from these early centuries. We can therefore safely assume that it was the same Japanese Mountain Cherry (*P. serrulata* var. *spontanea*) that entered the capital Kyoto after its founding.

The arrival of a Yoshino cherry to the imperial court went together with a growing appreciation: poems about the cherry increase in number. Memories of the glorious days of uncomplicated life in Nara became associated with the city's cherries, adding an air of nostalgia or even melancholy.[7] Although hardly scented, its fragrance entered the poetic mind to be enlarged as a nominator for glorious brilliance. The role of women inventing a native script that had the potential to express such subtle emotions is clear. Such a woman was Jōtō Mon'in Shōshi (988-1074), one of the emperor's wives. Concerned about the double-flowered cherry of the Nara period, Jōtō Mon'in took pity on the famed cherry, still in the uncertain hands of Kōfuku-ji's priests in the now deserted old capital Nara.[8] In private, a meeting was held with the official at court responsible for

5. 'Nara-no-yae-zakura' is a double form of *P. serrulata* var. *pubescens,* photo by author, cherry in author's collection Netherlands, 4/25/2004.

Kōfuku-ji, where it was decided that the cherry was to be moved to the imperial palace in Kyoto. Servants were sent out with an oxen-pulled cart to uproot and fetch the tree. They managed to root-ball and load it; but then priests and monks of Nara became alarmed and clamor and protest sprang up. Nothing else could be done than unload and replant the cherry to its old position in front of the Tōendō, a hall of Kōfuku-ji. When the report of this happening reached Jōtō Mon'in, her reaction was unexpected. Rather than being angry, she seemed happily surprised about the priests being so sincere in their love of the cherry.

Supported by other historic records we know that she decided to dispatch a yearly guardsmen service for the seven days that the tree was in flower, to protect it even better. As guardsmen she appointed the villagers of Yono (now in Ueno city, Mie prefecture), over the hills about twenty-five kilometers to the east of Nara. A strange move it seems at first, but Jōtō Mon'in must have been well informed: it was the villagers from Yono who had presented double-flowered cherries to the temple of Kōfuku-ji several times in the course of the Nara period. In 746 they even wanted to donate the double-flowered cherry to the emperor Shōmu himself. The village records state that the eighth-century emperor had rejected such a lowly present and returning home the villagers had just planted it at mount Mikasa.[9] It can hardly be a coincidence that the emperor "discovered" the cherry in the wild, so shortly after. Anyhow, three centuries later, in circles around Jōtō Mon'in, there was excitement about her strategic decision to appoint the villagers as guardsmen. A young woman named Ise-no-Taifu, in service of Jōtō Mon'in, was aspiring to a post as courtesan; for her approval, she wrote the following historic poem:

> *Inishieno Narano miyakono yahezakura*
> *kefu kokonoheni nihohi nurukana*
> Old capital Nara's double cherry
> Now in the court of Kyoto:
> brilliantly it blossoms

Ise-no-Taifu's poem is a beautiful and rhythmical play of words employing such tensions as between "old" and "new," a beauty of language hard to show in an English translation.[10] The poem suggests that the double-flowered cherry had arrived in the capital. It is likely that the villagers of Yono had brought it once more to the palace, this time in Kyoto. In centuries to follow, more records and poems refer to this double-flowered cherry in the capital.

Speaking of botany, the double-flowering cherry of Nara is a form of the Korean Mountain Cherry, *P. serrulata* var. *pubescens*. In its native habitat indeed double-flowered specimens are found, and apparently these are so stable that the double-flowered form of *P. serrulata* var. *pubescens* is well established in the wild, close to, exactly, the village Yono in Mie Prefecture.

6. In the wild one may come across specimens of P. *serrulata* var. *sportarea* with particularily deep-red coloring of young sprouts, photo by Kense Kuitert, cherry in Kyoto, Japan, 4/1/1997.

There is hardly any doubt now that the villagers donated this double-flowered cherry. The stability in the wild backs up the history that we are dealing with the same form found in the eighth century. It has small flowers with about thirty deeply bifid petals tightly set together, giving a precisely regular fringe of the flower (Fig. 5). At least three clones have been identified that in their botanic details are hard to distinguish.[11] The flower season is rather short, and also two weeks later than the Japanese Mountain Cherry, making chances for a natural hybridization very small.

In centuries to follow, cherries became the object of a wider interest of court nobles such as, for example, Fujiwara Teika (1162-1241). He is known as a literary man, commissioned by the emperor to help in compiling an important poetry collection of the days. But his personal diary is of even more interest: it speaks of cherries in quite some detail. We find remarks on an early cherry that flowered even before the plums (*P. mume*), and another late-flowering, pink, double cherry. He records the planting of cherry trees, including double-flowered ones, in gardens of various noblemen. Notes on propagation are found in 1226 on the twenty-seventh day of the first month. The weather is fine and the diary gives a short remark on dividing with a knife or scissors a branch of a double flowering cherry from a small tree, apparently to propagate. The scene takes place in the garden in front of a main hall. The twenty-seventh day of the first month would be late February in our modern calendar. Buds would already be out of their winter rest, therefore too late for preparing a graft, but for propagation by layering, late February should work.[12]

Cherries are fully used as garden plant in fashionable circles of the court entering the fourteenth century. A comment from about 1330 states:

> . . . Trees to plant at the house are pine and cherry. As pine the five-needled one is preferred (*Pinus parviflora* Sieb. & Zucc.) and for cherries single-flowered ones. Originally, double cherries were only found in the city of Nara, but recently these are found everywhere. However, of old, the cherries from Yoshino and the cherry in front of the main hall of the Imperial Palace were all single-flowered. Double cherries are grotesque, badly misshapen and distorted. Therefore it is better not to plant them. They are late in flowering, present a disastrous sight, and because of the bugs they are not to be preferred . . .

Conservative on the point of its appreciation of double-flowered cherries, this comment nevertheless shows us that the triumph of the cherry as a garden plant had taken a start. Double cherries are to be understood in this quote as the one from Nara but it also points to other cherries. The one from Nara is not particularly attacked by bugs.[13]

Political Symbolism: Ōshima Cherries Enter Kyoto

Political winds were turning worse and worse for the emperor and his court in the course of the twelfth and thirteenth centuries. Warlords, already largely in control of the country for more than a century from their government seat in eastern Japan, occupied the capital in 1336. Led by the shogun, they asserted power by choosing their own emperor. The displaced emperor fled to Yoshino, a place seen as an original homeland full of wistful memories for the imperial family, dramatically felt when Yoshino's cherries were in bloom. Also the shogun was well aware of the cultural values of the cherry. Urged by the clergy of the new Zen buddhism he founded a temple at Arashiyama in 1339 to soothe the soul of the exiled emperor who died in that year. Again cherries were brought from Yoshino, the emperor's last place of living, to be added to cherries planted earlier as part of an imperial garden.[14] Arashiyama had held the palaces of important court nobles such as Fujiwara Teika and several emperors. Therefore, it is obvious that not only were souls soothed here but also a claim to power was asserted.

What about the cherries coming in? Cherries were brought from the Nara region, like the double flowered one from Yono; single cherries came from Nara's Yoshino mountains. There, the villagers must have selected particularly nice specimens for cherry orders from the capital. But because of transport problems, trees would have been young and could only show their beauty at a young age. A young plant of the Japanese Mountain Cherry starts to flower after several years at a size when it is already more difficult to transport. Therefore, in selecting young cherry plants, the villagers of Yoshino would have chosen nice foliage colors rather than finding a particular beauty in flowers.[15] Sprouts of specimens in the wild at Yoshino are mostly brownish, few have yellowish, green, or red tints. The red ones are rare, but conspicuous and can develop their color most spectacularly (Fig. 6). At present, the Arashiyama-area in Kyoto is known for its red-sprouted cherries. The area brought such garden cherries as 'Arashiyama' and 'Tagui-arashi'; these are single-flowered forms that show a remarkably red coloring of leaf sprouts, but also of bud scales, bracts, calyx and sepals. A semidouble form 'Hōrinji' from this area has the same deep coloring. All are clearly selections from the Japanese Mountain Cherry (*P. serrulata* var. *spontanea*), and not hybrids.[16] Whether red sprouts were required in the capital or, on the contrary, promoted by Yoshino, we do not know, but the spectacular contrast of white blossom and deep red sprouts became the beauty standard for single-flowered cherries.

The planting of Yoshino's cherries at Arashiyama was a clear statement of power. But an even more evident cherry planting came in 1357 when the warlords put their own cherry in front of the main hall of the imperial palace in Kyoto.[17] Enforcing the symbolism of the deed, this new imperial cherry was brought from their native land, Kamakura in the east. Off the coast at Kamakura lie the Ōshima Islands, the new cherry was in fact a form of the Ōshima Cherry (*P. serrulata* var. *speciosa*). It has been identified as the cherry 'Kirigaya.'[18] It comes so close to the native Ōshima Cherry that it is judged as just a superb selection from the wild. 'Kirigaya' is a healthy form with single flowers, fragrant, and quite large. They measure up to five centimeters in diameter or even more when well manured and maintained. Indeed, from this basic quality as a garden plant, it is highly valued over just any wild specimen of the Ōshima Cherry.

Overt testimony to the symbolism of the cherry comes about a century later. The mid-fifteenth century is a time of political unrest and small skirmishes, leading to the ten-year Ōnin Wars (1467-1477). Most precise accounts of historical fact that include cherries are now found in diaries of priests in Kyoto. They had their temples sponsored by the military rulers. At the time a cherry 'Fugenzō,' famous from a tree in a Kamakura temple, had arrived in Kyoto as recorded by a Buddhist priest of the Zen sect, Ōsen Keisan (1429-1493). Keisan used to take his guests to the cherry when it was in flower. It stood close by his residence in the temple Shōkoku-ji, just north of the imperial palace. During the warfare he could not visit it, and when he returned to enjoy the spring show for the first time again, he felt as if ten-year-old buds were bursting into bloom. The intensity of the experience after a period of political stress made him rush to a remarkable conclusion. Although hardly an established nation when compared with neighboring China he noted:

> Our country is the country of the cherry (*sakura*). However, one rather simply says "the flowers" (*hana* in Japanese) rather than saying cherries. It has the same precious meaning as the Peony has for Chinese Loyang, or the Crab Apple for Szechuan in China.

Then he turns to 'Fugenzō':

> People say that there is a temple hall in Kamakura where the saint Fugen is enshrined, but there is also a cherry in the compound and it is called therefore 'Fugendō' (as the name of the hall (*dō*), for Fugen). Others call this cherry 'Fugenzō,' (meaning Fugen's elephant (*zō*)), because the words for "flowers" and for "nose" in Japanese both have the same pronunciation (*hana* for both). The big and white flowers resemble the nose of the white elephant on which Fugen rides. Both explanations are correct. This cherry is now found in the west of Kyoto, and it really is an excellent flowering cherry.[19]

This Kamakura cherry is even more spectacular, having all the qualities of 'Kirigaya,' but on top of that, the flowers are double, and stay much longer on the tree. Also, for a double-flowered cherry the tree's height is impressive, reaching up to eighteen meters with proper care. Since the planting of 'Kirigaya' in the sacred palace grounds a second spectacular garden cherry 'Fugenzō' had arrived from the eastern provinces of Kamakura, about four hundred kilometers to the east of Kyoto. Cherries were therefore moved about the country over long distances. Hitherto isolated plant material of the highly variable Ōshima Cherry was brought to Kyoto, whereas Japanese Mountain Cherries had been brought to the city ever since the centuries of its founding. Cherry populations that had developed separate identities in the course of their evolution were brought together and could start to hybridize, a process that would lead to a number of distinctly different forms in the following centuries. Contrary to the double flowered cherry from Nara, which is a form of the Korean Mountain Cherry, the Ōshima Cherry and the Japanese Mountain Cherry flower at the same time, giving full and free play to the variabilities inherent to both. But before turning to the hybrids, it should be explained how the cherries were used in the gardens of the period. Named forms of the cherry can only be understood from the way they were discovered and enjoyed as garden plants.

Cherry Aesthetics in Garden Design

Cherries form a part of the garden world in literature of the time. Setting the tone is *The Tale of Genji*, written in the years around 1000 A.D., and a famous classic in courtly circles ever since. *The Tale* describes to us the fictional world of a courtly prince engaged in endless love affairs in a setting of palaces and gardens. One garden designed as an arrangement of lakes and hills is the stage for a boating party with music: " . . . the hills were high in the south-east quarter, where cherry trees were planted in large numbers. The pond was most attractively designed. Among the plantings in the forward parts of the garden were cinquefoil pine, red plum, cherry, wisteria, Kerria, and rock azalea, most of them trees and shrubs that are enjoyed in spring." Or in a later section: " . . . The branches caught in mists from either side were like a tapestry, and far away in

Murasaki's private gardens a willow trailed its branches in a deepening green and cherry blossoms were rich and sensuous . . ."[20]

The distinction between the forward planting and the background is described impressionistic and convincing. The forward part of the garden is in fact the open area just in front of the veranda of the main hall of a typical nobleman's mansion. The usual "forward garden" (*senzai*) could have solitary plants, including a cherry. The hills in the south-east are likely to be situated at the far end of the garden, these have cherries in large numbers as a background scenery to the pond that lies in front of it when seen from the main hall. The second quote has cherries and willows as background, a combination that heralds a virtuous ruler in the classic, Chinese manner. The cherries here were probably even standing outside in other gardens.

The Tale of Genji inspired many a courtly garden, even up to the design of cherry planting. For example, an emperor's residence known as Kameyama-dono was laid out on the site of Fujiwara Teika's retreat at Arashiyama. Work started in 1255. A spacious garden was made with a pond fed with water from the Ōi River that runs between the hills of Arashiyama. In front of the hall of Kameyama-dono stood a cherry, the opening of its flower buds were the reason for an imperial visit in the spring of 1263, the sight was more beautiful than ever, as a source mentions. Outside the residence, large numbers of cherries brought from Yoshino were planted on the foothills of Arashiyama. Seen from the imperial pond garden, the cherries on the hillside opposite the estate formed a visual background beyond the garden.[21] In close-up, by contrast, it is the beauty of the botanical details of the budding blossom that forms the joy of the cherry in front of the residence itself. Cherries are employed in foreground and in background design.

Apart from such aesthetics, the cherry was also used in a more formal design. From records but also sources such as *The Tale of Genji* we know that a cherry was typically found in the small yard used for imperial football (*kemari*); it was either planted on one, or on all four corners of the field as a corner tree.[22]

Strikingly, the eleventh-century famous garden text *Sakuteiki* does not mention principles of cherry planting, although in poetry, fiction, and historic fact it is clearly part of the garden world. On the whole, the *Sakuteiki* is remarkably scarce on aesthetic principles of planting design, mentioning only twenty or so names. Hardly any of the woody plants in the above quotations from *The Tale of Genji* features in this garden text. Only when treating various styles of setting stones, or designing islands it gives some scattered comments on what kind of plant would associate best with certain rock styles; these are mostly perennial rather than woody plants. Only at the section on tree planting is there mention of "flowering trees," to be understood as cherries. These should be planted in the east, whereas trees with fall colors must be in the west. Referring to the classic zodiac the *Sakuteiki* actually repeats what is said in the orthodox Chinese theory of the Five Elements. It resembles the planting design in the four quarters of the football field. As such it is stating a fact, rather than giving advice or an idea on garden design.

The *Sakuteiki* was written by a nobleman to present a personal statement to his public. Given the overt indications and symbolism of plants in poetry and literature, it is obvious that advice on plant design could only repeat things that had been said already many times before. It would be so plain and blatant in a treatise specializing on gardens that it would turn the whole effort into just one banal demonstration of writing down what any noble knew already. So the reason for absence of advice on the design of planting is obvious; at one point the writer indeed condemns superficial garden wisdom.[23]

Centuries later a division in appreciating cherries in the fore- and background is found in a second treatise on gardens, this time by gardeners who were not very well versed in classic literature. Nevertheless, they were well aware of the proper

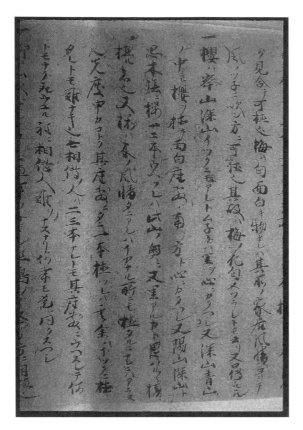

7. A detailed design strategy for cherry planting appeared in the fifteenth century handwritten manual *Sansui narabini yakeizu*. The rather unpolished text and writing betray the practical purpose of the text (reproduced from facsimile, *Sonkeikaku Sōkan,* Tokyo: Ikutokuzaidan, 1930).

aesthetic of cherries in the garden, and had even developed a design strategy to reproduce it as a man-made work of art. From olden times in Kyoto, gardeners are living above all in the north-western parts of the city. At the founding of the capital it was here that naturalized Koreans and Chinese settled. These men had come as civil engineers, architects, and craftsmen to help build the new capital. Civil engineering meant placing foundation rocks for buildings, arranging the water system, building ponds, or making gardens. In the course of time, a temple in the area, Ninna-ji, close to Arashiyama, became the center of professional gardening: it forms the origin of the fame that gardeners of this part of Kyoto still have. Ninna-ji itself was organized as a temple compound sponsored by the imperial court from the end of the ninth century onwards. In fact it was a residential area in which courtly residences were developed as small temples, all belonging to the Ninna-ji compound. From the end of the ninth century until the late twelfth, the area had about seventy of such courtly temple residences.[24] In the course of these centuries, the typical residence came to have one main hall (*shinden*), that was styled after the imperial palace itself. It faced a garden or yard to the south of it, and one may imagine that many a hall had its cherry as in the palace. Entering the fifteenth century, fashions in architecture changed. Instead of one main hall, a less rigid arrangement of smaller buildings had the most luxurious room to receive guests as the center piece of the temple residence. Again, this main room faced a garden south of it. The area around Ninna-ji remained upper-class residential and had still more than sixty temple residences at the outbreak of the Ōnin Wars in 1467.

The gardeners of Ninna-ji were well aware of the courtly appreciation of the cherry, which is clear from a text that stems from their circles. The manuscript *Sansui narabini yakeizu* (dated 1448 as well as 1466) is a manual, written and compiled by low-ranking priests of esoteric buddhism as it was practiced in Ninna-ji. The rather unpolished text must have served as an instruction manual, listing all possible information on the subject as an aid to the memory when teaching others that probably could not read. It holds a wealth of practical hints and technical ideas that makes this source into a veritable gold mine on medieval gardening. The manual's advice is quite detailed on the planting design of cherries, for which it gives a separate section (Fig. 7). Departing from the typical situation of the main room of a residence, facing south, it describes an advanced planting principle:

> Cherries may be found on peaks and deep in the mountains, but only if the countryside is beyond, and that is what you have to keep in mind. Thus, it is interesting to plant cherries in between the deep, dark green (garden) hills, keeping in mind that the main room (*zashiki*) has south as its direction. To give the impression of deep

mountains to the garden hills at the south, you may plant one or two cherries in the shade of the trees, giving the illusion that beyond the mountains there is a countryside village. Also, the cherry has a poetic feeling as a tree, and can be planted just about anywhere without problems. As said above, at the main room you plant (only) one cherry, some others could be elsewhere, that it is no problem. Somebody who was not given instructions on the traditions would plant perhaps two, or three in front of the main room, certainly to be disapproved by someone who has received traditional instructions. In any case, you have to do it in the same way.[25]

Although perhaps a little cryptic at first reading, the idea is clear. It departs from the mental image of cherries in the landscape that everybody knew. In the natural landscape, cherries inevitably betrayed the secondary forest of a village in the countryside (*sato*). They were to be planted in between the garden hills at the far back of the garden, where they would stand in the shadow of higher trees at the back. With the sunlight shining through from the back, they would glow when in blossom. There, in the far background, high trees would suggest the deep and dark green hills; the cherries, planted in between the sloping lines of hillocks would suggest the entry of a mountain valley where a gentle countryside would be. Thus, the garden intended to recreate a reference to the domesticated landscape bordered by cherries where the primeval forest began.

In earlier and greater gardens, such as at Arashiyama, cherries functioned as a backdrop to man's cultured environment; there they were planted even outside the garden. Clouds of cherry blossom as a background to a village scenery must for the noblemen have been a nostalgic landscape imagery as found in the earlier poems where far cherries attracted the attention of the poet. The Ninna-ji manual makes clear that cherries recalled the homely countryside where man was comfortable with himself. On a smaller scale the archetype was recreated in the planting design of their private gardens as an indicator for the landscape under human control, psychologically a safe and friendly place to be.

The quote continues in the tradition of the main hall of the imperial palace facing south, where one cherry was to be planted close to the hall, providing a foreground. Also for the best room, two or three are too much, you should keep to one cherry. Here the best of its poetic qualities found in botanic details, such as budding, fragrance, or the shedding of its petals, could be appreciated at best.[26] Later in the manual, the cherry is mentioned once more where it is listed as one of the plants in design of deep mountains. At this point the manual speaks of *yama-zakura;* we can interpret this as Japanese Mountain Cherries (*P. serrulata* var. *spontanea*).

Without much doubt gardeners were growing cherries for their clients. Cherries are rather easily grown from seed, at that time likely collected at nearby Arashiyama or from garden trees. An even easier method is simply digging up young plants found as seedling. Seeds in a bird dropping easily germinate and such seedlings are often found under places where birds rest, like big trees or along a garden fence. The parentage of such cherry finds is not traceable: they could be wild or random hybrids, adding expectation and surprise to the joy of growing cherries in the nursery and pride to the owner of a unique specimen.[27]

Although the manual speaks of cherries, and later more precisely of Mountain Cherries, it does not prove the existence of garden hybrids at this point. But we may imagine that better seedling specimens were reserved for better clients. Gardeners must have planted them in the gardens of the upper-class living in the area around Ninna-ji, where the best cherry went to the best position in the garden, such as the solitary position at the veranda of the main room. The admonition to refrain from planting more than one could point to such an awareness of unique specimens. From the manual we must conclude that gardeners were well aware of the feelings that properly planted and maintained cherries could evoke in the garden. In these days, 'Kirigaya' and 'Fugenzō' were clearly identified, not only as one peculiar garden tree but also as a garden form. Singular, proper names attached to cherries, came to identify in extension the propagated offspring as well. The role of gardeners is clear.

Later Cherry History of the Ninna-Ji Area

Several cherry gardens and collections still exist in the Ninna-ji area.[28] How do these relate to this earlier cherry history?

Ninna-ji, as well as the residences around it, were not spared from the Ōnin Wars with its disastrous fires that broke out shortly after the last date at the end of the manual. Many a cherry must have perished in the fires as well. After this period of serious trouble Ninna-ji was not rebuilt, although religious services continued in a nearby retreat. Only in the seventeenth century was it set up again. The role in cherry history of a neighboring temple compound Myōshin-ji has yet to be established. In the late fifteenth century, just after the Ōnin Wars were over, Myōshin-ji was made to flourish again by priest Sekkō Sōjin (1408-1486). At that time many cherries were planted on the "banks" of the temple, probably to be understood as a planting on the surrounding slopes after grading the site.[29] It is a guess that these cherries were planted by Ninna-ji gardeners, and that these were propagated or transplanted garden trees, previous cherry prides of courtly residences now destroyed.

Ninna-ji is found along a road that led to the center of the city. Along a similar route connecting Arashiyama and the inner city of Kyoto lies the Hirano Shrine of the Shinto religion, known at present for its splendid cherry garden. An early thirteenth-century record speaks of Ninna-ji gardeners working in this shrine. Anyhow, Hirano Shrine was also devastated in the Ōnin Wars; it became a place known for infestations of termites, and human bones lying around. Only in the early seventeenth century was its main hall rebuilt. The head of the shrine became a court noble Nishi-no-tōin Tokiyoshi (1552-1639). He was an aesthete and leading poet of the emperor's salon. In quite some detail he records the assembling and planting of cherries in his logbook; in 1629 and 1630 more than thirty trees are brought in. The planting starts only a few years after the rebuilding works had begun; trees must have been nursed for some years for preparation, showing that it was a well-planned project. Rather than speaking of "trees" in a general sense, the text speaks of "propagations," suggesting that one wanted to preserve existing specimens.[30] Indeed, some well-known garden forms are coming in: 'Kirigaya,' a name that relates to the cherry that had been planted in the imperial palace in 1357, comes to the Hirano Shrine in 1629. The cherry 'Fugenzō' is in full flower in 1638, suggesting that it was already a tree of quite some size. It had entered Kyoto in the mid-fifteenth century, so that it had survived about two centuries by now. This means several generations, for instance ten or so, of propagated trees. 'Fugenzō' is not fertile, having a completely deformed pistil and ovary, so vegetative propagation is indeed the only way of keeping it. No doubt the clone was preserved exactly because it had caught the attention of the literati and had appeared in the writings of well-known men.[31] Once the cherry garden was established, Tokiyoshi's logbook mentions that cherry twigs in flower were presented to the empress, for instance in spring 1637 and 1638.[32] The garden must have been of quite some standing and scale, clearly documented in his diary. Nevertheless, no mention of it is made in any of the official Kyoto chronicles until the mid-eighteenth century. The timing of court noble Tokiyoshi's activities to assemble an off-the-record collection seems significant in light of the political situation of the time.

Japan had since the early years of the seventeenth century entered a period of centralized rule under shogun Tokugawa's government now residing in Edo. For the first time in its long history, political power was drained from Kyoto completely, shifting cultural hegemony to Edo as well. The central government of the shogun took measures intended to curtail the power and engage the loyalty of the imperial court, with the extended intention to consolidate control over Kyoto. From 1613 on, for instance, the shogunate began to require its approval in the old right that the imperial court could bestow ranks on the buddhist temple clergy. This affected not only the political power of the court, but in fact also its income, as the priesthood paid substantially for getting such titles. Even the emperor himself became involved in political intrigues. Against his will, emperor Go Mizunoo (1596-1680) was forced to marry a shogun's daughter Tokugawa Masako (1607-1678). The marriage took place in 1620, after which she became an official empress in 1624. The shogun was thus, in the traditional way, ensured of family ties with the court. Self-determined and probably already angered by the increasing pressure of the shogun's government, the emperor bestowed high clergy ranks on some priests of the temples Myōshin-ji and Daitoku-ji in 1627. These titles were immediately pronounced invalid by the shogunate. Some persons were exiled and emperor Go Mizunoo abdicated angrily in favor of his five-year-old daughter, first child of Masako. In 1629, this little girl became the first reigning female since the eighth century; this was the year in which the cherry planting in Hirano Shrine starts. Apart from the personal diary of the shrine's head, no records are found elsewhere of his planting of 'Kirigaya' and 'Fugenzō,' two cherries that had such an overt symbolic role in asserting the power of the shogun in centuries before. One gets the impression that the court wanted to secretly preserve a

8. 'Kirigaya' became one of the standards in cherry botany after publication of Matsuoka Gentatsu's *Igansai Ōhin,* where it plays a major role as a bench mark in comparing cherries (1758, facs. reprint Tokyo: Bunkyūdō, 1891).

political and cultural heritage familiar to the cultured elite of Kyoto, but hardly known to the new shogun and his men.[33] At the same time, however, Tokiyoshi presented flowering twigs of these cherries not only to empress Masako, the shogun's daughter, but also to the shogun's deputy Itakura Shigemune (1586-1656), perhaps to mitigate the deed. Cherries entering Hirano Shrine could have been propagated from the collection that was standing in Myōshin-ji, now also victim of political intrigue. But Tokiyoshi's logbook gives only three names of garden forms: 'Kirigaya,' 'Fugenzō,' and 'Beni-zakura.' [34]

Ninna-ji only came to flourish again in the 1640s. Works were financed by the third Tokugawa shogun Iemitsu (1604-1651). He was the older brother of Masako, forcefully married to Go Mizunoo in 1620, giving him reasons for a shogunal donation of 1634. In the 1660s a large number of cherries, without a clear date, nor a documented origin of the planting material, were planted in Ninna-ji. It attracted the attention of several scribes from the 1680s on; they mention double-flowering cherries as well.[35]

There is no information on names of garden forms at Ninna-ji for the next century or so to come. We have to wait until 1793 for a first detailed list of names that gives: 'Kirigaya,' 'Edo,' 'Shiogama,' 'Roma,' 'Shibayama,' 'Akebono,' followed by an "et cetera." 'Goma-zakura' and a 'Shidare-ito-zakura' are standing at the main hall.[36] Interestingly, 'Kirigaya,' 'Roma,' and

'Akebono' have a relation with emperor Go Mizunoo. 'Roma,' according to one source, was grown by the emperor himself from a seed of 'Kirigaya.' These cherries must have been among the seven that the emperor had commemorated by imperial order. This imperial order has to be understood as the selecting and naming of new garden forms to be retained by propagation, for which the emperor clearly sets a pattern.[37] We can only wonder where the Ninna-ji cherries had come from, and guess that some of them were originally courtly plants, perhaps propagated from trees in Hirano Shrine. Anyhow, it is significant that with increasing stability and nationwide peace, cherished cherries could also come to the public and become named garden forms to be preserved and propagated. At present, 'Kirigaya' is still around, but most of the names of 1793 are no longer found.[38] (Fig. 8.)

One can not conclude that cherry hybrids in this area are all garden trees from previous gardens of the nobility. Later cherry collectors have been, and are still active in this area. But some spectacular forms in the collections at Ninna-ji and Hirano Shrine are still found only here, as a unique specimen with only some duplicates in modern research collections. Others that are wider spread as a clone are clearly traced to a parent tree in Hirano Shrine or Ninna-ji temple. These peculiar forms could be old, above all if they have a typical beauty or unique characteristic, and of course if they are not too difficult to grow. Hirano Shrine has quite a few of these, such as 'Tsukubane,' 'Imose,' 'Kinugasa,' or 'Nezame.'[39] Ninna-ji also has such peculiar forms, like its 'Kuruma-gaeshi' and 'Ariake.' Records from the late eighteenth century on Ninna-ji mention the yellow and greenish flowered forms 'Asagi,' 'Ukon,' and 'Kizakura,' all three still present in the temple's cherry garden.[40]

When it comes to the botany of these peculiar cherries in both gardens, it is mostly hybrids of the Japanese Mountain Cherry (*P. serrulata* var. *spontanea*) with a rather strong influence of the Ōshima Cherry (*P. serrulata* var. *speciosa*). But also Korean Mountain Cherry (*P. serrulata* var. *pubescens*) influence is seen in Hirano's 'Shibayama.'

Much still remains unclear and the only thing we can do is pose a hypothesis. The region of Ninna-ji and Hirano Shrine, in between the Mountain Cherries from Yoshino at Arashiyama and the center of the city, with its gardeners living close by, must have been an increasingly interesting cherry-hybrid area from the fourteenth century when Ōshima Cherries were planted in the imperial palace. The hybrids' parents were not simple wild plants, but were selected in their natural habitats, Yoshino and Kamakura, to serve a courtly taste and gain heavy cultural values in the capital. They were already of an extreme aesthetic quality, and most of them very fertile as well, easily leading to spectacular offspring. Hybrids enter the nearby Ninna-ji and Hirano Shrine from the seventeenth century to be preserved as named forms. They must have been found, selected or generated by gardeners working around Ninna-ji, catering to the tastes of a courtly and military elite. Indeed, they are spectacular garden plants that have stood the ages.

NOTES

[1.] On the botany and wider cultural context of Japan's cherries, see Wybe Kuitert, *Japanese Flowering Cherries* (Portland, Ore.: Timber Press, 1999) with bibliography and further references.

[2.] See, for example, *Man'yōshū*, 1872: *Miwataseba Kasuga no nohe ni kasumitachi sakinihoheru ha sakura bana kamo.* As far as the eye can see, the fields along Mount Kasuga are as in mist: can it be the brightly blossoming flowers of the cherry blossom? Transcriptions here follow Sadake Akihiro, et al. (ed.), *Man'yōshū honbunhen* (Tokyo: Hanawa Shobō, 1993).

[3.] *Man'yōshū*,1776, with comment in Yamada Takamitsu and Nakajima Shintaro, *Man'yō shokubutsu jiten* (Tokyo: Hokuryūkan, 1995), 254–261.

[4.] For example *Man'yōshū*, 1047. In my previous *Japanese Flowering Cherries,* I have overstressed this assuring quality of the cherry, proposing that it was felt as a "protective belt"; rightly criticized by Robin D. Gill in a personal message.

[5.] Here the story follows *Nara meisho yae-zakura,* a source on local geography dated 1678, written by Ōkubo Hidefusa and Motobayashi Koresachi, and illustrated by Hishigawa Moronobu; see Koshimizu Takuji, *Meizakura-Nara yaezakura* in *Sakura* (Kyōto: Kyōto Engei Kurabu, 1968), 27–29. Of olden times mount Mikasa is owned by the nearby Kasuga Shrine, and not by Kōfuku-ji; the story has fictional points.

[6.] Emperor Ninmei, had a cherry planted in front of the palace hall in stead of the usual plum in the Jōwa period (834–848). See: *Kojidan* (1212–1215), by Minamoto Akikane (1160–1215), quoted in Yamada Takao, *Ōshi, chūko no maki,* in *Sakura,* no. 3. (Tokyo: Sakura-no-kwai, 1920), 30. According to Sato Taihei, *Sakura to Nihonminzoku* (Tokyo: Daitō shuppansha, 1937), 44–45, 121, and 123 a cherry was planted in stead of a plum already at the founding of the capital by emperor Kanmu in 794, pointing to a source *Hana-tachibana no ki,* written in 869 by a priest Shuntō.

[7.] Basic is a series of earlier *Man'yōshū* poems (1044–1049) that describe the desertion, overlooking the once so flourishing valley and city of Nara. These were perhaps written by persons that had to flee the city after political troubles, rather than after the 784 moving of the capital. See also: Niels Gülberg, "Japanische Kirschblüten duften nicht," (Bochum/Heidelberg: *Hefte für Ostasiatische Literatur,* 12, März 1992), 99–106, seeing a change of "paradigm" in the treatment of plum and cherry in poetry around the same time. I thank Ivo Smits for bringing this source to my attention.

[8.] Jōtō Mon'in Shōshi was a consort and later empress of emperor Ichijō. She was the eldest daughter of powerful politician Fujiwara Michinaga (966–1027). The story is given in *Shasekishū,* by Mujū Dōkyō compiled in 1279–1283. See: Watanabe Tsunaya, *Shasekishū,* (*Nihon koten bungaku taikei,* Vol. 85, Tokyo: Iwanami Shoten, 1966), 376–377.

[9.] See: *Mieken chimei daijiten,* (Tokyo: *Nihon chimei daijiten,* vol. 24 Kadogawa shoten, 1993) 1111–1112. See also: Koshimizu Takuji, *Meizakura-Nara yaezakura* in *Sakura,* (Kyōto: Kyōto Engei Kurabu, 1968), 27–29; the records of the village studied by Yoshizumi Kangen, a local historian, even give a name of an eighth-century village headman. The village temple dedicated to Kannon had a double-flowered cherry standing in front of it.

[10.] See Mukōgaku Tosho (ed.) *Hyakunin isshu no techō* (Tokyo: Shogakkan 1989) 122–123. The poem was published in: *Shikawakashū* (compiled 1151–1154) by Akisuke Fujiwara and became a classic. "Nihohi" (*nioi*) not to be translated as "fragrance" as Niels Gülberg rightly remarks, but rather as "brilliance," or "glory."

[11.] The clonal cultivar 'Nara-no-yae-zakura' from the temple Chisoku-in, Nara, was described as *Prunus antiqua* by Miyoshi Manabu in *Shokubutsugaku zasshi,* Vol. 36, 1922. It is given as *Prunus leveilleana* Koehne cv. Nara-zakura, in *Idenken no sakura* (Mishima: Kokuritsu Idengaku Kenkyūjo, 1995), 58; and as *Prunus verecunda* cv. Antiqua in Kawasaki Tetsuya, *Nihon no sakura* (Tokyo: Yama to keikokusha, 1993), 204. A selection with an excellent, erect tree shape, is known as the clone 'Yono-no-yae-zakura' classified as *Prunus leveilleana* Koehne cv. Nara-zakura in *Idenken no sakura*, 82 where a third clone and some history are shortly discussed as well.

[12.] The two Chinese characters that Teika uses for propagation are ZOKU and KEI. "ZOKU" also reads as *tsuzu(ku/keru),* and stands for the verb to continue (intr./tr.). "KEI," or *tsu(gu),* means: succeed to, inherit; follow; patch, join. The latter could be interpreted as grafting, close to the presently used character "SETSU," or *tsu(gu)* join; graft. Since "SETSU" had been used as the Chinese term for grafting from the eighth century (see communication by Georges Métailié in this volume) it should have been known by Teika, but he does not use it. It is very questionable that Teika speaks about grafting as is usually assumed, when reconstructing the time of the year. He must be speaking of propagation by layering, probably of a side branch sitting on the stem of a tree with moss as substrate, as I have seen farmers doing to preserve precious village cherries in Japan's countryside. It is a simple and sure method, hardly requiring the dexterity needed for grafting. The character "KEI" reads therefore rather as *tsugu* in *uetsugu*: to plant a propagation, and not as grafting. Teika's diary *Meigetsuki,* edition Hayakawa Junsarō, 1912. Later records like the one by Tachibana Narisue in his *Kokon chomonjū,* Vol. 19 are often taken as confirmation that Teika speaks of grafting, but again dates are too late for grafting, and perfect for layering.

[13.] Yoshida Kenkō in his scribbles of a literary man, *Tsurezure gusa,* the section translated here from the edition with annotations: Inamura Toku, *Tsurezure gusa yōkai,* (Tokyo: Yūseidō Shuppan, 1981). Sato Taihei, *Sakura to Nihonminzoku* (Tokyo: Daitō shuppansha, 1937), 207–208 mentions that Kenkō was a lover of cherries and selected garden trees himself. He built a small retreat (*mujōjo*) at Narabiga-oka, close to Ninna-ji, and planted a cherry next to it.

[14.] A thousand cherries from Yoshino were planted, adding a small shrine for Zaō Gongen, the cherry god of Yoshino. See Toyama Eisaku, *Muromachi jidai teienshi* (Tokyo: Iwanami Shoten, 1934; reprint Kyōto: Shibunkaku, 1973), 403, referring to sources as *Niwatakeki.* See also note 21.

[15.] On the nursery, a young Japanese Mountain Cherry typically gives its first hesitating flowers one or two years later than the Oshima Cherry. Any young cherry's first flowers are few and not very well developed, because of strong growth. With slower growth on older trees, flowers develop better. Double-flowering specimens have the habit of showing more petals per flower with increasing age, up to a record in chrysanthemum-forms of 360 counted in flowers of an old 'Kenrokuen-kiku-zakura.'

[16.] At the time when Nagaoka-kyō (close to Kyoto) had served as capital from 784–794 A.D., cherries had also been brought from Yoshino and planted on Oshio-no-yama Mountain a few kilometers south of Arashiyama, to ban evil spirits. See Kayama Masuhiko, *Kyōto no Sakura, daiisshū* (Kyōto: Kyōto engei kurabu, 1938), 75–77, referring to local temple records. A cultivar 'Koshioyama' (written with the same characters as Oshio-no-yama but in a different pronunciation) resembles 'Tagui-arashi' and can be understood as another good selection from the wild material at Oshio-no-yama. See Kawasaki Tetsuya, *Nihon no sakura,* (Tokyo: Yama to keikokusha, 1993), 222.

[17.] See *Entairyaku,* a logbook of historical records from 1308–1360, quoted in: Yamada Takao, *Ōshi,* (Tokyo: Sakura shobō, 1941), 121. The imperial palace was the Tsuchi-mikado-dono palace, largely at the site of the present imperial palace, see Hayashiya Tatsusaburō, ed., *Kyōto no rekishi* ((Kyōto: Kyōto-shihen, 1971, vol.2, and 3).

[18.] See Kayama Masuhiko, *Kyōto no Sakura, daiisshū* (Kyōto: Kyōto engei kurabu, 1938), 6, and 16. 'Kirigaya' reappears in fifteenth century records of priests; see Yamada Takao, *Ōshi* (Tokyo: Sakura shobō, 1941), 122–123 with some quotes. The *Inryōken-nichiroku* a fifteenth-century log kept by priests of one of the subtemples of the monastery Shōkokuji gives 'Kirigaya' alongside 'Fugenzō,' and an otherwise unknown 'Shinshū-zakura'; see Hida Norio, *Nihon teien no shokusaishi* (Tokyo: *Randosuke-pu kenkyū,* 68/1, 44–51, 2004), 47. 'Mikuruma-gaeshi' is a synonym for 'Kirigaya.'

[19.] Osen Keisan's poem quoted in: Yamada Takao, *Ōshi* (Tokyo: Sakura shobō, 1941), 137. Osen Keisan could visit the cherry in the spring of 1474, about the date of his recording. His idea of the cherry as Japan's flower was taken up with the same comparisons with China in Matsuoka Gentatsu, *Igansai ōhin* (1758, facs. reprint Tokyo: Bunkyūdō, 1891), laying foundations for later nationalistic symbolism. Records in the *Hekizan-nichi-roku* by a fifteenth-century priest Taikyoku suggest the presence in Kyoto of 'Fugenzō' already in 1459.

[20.] *The Tale of Genji* was written by a courtly lady Murasaki Shikibu. She was in service of Jōtō Mon'in and is known to have associated with the poetess Ise-no-Taifu. The English quote largely follows Seidensticker, E.G. *Murasaki Shikibu, The Tale of Genji* (Tokyo: Rutland, 1982), 384, and 418–419; checked on Yamagishi Tokuhei, *Genji monogatari II* (*Nihon koten bungaku taikei,* Vol. 15, Tokyo: Iwanami Shoten, 1959), 322, and 396.

[21.] The cherry party of 1263 was held at Kameyama-dono, the palace of emperor Go Saga (1220–1272). The main hall of Kameyama-dono faced south, in front of it was a pond. Over the pond one could see the river in front of the north-east facing foothill of Arashiyama. A terrace (*sajiki*) was provided to overlook the garden and the landscape beyond. See Hisatsune Shūji, *Kyōto meienki gekan* (Tokyo: Seibundō Shinkōsha, 1969), 59, referring to *Masu-kagami,* a fourteenth-century compilation of historical records, and *Godai teiō monogatari,* a late fourteenth century history book. About eighty years after this cherry party, shogun Ashikaga Takauji would add his cherries on the same hill. A similar planting of one cherry in front of the main hall, and mass planting in the background was envisioned at Kitayama-dono (now Kinkaku-ji) Toyama Eisaku, *Muromachi jidai teienshi* (Tokyo: Iwanami Shoten, 1934; reprint Kyōto: Shibunkaku, 1973), 508–511.

[22.] A football party under a shower of cherry petals shed in Kitayama-dono is described in Yamada Takao, *Oshi, chūko no maki,* in *Sakura,* no. 4. (Tokyo: Sakura-no-kwai, 1921), 44–45, with a quote from *Masu-kagami.* As a standard, a willow was planted in the south-east, a maple in the south-west, a pine tree in the north-west, and a cherry in the north-east corner of the field. See Yamada Takao, *Oshi, kinko no maki,* in *Sakura,* no. 4. (Tokyo: Sakura-no-kwai, 1921), 41–45.

[23.] See also my *Themes in the history of Japanese Garden Art,* 51, 52 on principles of design in this period. *Sakuteiki*'s author makes fun of people who want to judge the setting of stones according to a certain style. See Michel Vieillard-Baron , *De La Création Des Jardins—Traduction du Sakuteiki* (Tokyo: Maison Franco-Japonaise, 1997), 32; or Takei Jirō., and Marc Peter Keane, *Sakuteiki, Visions of the Japanese Garden* (Boston: Tuttle, 2001), 166, though mistakenly speaking of garden styles, whereas it only concerns the stones.

[24.] Murayama Shūichi. *Heiankyō, Kugekizoku no seikatsu to bunka* (Tokyo: Shibundō, 1957), 14–26. The descent of the gardeners living around Ninna-ji can be traced in the records to lineages and names found in the eleventh-century *Sakuteiki.* Other evidence shows that monks of this temple were engaged in garden works in the thirteenth century. See also Naka Takahiro, *Ninna-ji shinden teien,* in *Nihon teien kenkyū* (Kyōto: Nihon teien kenkyū senta-, 2002), 59–63. See also *Kyōtofu no chimei,* vol..26, *Nihon chimei daijiten* (Tokyo: Kadogawa shoten, 1991).

[25.] See the text edition Egami Yasushi. "*Dōji kudensho tsuki sansui narabini yakeizu—kōkan ge*" (Bijutsu kenkyū 250, 1967), 25. The English translation in David A. Slawson, *Secret Teachings in the Art of Japanese Gardens* (Tokyo, New York, London: Kodansha, 1987), unnumbered page 164, misses the point, unforgivably translating *zashiki* as "home site," suggesting the native habitat of the cherry. *Zashiki* is the best room of a residence, where guests were received, often with a garden in front. Slawson's translation does not catch the idea of depth of perspective, missing *mukō,* "beyond."

[26.] For example, the falling of petals of the cherry at the main hall was the highlight of a day with poetry and music at Kitayama-dono in 1259, see Toyama Eisaku, *Muromachi jidai teienshi* (Tokyo: Iwanami Shoten, 1934; reprint Kyōto: Shibunkaku, 1973), 510–511 after *Shōka san'nen kitayama gyōkō waka* by Fujiwara Michiyoshi. A party was held at the budding of the same cherry in 1263, see Toyama Eisaku, *Muromachi jidai teienshi* (Tokyo: Iwanami Shoten, 1934; reprint Kyōto: Shibunkaku, 1973), 390 after *Masu-kagami* Vol. 8 *Yama-no-momijiba.*

[27.] The forests surrounding Kyoto are still stripped from seedlings, not only cherries, for use as garden plant. Sano Tōeimon V, famous Kyoto cherry lover, having his nursery close to Ninna-ji, told me once that cherries should not be preserved as fixed cultivars under a given name: "cherries are not like that" (*sakura ha soiu mono ja nai*). Given the extreme variability and the unusually high chance to get an improved cherry from your random seedlings, his words show what the gardeners of Japan have done for centuries: just enjoy playing with the gorgeous and fickle genes of cherries.

[28.] Apart from cherry gardens at Ninna-ji and Hirano-jinja, one may find cherry collections at the nearby nursery Uetō of Sano Tōemon and at the Utano Hospital.

[29.] See Tsuneo Kajūji *Kyōto no sakura* in *Sakura,* no. 8. (Tokyo: Sakura-no-kwai, 1925). Cherry connoisseur Tsuneo Kajūji (1882–1936) was a prominent nobleman who prepared an illustrated cherry catalogue *Koto-meibokuki,* 1925. Such planting on banks is seen in Ninna-ji in later history, for example illustrated in *Miyako miyage* (1677); see Kayama Masuhiko, *Kyōto no Sakura, daiisshū* (Kyōto: Kyōto engei kurabu, 1938), 4–5, and 8.

[30.] See Kayama Masuhiko, *Hirano no sakura* (Kyōto: Hirano-jinja Samusho, 1933), 2 and 3, quoting Nishi-no-tōin Tokiyoshi's log *Tokiyoshi kyōki,* kept in the shrine. I thank Machida Kaori for helping me with reading the text. Again the character "KEI" is used for planting a propagation. It concerns duplicated plants, see also the way the character KEI is used in Kayama Masuhiko, *Hirano no sakura* (Kyōto: Hirano-jinja Samusho, 1933), 36 and 123. Neighboring Kitano Shrine is also known as a medieval center of gardeners.

[31] A most complete treatment of the history of the cultivar 'Fugenzō' given in Kayama Masuhiko, *Kyōto no Sakura, daiisshū* (Kyōto: Kyōto engei kurabu, 1938), 122–148, leaving no doubt that the present cultivar 'Fugenzō' is the same as the fifteenth-century one on a clonal level.

[32] Flowering twigs were presented to Masako (1607–1678), second wife of Emperor Go Mizunoo, daughter of second Tokugawa shogun Hidetada. Empress Meishō (1623–1696, r. 1629–1643) was the eldest daughter of Masako and Go Mizunoo.

[33] See for similar cultural expressions, negating shogun culture, my *Themes in the history of Japanese Garden Art*, 166–168, 204.

[34] Tokiyoshi gives 'Fugendō,' synonym for 'Fugenzō.' See on its deformed pistils my *Japanese Flowering Cherries*, 236–244, with photo 121. In 1639 the logbook mentions a *beni-zakura*, a name often used for the deep pink *P. sargentii*, but too vague here to draw conclusions.

[35] Kurokawa Dōyū's *Yōshū fushi* (1682) mentions that large amounts of cherries had been planted in recent years in Ninna-ji, making it a cherry site comparable to some other famous sites, see Kayama Masuhiko, *Ōmuro no sakura* (Kyōto: Dai-honzan Ninnaji, 1931), 5 with other evidence that the planting must have taken place around the 1660s. In front of the temple a horse-racing course was laid out, decorated with cherries, planted on the earthen embankment on which the gate still stands.

[36] Akisato Ritō, *Miyako kagetsu meisho* (1793) gives the list, quoted in Kayama Masuhiko, *Ōmuro no sakura* (Kyōto: Dai-honzan Ninnaji, 1931), 7.

[37] Emperor Go Mizunoo grew 'Roma-zakura' from a seed of 'Kirigaya.' 'Roma' derives from the place where it was growing: the room (*ma*) of one of the pavilions (*ro*) in the garden of the palace for the retired emperor (Sento Gosho). It must have been one of the cherries he recommended by imperial order (*chokumei*), like the 'Akatsuki-zakura' (named 'Myōjō-zakura' by the emperor) with remarkably large, single flowers, over six centimeters in diameter; see on both cherries Matsuoka Gentatsu, *Igansai ōhin* (1758; Tokyo: Bunkyūdō, reprint 1891). 'Akebono-zakura,' originally in the imperial palace, was later a famous cherry in Kanga-an and was also recommended by Go Mizunoo (*Kokon Yōrankō* after *Rokuroku sakura shurui*). On Go Mizunoo recommending seven cherries: see Hirose Kain *Sanjuroku-ōfu* (1824). 'Edo-zakura' stood in Ninna-ji according to Matsuoka Gentatsu, *Igansai ōhin* (1758; Tokyo: Bunkyūdō, reprint 1891) an important source for cherry botany, written in 1711–1716, and published in 1758. See also Hiroe Minosuke, *Sakura to jinsei* (Tokyo: Meigen Shobo, 1976), 16, 56, 240, 241, 243.

[38] 'Kirigaya' became one of the standards in Japanese cherry botany and was used as a benchmark for comparing other cherries in Matsuoka Gentatsu, *Igansai ōhin* (1758; Tokyo: Bunkyūdō, reprint 1891), and later sources. The plant that Carrière received and used as a type specimen to describe *Prunus lannesiana* (1872) was likely a potted 'Kirigaya'(syn. 'Mikuruma-gaeshi'); see my *Japanese Flowering Cherries*, 87, 88.

[39] 'Imose' is a clone with its parent tree in Hirano Shrine. Other cherries in Hirano are now famous but known here under different names and also could have originated here. 'Shōgetsu' is found as 'Nadeshiko,' and a form that appears to be 'Kanzan' is called 'Okame' in Hirano Shrine.

[40] 'Kuruma-gaeshi' and 'Ariake' (in Ninna-ji) are not very original names. The Ninna-ji collections are less convincingly preserved than at Hirano Shrine. But the green or cream-flowered cherries at Ninna-ji, such as 'Ukon,' 'Gyoikō,' and 'Asagi' appear in a description of Ninna-ji in Akisato Ritō, *Miyako meisho zue* (Takehara Shunchōsai, 1780). That there are three of such greenish/cream forms in the 1780s points to a rather long variability history already at that time.

Botanical Progress,
Horticultural Innovations
and Cultural Changes

Botanical Progress,
Horticultural Innovations
and Cultural Changes

Botanical Progress,
Horticultural Innovations
and Cultural Changes

Botanical Progress,
Horticultural Innovations
and Cultural Changes

Georges Métailié

Every author who, over the past fifty years, has written about the history of grafting in China has considered the Yuan dynasty (1271-1280-1368) of particular importance. I questioned this fact and wondered what could be the reason(s). First, I considered this history and compared the situation during this period with what had happened previously and after. Obviously, the first answer was an economic one. After the wars that had led to the establishment of the new dynasty, the agriculture of the country was in a terrible state of desolation and reconstruction was necessary. The new rulers did not choose to adapt a pastoral regime to Northern China, but decided instead to develop the traditional Chinese farming methods, particularly sericulture. In this process, grafting of new and better varieties of mulberry trees was recommended. Compared with the previous literature dealing with tree grafting, the remarkable point is that the two main agricultural books written during the Yuan dynasty, one imperially commissioned and the other by private initiative, made a systematic presentation of all the various techniques that were known. Eventually, this classification became the reference for the horticultural and agricultural texts written afterward.

Through a brief historical presentation of spontaneous and artificial grafting in China, the social impact of this horticultural practice is briefly analyzed. Grafting, consisting "in placing two cut surfaces of one or of different plants under conditions which cause them to unite and grow together"[1] shows that the vocabulary used to name the various actions and material playing a part in this process can give information on the symbolic representation that it can create. On this basis, a tentative answer is proposed both from an agricultural and a cultural point of view to the question "why did grafting appear to have played a particularly significant part in the Chinese society during the Yuan dynasty?".

History of Grafting in China

Spontaneous "interlocking of principle of trees," mulianli
References of grafting of trees as a common practice in horticulture appear rather late in Chinese literature and the various authors who have written over the last fifty years on the history of this practice have all considered another fact as a forerunner of this cultural technique in China. It is what is called *mulianli*, "interlocking of principle of trees," a kind of spontaneous grafting.

This phenomenon—which can be observed rather often,[2] at least in temperate forests—was considered in ancient China as particularly auspicious. When it happened it was considered proof of the acknowledgement by heaven of the virtue of the

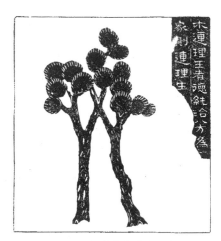

生為治者木作從字明此
一八德連治符侶顯榜字
則方澤理志瑞治其治畫
為合純王云志字治畫

Figure 1: The intertwining trees. Wu Liang Ci ceiling carving. Reconstruction (Feng & Feng, Shi suo, 4.32). Translation of the text (following Wu Hung, 1989): "The intertwining trees. Intertwining branches grow when a ruler's virtue is pure and harmonious, and when the eight directions are unified into a single family".

sovereign. Between the third and the eighth century A.D., the dynastic annals recorded not less than 256 reports made to the various rulers by local authorities. It is interesting to notice that these notations were particularly numerous during the periods of internal troubles. This fact may not be purely coincidental and, as suggested by Tan Bi'an[3] there could have been artificial grafting among the ones recorded as spontaneous. Probably it was not useless for the officials of the various kings in competition to help them to show in this way their legitimacy. One of the most ancient representations of such interlocking trees can be seen on the stone engravings of the famous Wu Liang Shrine of the Han Dynasty (206 B.C.-220 A.D.) as reproduced in the nineteenth-century *Jin shi suo* (An index to bronze and stone carvings) by the Feng brothers[4] (Fig. 1). Following again Tan Bi'an (1956) who is the first author to have analyzed these data, this kind of natural grafting concerned trees of the same botanical genus or of different genera and families. We also have found various texts relating the same thing to a fungus of a great importance in Chinese Taoist tradition, the *lingzhi (Ganoderma lucidum)*, a symbol of immortality. All Chinese authors writing about the history of grafting in China considered the observation of *mulianli* to be one of the factors at the origin of the practice of grafting. This does not seem sufficient evidence, however, and an argument can be made that the vocabulary for the two is completely different. For this reason, I prefer to translate the Chinese expression as "interlocking principle of trees" than as "spontaneous grafting" to forbid any anachronistic deduction. However, an important fact to keep in mind is the strong symbolic value associating the occurrence of the phenomenon to the moral attitude of the ruler.

Artificial grafting

When we come to grafting proper, the first mention in the literature is to be found in a book on agricultural practices written during the first century B.C. by Fan Shengzhi, a eunuch serving the Emperor Cheng Di, who reigned from 32 to 7 B.C. This text has been partially reconstructed using excerpts quoted in various other books. The method for harvesting particularly big gourds recommends a kind of grafting[5].

Shallow pits cultivation of calabash gourds (*Lagenaria vulgaris* Seringe):

Collect seeds from big gourds. If one gets seeds from a gourd one bushel (10 1.) in size, one harvests gourds ten times this size, one *dan*. With seeds of one *dan* gourds, one gets ten *dan* gourds.

Firstly, dig the ground in order to make pits with a square perimeter, three feet across and deep. Mix silkworm's excrements with the earth, then pour down into the pit, tramp down firm. Water, when the water has seeped away, put ten seeds and cover with the previous manure.

When they have grown up to more than two feet, put together the ten stems and bind them with a cloth on five inches and then seal them with mud. After a few days the bound part unite into one. Keep the strongest stem and eliminate the others. Later pinch off all the branches without fruits.

To keep the best fruits: The first three fruits are not good, eliminate them. Keep the fourth, fifth and sixth. Just leave three fruits in each pit. By dry weather it is necessary to irrigate. Dig a small ditch around the pit, four or five inches deep, let water in to moisten the vine. Never pour water into the pit.

Even if this passage is a clear mention of grafting by approach, there is no specific name given for this technique. Up to now, it is the first description known of grafting in Chinese history. However, some earlier texts can be interpreted as indirect witnesses of grafting techniques. Probably the most famous and most often quoted is the following sentence found in two early texts, the Rites of *Zhou Zhou li* (second century B.C.) and the Master of Huainan *Huainanzi* (ca. 139 B.C.) "when the orange tree crosses the Huai River [which means "go North"] it turns into the thorny lime bush."[6] Among the various hypothesis—which have been explored by Joseph Needham and Lu Gwei-djen[7]—raised to explain this transformation, the most plausible is the grafting of sweet orange on thorny lime bush, this very hardy tree having been and being still used as stock for citrus fruits. Under more severe climatic conditions, the most fragile grafted part would have died and the stock would have carried on growing, which is often the case in arboriculture.

The fact that the first Chinese encyclopedia *Er ya* compiled during the third century B.C. mentions some "seedless" varieties of fruits *wushi*, jujube and prune, attests to a vegetative propagation of cultivated trees, by cuttings or grafting. A rather difficult passage of this text can be understood as "[if] one grafts the seedless prune tree *xiu* on the *lu* prune tree, the fruits are red and variegated."[8] Anyway, in the first dictionary of characters *Shuo wen jie zi* achieved in 100 A.D. by Xu Shen's, and presented to the emperor in 121 A.D., one finds an entry *jie* defined as "[to] link trees"[9]. A commentator of the Qing dynasty Duan Yuzai (1735-1815) explains this definition in the following way "nowadays those who cultivate flowers or fruits if they transport this branch and graft[10] it on that tree, the flowers and fruits are similar to those of this tree." The entries of this dictionary are archaic characters of the third century B.C. These various data allow us to assume that grafting already existed in northern parts of China, at least since the third century B.C.

Later on, two poems, one by Fu Xuan (217-278), the other by Sou Xin (502-557), mentioned a grafted prune tree with three different varieties for the first, and grafted pear trees with more perfumed and less easy dropping fruits for the second.

The first clear mention of grafting of trees with a precise description of the process appeared in another agricultural treatise, the *Qi Min Yao Shu* (Essential Techniques for the Peasantry), achieved between 533 and 544 A.D. by Jia Sixie a man involved in agricultural production. This book gives a complete survey of the various techniques from the cultivation to the transformation of the agricultural products, which were in use in northern China during the sixth century.[11] Lost as a whole, the text has been reconstructed thanks to the many quotations, which had been used from this fundamental manual later on. Chapter 37, in the fourth part of the book, devoted to the culture of fruit trees, is entitled "Inserting pear trees"[12] *cha li* 插梨, where "inserting" must be understood for "grafting." In a previous chapter, the same term means "to propagate by cuttings." The text indicates that the first benefit to arise from grafting versus sowing is the hastening of production. Precise details are given concerning the choice of trees for stocks. With two wild species of pear trees that the author recommended,[13] he assumed that the pears would be big and smooth. On the contrary, grafting peach tree on mulberry would give the worst pears. On pomegranate or jujube, the fruits would be of good quality but the grafting would fail in eighty to ninety percent of the cases. The best period for grafting was when the leaves began to move slightly, the worst when the buds were opening. Jia Sixie gives precise details:

> Stocks up to the thickness of a forearm will do.
>
> Have a rope made of hemp, wrap the trunk. Saw the stock at five or six inches from the soil [note: without wrapping, the skin may be damaged when sawing. If one leaves a tall stock, the scions are going to be luxuriant and, in case of wind, will be broken].
>
> Cut slantwise a piece of bamboo to make a wedge. Insert it between the skin and the wood one inch deep.
>
> Choose a twig of a good pear tree among the branches on the sunny side [note: those of the shady side give few fruits], five or six inches long, cut it slantwise, same size as the bamboo wedge. Gently incise above the slanting part to remove the dark skin [note: don't injure the dark green skin. If it is injured, the scion dies]. Pull out the bamboo wedge and insert the pear up to the peeled part. The side with wood against the wood, the rounded part with skin against the skin. Once it is well fixed, wrap up with silk batting the head of the stock, and seal upon with mud, cover with earth leaving alone the head of the scion. Water on the scion twig. When all water has seeped away, cover with some dry earth and do not let harden by lack of moisture. You will not lose a single one among one hundred [. . .]. If the stock is split in four; among ten there will be not a single success [note: This is natural, the wood is split and the skin separates due to emptiness and drying].
>
> Once the pear tree grows, eliminate all the leaves which can grow on the stock [note: otherwise the 'potential strength' is divided and the growth of the pear retarded].[14]

The author insists on the crucial points for the success of the operation, dark green skin against skin, that is, cambium against cambium, and sealing with mud of the grafted part to prevent any desiccation. He describes precisely the grafting process, though there is not yet a specific terminology. The terms used are all borrowed from the general vocabulary. The stock is named by the name of the tree "crab-apple" *tu* and the scion is just called "pear" *li*. However, in one sentence, the first is referred to as *zhu*, the "host," the second as *ke*, the guest.

We learn also that for a good development of the grafted tree, one needs to eliminate all the leaves that may grow on the stock because if not, the *shi* "potential strength" of the stock is divided and the growth of the pear tree slowed down. This point is very important because the term *shi* is crucial in Chinese thought.[15] Its more general meaning is "propensity," "potential strength," but it also means "genitalia," more precisely male genitalia.

Up to this point, in the process of grafting, the stock can be understood symbolically as the host, and perhaps less consciously, the male lineage of a family, and the scion, the guest who enters one's family.

Of the succeeding five centuries, two texts with references to grafting trees written during the Tang dynasty (618-907 A.D.) have remained. In "Necessary for the four seasons" *Si shi zuan yao* by Han E (probably eighth century A.D.) the author indicates a technical term for the stock *shuzhen* "tree-anvil." Compared to the Qi Min Yao Shu, a basic principle is indicated, one can graft trees whose seeds inside the fruit belong to the same "categories" *lei*. For instance, an apple tree (*Malus pumila* Berkh. var. *rinki* Rehder) or a pear tree on a flowering quince, and a chestnut on an oak. Something new appeared during the tenth and eleventh centuries. The aristocracy of the Song dynasty (960-1279) was progressively very fond of peonies, which became expensive commodities, and this fact gave birth to professional breeders. Grafting played an important part in their activities as it was used as a way to produce or to help to produce new varieties.

In "Notes about the flowers and trees of Luoyang" *Luoyang huamu ji*, composed between 1068 and 1077, in which several hundred different kinds of flowers are described, the author Zhou Xu or Zhou Shihou recommends fourteen species to be used as stock for some twenty generic kinds of flowering plants, in Luoyang (in actual Honan province). For instance, a flowering quince *mugua (Chaenomeles lagenaria* Koidz.= *Cydonia lagenaria* Lois.) is recommended for several varieties of the same species, quince *wenbo (C. vulgaris* Pers.), for the Chinese flowering quince *mingzha (Chaenomeles sinensis* Khoene = *Cydonia sinensis* Thouin), cherry tree for peach tree, *Magnolia liliflora* Desr. *mubi* for the same species and *Magnolia denudata* Desr. *mulan*, wild rose for double yellow roses and all thorny flowers . . .

The aesthete Emperor Hui Zong (r. 1101-1126) who had created a special office called "Net of flowers and stones" *hua shi gang,* to collect the strangest and most beautiful things in the empire, ordered a man named Liu You, from what is today Sichuan province, to come and live several months at the capital to work in the imperial garden because of his fame as an extraordinary grafter.[16] Among the marvels he produced there were a five-colored blackberry rose (*Rosa rubus* Lev. et Van. = *Rubus commersonii* Poir); hundreds of different kinds of tree peonies *mudan (Paeonia suffruticosa* Andr. = *P. moutan* Sims.), and herbaceous peonies *shaoyao (Paeonia albiflora* Pall.), some with different kinds of flowers on the same plant, some with the same flowers but of different colors; and golden lotuses with double flowers, an indescribable luxuriance of fragrance and beauty.

As we can see, the main purpose of grafting seems to have changed from the production of better fruits towards the propagation of more beautiful flowers. Several monographs were devoted to the tree peony, which was really the flower-king. The growers used to go and pick seeds of wild peonies not too far from the town of Luoyang. The young plants raised from these seeds, after two or three years, were used as stock to graft precious varieties. Another ornamental, the Japanese apricot *mei (Prunus mume* Sieb. et Zucc.), often ambiguously called "prune tree" in sinological publications, became highly valued by the scholars for its flowers at that time. Grafting offered a good possibility for the owner of a rare or particularly beautiful variety to partake this pleasure with a visiting friend who would leave with a cutting that he could graft on a tree of his own garden.[17] Fan Chengda (1126-1193) in his *Mei pu* "Treatise on Japanese apricot" wrote that during such a visit he "had got a twig to graft separately" and he added that now this famous variety grew in two different places in the town.[18] He noted also that one spontaneous kind named "wild apricot" *yemei,* very common, was grown from seeds and never grafted. It was probably used as stock. Another scholar, Chen Si, active between 1225 and 1265 wrote a "Treatise on the crab apple" (*Malus spectabilis* Borkh.= *Pyrus spectabilis* Aiton.) *Haitang pu* where grafting of young crab apple branches on a pear tree is recommended to strengthen the vegetation of the crab tree. As for the technical vocabulary the standard term now for grafting is *jie*, which, in its first acceptation, means "to unite." In the various treatises on peonies the standard term for stock is *zuzi* "(male) ancestor" which confirms our hypothesis of an analogy between stock and family.

Up to the end of 1260, if grafting appears to have become very popular, so far there is not yet any single treatise that would summarize the knowledge on the subject. Things change, obviously, with the conquest of China by the Mongols and the establishment of a foreign new dynasty over the whole empire in 1280. This dynasty had been founded in 1206 under the name "Mongol" with Karakorum as capital. The transfer of the capital some fifteen hundred kilometers eastward in a new city called Khanbalik (today Beijing) built in 1267, was followed in 1271 by the change of the Mongol name of the dynasty for a Chinese one, Yuan. The Northern part of China had been devastated by the war and one would have thought that the new rulers would have taken this opportunity to install, as they had done previously elsewhere, a pastoral farming following their own cultural habits. Actually, this did not happen at all and the new rulers decided, on the contrary, to develop agriculture. Already in 1273, before the complete victory over the Song dynasty, the Mongol emperor Qubilai Khan had ordered the compilation of a book in Chinese, "Essentials of agriculture and sericulture" *Nong sang ji yao*, to enhance the reconstruction of agriculture and stimulate in particular the production of silk in southern parts of China. Two chapters of the book are devoted to grafting, which is called "graft-exchange" *jiehuan* or simply *jie*. One concerns the culture of mulberry trees, the other the grafting of fruit trees. The first one is actually a small treatise on grafting. It summarizes the experiences and describes four different kinds of grafting (Fig. 2). Apparently the first problem was to regenerate old trees. Two methods were recommended for that. One is named *chajie* "grafting by insertion," and its precise description corresponds to a crown or rind-grafting.[19, 20] The other is named *pijie* "grafting by splitting." It is a kind of cleft-grafting, called 'inlay graft' (Gardner, 1958, p. 130, fig. 49). Triangular incisions are done on the side of the stock where the scions are inserted. Both can be used on trunks or roots of old trees. The third one is a kind of shield budding, called *yejie* "dimple" graft. It is proper to big or small trees. The fourth method is called *dajie* "reciprocal graft." It consists of grafting on a young budding stock of one year a scion of the same size, both being cut in slope and bound together. It is called 'splice graft' or 'whip graft' (Gardner, 1958, p. 140, fig. 57). The content of the book relies mainly on the *Qi min yao shu* and the *Si shi zuan yao*. However, the shield budding appears here for the first time. Sixty years later, the "Book of agriculture" *Nong shu* was written by Wang Zhen, who signed the preface in 1313. The book was initially intended to help the local civil servants to understand agricultural practices and let them be able to properly manage the farmers under their jurisdiction[21]. In the chapter "Nong sang tong jue,"[22] there is a presentation of six grafting techniques under the title *jiebo* "unite-exchange." The first one is called "body grafting" *shenjie*[23]; the second is "root grafting" *genjie*; the third, "skin grafting" *pijie;* the fourth "branch grafting" *zhijie*; the fifth "dimple grafting" *yejie*; and the last "reciprocal grafting" *dajie*. Because of the importance of the book, it will remain the definitive reference quoted until modern times. In reality, it is just a new formulation of what had been indicated in the previous text. Grafting on "root" (i.e., very low stock) and on trunk are two variants of *chajie* "grafting by insertion"; "skin grafting" *pijie* is the same thing as "grafting by splitting" *pijie* and "branch grafting" *zhijie* simply the same kind of graft but on branches and not on the trunk. The last two, "dimple grafting" *yejie* and "reciprocal grafting" *dajie* have the same names in the two texts. The next step is to be found in the *Treatise of the beautiful plants Qun fang pu,* achieved by around 1620 by Wang Xiangjin. After having quoted extensively the text of Wang Zhen, the author adds that when these methods don't work, there is another possibility, called "sticking" *guotie* which is grafting by approach.

Grafting for What Purpose?

After this brief bird's-eye view of the evolution of grafting techniques, now let us consider what was the purpose for grafting as expressed by the authors of the different texts.

It could hasten the growth of good varieties and give the opportunity to get bigger, sweeter, and smoother fruits. Actually, there were a great number of wild species of fruit trees that could be used as stock. We found several times the remark that after three graftings of cultivated varieties the stones in the fruits became smaller. This may explain why several varieties without a stone *wu shi*, literally "fruitless," are mentioned for jujube, prune, and lichee, for instance. Coming to flowers, grafting was considered as a way to create new varieties. In the case of tree peonies, this was obtained by grafting old famous varieties on wild herbaceous peonies, or by multiple grafting using different famous cultivated varieties. Examples of new varieties, different from both stock and scion, which had appeared after grafting, can be found quoted in various texts.[24] It is striking that this fact, which is considered impossible by geneticists, is taken for granted by modern historians in China, following the michurinian views on grafting.[25] However, it is remarkable that the wonderful results obtained by grafting peonies during the Song dynasty were actually considered with caution by the professionals. There is one example, reported by Zhang Bangji in a monograph on tree peonies,[26] about an extraordinary pale yellow *mudan* flower which appeared in the nursery of the Niu family in 1112. Here is an excerpt of the translation of this text by Joseph Needham and Lu Gwei-djen (406-407):

> Grower Niu named it 'Gold-thread yellow' *lujinhuang*. He built a marquee of bamboo matting around it, with suitable barriers, and spread caerulean silk decorations at the gate, after which guards were set so that only those who paid 1000 cash were allowed to see it. In ten days the family made hundreds of thousands of cash. I myself was one of those who managed to get it and see it.
>
> Later the governor of the prefecture heard about it, and wanted to send up a cutting in presentation to the imperial court, but all the nursery men maintained that he must not do so, saying that this was no ordinary flower and might only too easily change. After a while the governor renewed his suggestion, wondering how best to react to the phenomenon, and proposing that a divided root should be presented but the nurserymen replied politely but firmly the same way as before. Next year when the flowers opened they had all reverted to the previous (common) variety. This was truly a queer marvel of the plant world.

This "Gold-thread yellow" was probably a chimera or a mosaic.

In the "Book of Agriculture" *Nong Shu* (preface 1313) we learn that "mulberry trees and fruit trees are all ameliorated by grafting" in a way that is compared with the transformation of caterpillars into butterflies. The scions must be beautiful, chosen from south-facing branches because they have a strong *qi* and they grow easily. Grafting is possible with plants of the same categories. In addition to the previous remarks taken from ancient horticultural books, perhaps the best summary of what grafting could have represented during the Song dynasty is expressed in a poem by Chen Guan[27] (1059-1124), titled "Grafting flowers"[28] *jie hua*.

> From red I can make purple royal.
> From a petal I can make a thousand.
> From a small flower I can make a larger one.
> Where there are few fruits I can make many.
> Heaven endows [everything] with a fixed disposition.
> I have the power to transcend that.
> I boast that in grafting flowers
> My own hand grasps the power of creation.
> People who learn of that are taken aback.

And sigh dubiously.

With his cleverness would he do marvels?

Is it possible to transform what Heaven has done?

I long to pluck chrysanthemums in Spring time.

I long to admire peach flowers in Winter.

If you cannot cultivate with grafting

Your talent is just the fruit of labor.

With rain and dew grass must grow.

With snow and frost pine does not die.

If it does not die, it is because of its proper nature.

If grass must grow, it is to follow the weather.

Oh! You whose transformations

Are really affected by the seasons.

Are you capable to master [the work of] the seasons?

What is there that does not follow the seasons?

Two centuries later, Wang Zhen appears to confirm the affirmation of Chen Guan. He wrote:

[O]nce the graft has succeeded the *qi* circulate. With ugliness one makes beauty, one is transformed into the other, no word can express the advantages. Since, nowadays, farming gives poor benefits, one can do better by grafting. It is the same as if one could change weeds into rich crops, rough stone into precious jade. He who wants to live from his labour, how couldn't he use it?

In addition to these obvious economical aspects, we have also found more social side effects: recognition of the ambiguous skill of the grafter, on the one hand, and an elegant way to maintain distinguished exchanges between scholars gardeners, on the other.

Now, I would like to come to a more speculative point that may have contributed, with economic reasons, to explain why it is during the relatively short foreign Mongol dynasty that grafting appeared to have reached a prominent position in agricultural books and possibly in orchards. For this purpose, my only allies will be the terms and their meanings.

The Technical Vocabulary for Grafting and its Symbolic Value

We have already met various terms for the different objects and actions involved in the grafting process. We may notice that, beside *cha* "insert" used in the *Qi min yao shu,* which translates the actual action of grafting, all the other terms mean "unite" *jie,* "exchange" *huan, bo.* We remember that there had been since Antiquity a strong symbolic relationship between trees and the virtue of the sovereign. When we consider the description of grafting in the books published during the Yuan dynasty, one cannot but acknowledge the purely anthropocentric vision of the tree body. The wood is called *gu* "bone," the dark green layer under the skin is *jirou* "tendon and flesh" where circulate *jinye,* saliva; the periphery is *pi* "skin."

In the first mode of grafting, analogous to what was described in the *Qi min yao shu,* a crown graft, before inserting the scion into "tendon and flesh," against the bone, the grafter must put it into his mouth in order to warm it up and let it "borrow human *qi* to facilitate its survival." A note specifies that "on this occasion one cannot drink alcoholic beverage or eat things with a strong taste." Putting the scion into one's mouth before inserting it into the stock is also indicated in the *Nong shu*, where one reads that in doing so "the saliva is used to help its *qi*." These details remind us that in the Chinese conception of the natural world, particularly with the neo-Confucian school, living things, plants, animals, and man do not have differences of nature between them, but of "degree of consciousness".[29] All possess the *li* "basic structural principle" and are animated by the *qi* this "force" resulting from the yin–yang activity. It is the way in which the two react to each other that produces the differences between the "ten thousands things" *wan wu*. This may help to better appreciate the symbolic meaning of the spontaneous grafting of plants. This linkage of their *li* actually concerns the fundamental principle of organization of all the things on one hand. By contrast, the fundamental function of the ruler in imperial China was to maintain, by a proper personal and social attitude, the equilibrium between Heaven (Yang) and Earth (Yin), which brought harmony to the whole kingdom.

We have already mentioned that the first designation of the stock had been the "host," which possessed the potential vital strength that could properly nourish the scion considered as a "guest;" then the stock became, among nursery men of the eleventh and twelfth centuries an "ancestor." In this respect, grafting can be seen as the image of the entry of somebody into a clan or a family. Actually, it is what happened at a later period since the modern term for grafting is a binomial associating the standard term since the eighth century *jie* with *jia* which means "to marry (for a woman)," the character representing a woman and a house/family. The term is not used to name grafting, at least before the seventeenth century. We have better evidence, however. In the *Treatise of the crab apple* (mid–eleventh century), Chen Si expresses the following sentence "bring close to the pear tree the young twig and graft it" using for what I have translated by "graft" a very particular term *zhui* meaning "to be adopted after marrying as a son by the parents of his wife" when there is no male heir in the wife's family in order to assume the filial duties and particularly the cult to the ancestors. In the same period, grafting of mulberry was recommended to rejuvenate old trees. In order to get good results, the scion had to be chosen full of strong *qi*, because grafting was considered to imply a mutual transformation that produced positive effects. Actually, extraordinary results of grafts had to be reported to the Court. In view of this, I am tempted to consider that it could not be by chance that this technique became officially widespread over an empire when its foreign ruler had just decided to adopt a highly symbolic Chinese name "Yuan," meaning "the origin," and had also chosen to enhance the traditional Chinese agricultural system of production but carried on governing China, mainly with non-Chinese as high officials, among them, possibly Marco Polo, the emperor, and his officials all like new scions full of life on the old trunk of the exhausted Chinese empire.

[1.] Georges Nicholson, *The Illustrated Dictionary of Gardening, A Practical and Scientific Enciclopædia of Horticulture for Gardeners and Botanists* (London: L. Upcot Gill, 1885), vol. 2:87.

[2.] See, for instance, R.J. Garner, *The Grafters' Handbook* (London: Faber & Faber, 1958), 29-35; and Henry I. Baldwin, "Trees that Unite with Each Other", *The Scientific Monthly* 47, no. 1 (1938): 80–85.

[3.] Tan Bi'an, "Zhongguo gudai de jiemu jishu" ("Techniques for grafting trees in Ancient China"), *Nongye xuebao* 7, no. 4 (1956): 419–450 (see 446).

[4.] Feng Yunpeng, Feng Yunyuan, *Jin shi suo* (1821), vol. 4:16b; See also Wu Hung, *The Wu Liang Shrine. The Ideology of Early Chinese Pictorial Art* (Stanford, California: Stanford University Press, 1989), 240.

[5.] Shih Sheng-Han, *On "Fan Sheng-Zhi shu" – An agriculturistic book of China written by Fan Sheng-Chih in the first century B.C.* (Peking: Science Press, 1959, 68), 37–38.

[6.] Joseph Needham and Lu Gwei–djen, *Science and Civilisation in China* Vol. 6: 1: *Botany* (Cambridge: Cambridge University Press, 1986), 103; see *Zhou li. Zhou rites* (2nd century B.C.) French translation: Edouard Biot, *Le Tcheou-Li ou Rites des Tcheou* (Paris: 1851), ch.11:3a ; and Huainan zi, *Master of Huainan* (c. 139 B.C.) French translation: Charles Le Blanc, Rémi Mathieu, eds. (Paris: Gallimard, 2003), ch.1:6b.

[7.] Joseph Needham and Lu Gwei–djen, *Science and Civilisation in China* Vol. 6: 1: *Botany* (Cambridge: Cambridge University Press, 1986), 103-108.

[8.] Emil Bretschneider, "Botanicon Sinicum," *Journal of the China Branch of the Royal Asiatic Society for the year 1890–91* II (1893): 269-271 [Reprint : Nendeln/Liechtenstein: Kraus Reprint Limited, 1967]

[9.] Ding Fubao, *Shuowen jiezi gu lin* (Forest of commentaries on the *Shuowen jiezi*), (Shanghai : Yixue shuju, 1928), 2548a.

[10.] We may notice that the standard modern term used by Duan Yuzai meaning "to graft" and the entry of the ancient dictionary are homophonous.

[11.] For a presentation of the book and the translation of some passages, see Shih Sheng-Han, *A preliminary survey of the book "Ch'I min yao shu" - An agricultural encyclopaedia of the 6th century* (Peking: Science Press, 1982); and Francesca Bray, *Agriculture*, vol. 1.6:2 in Joseph Needham and Lu Gwei–djen, *Science and Civilisation in China* (Cambridge: Cambridge University Press, 1984). In this last book, after a general presentation (55–59), many concrete references to the *Qi min yao shu* are done in various parts of the text.

[12.] *Pyrus serotina* Rehder = nashi

[13.] *tang*: *Pyrus phaecarpa*, *du*: *Pyrus betulifolia*

[14.] Trans. pp. 58–59, in Shih Sheng-Han, *A preliminary survey of the book "Ch'I min yao shu" – An Agricultural encyclopaedia of the 6th century* (Peking: Science Press, 1982), modified.

[15.] Francois Jullien, La propension des choses. Pour une histoire de l'efficacité en Chine, (Paris : Editions du Seuil, 1992); in English: The propensity of things : toward a history of efficacy in China, translated by Janet Lloyd, (New York : Zone Books, 1995).

[16.] Li Lian, Bianjing jiu yi ji (Notes on strange things of the capital Kaifeng) (Shanghai:Shangwu yinshuguan, 1937, 86), 56-57.

[17.] This garden was very often mainly composed of potted trees that could be transported, following the scholar civil servants in their numerous transfers along their administrative carriers. On the scholar gardeners, see George Métailié, "Some Hints on 'Scholar Gardens' and Plants in Traditional China," *Studies in the History of Gardens & Designed Landscapes* 18, no. 3 (1998): 248-256.

[18.] Fan, in Anon., *Shenghuo yu bowu congshu (Collectanea on Life and natural sciences)* (Shanghai: Shanghai guji chubanshe, 1993), 2.

[19.] Georges Nicholson, *The Illustrated Dictionary of Gardening, A Practical and Scientific Enciclopædia of Horticulture for Gardeners and Botanists* (London: L. Upcot Gill, 1885), vol. 2:87-88.

[20.] For a more modern reference, see R.J. Garner, *The Grafters' Handbook* (London: Faber & Faber, 1958).

[21.] Francesca Bray, *Agriculture*, vol. 1.6:2 in Joseph Needham and Lu Gwei–djen, *Science and Civilisation in China* (Cambridge: Cambridge University Press, 1984), 60.

[22.] Wang Yuhu, *Wang Zhen Nong Shu (Book of agriculture by Wang Zhen)* (Beijing: Nongye chubanshe, 1981), 54–55.

[23.] Georges Nicholson, *The Illustrated Dictionary of Gardening, A Practical and Scientific Enciclopædia of Horticulture for Gardeners and Botanists* (London: L. Upcot Gill, 1885), vol. 2:89.

[24.] Chenzhou mudan ji, p. 10, *juan 289, ce 554*, in Chen Menglei, Jiang Tingxi and als. (ed.) [1725]. *Gujin tushu jicheng* (Compilation of ancient and modern texts and images) Zhonghua shuju, 1934.

[25.] Sheng Chenggui, "Woguo gudai wenxian zhong youguan jiemu de jizai" ("Records concerning grafting of trees in ancient Chinese literature"), *Shengwuxue tongbao*, no. 4 (1954): 6–9; Tan Bi'an, "Zhongguo gudai de jiemu jishu" ("Techniques for grafting trees in Ancient China"), *Nongye xuebao* 7, no. 4 (1956): 419–450; Zhang Zongzi, "Wogu gudai huamu jiajie tanyuan" ("Preliminary discussion on grafting flowers and trees in Ancient China"), *Nongye kaogu* , no. 2 (1988): 323–327; Zhou Zhaoji, "Zhongguo jiajie jishu de qiyuan he yanjiu" ("Researches on and origins of the grafting techniques in China"), *Ziran kexue shi yanjiu* 13, no. 3 (1994): 264–72. Even a modern small practical manual indicates that new varieties are obtained by grafting. See Li Jihua, *Zhiwu de jiajie (Plant grafting)* (Shanghai: Shanghai renmin chubanshe, 1977), vol. 1:92. If we cannot negate the reality of these new products, there are not new genetic varieties but probably the result of a chimera. A recent discovery by researchers of a laboratory of Strasbourg University–UMR "Vigne et Vin d'Alsace"–Inra Colmar, France - confirms the possibility of this point. They have found that "Pinot gris" is a natural chimera resulting of the association of an epidermis of "gris" ("grey"), which confers its color and probably its aromatic properties, with internal tissues of "blanc" ("white") type. These characteristics are transmitted through vegetative propagation of vine plants – see Didier Merdinoglu, "Le pinot gris a deux genomes," *INRA-la lettre*, no. 7 (2004): 5.

[26.] Tan, Bi'an, "Zhongguo gudai jiemu jishu" (Grafting techniques in Ancient China), *Nongye xuebao*, 7 no. 4 (1956): 419-450 (with an English abstract) ;

also p. 61, *juan* 12, *ce* 531 in Chen Menglei, Jiang Tingxi and als. (ed.) [1725]. *Gujin tushu jicheng* (Compilation of ancient and modern texts and images) Zhonghua shuju, 1934.

[27.] See Tan Bi'an, "Zhongguo gudai de jiemu jishu" ("Techniques for grafting trees in Ancient China"), *Nongye xuebao* 7, no. 4 (1956): 427.

[28.] I would like to thank Maria Eva Subtelny who kindly helped me to improve my first translation.

[29.] Derk Bodde, *Chinese Thought, Society, and Science* (Honolulu: University of Hawaii Press, 1991), 27.

Botanical Progress,
Horticultural Innovations
and Cultural Changes

Botanical Progress,
Horticultural Innovations
and Cultural Changes

Botanical Progress,
Horticultural Innovations
and Cultural Changes

Botanical Progress,
Horticultural Innovations
and Cultural Changes

The Chinampas Before and After the Conquest

Saúl Alcántara Onofre

When the Spaniards arrived in the Basin of Mexico in 1519, lacustrine zones were playing important roles in the lives of the Mexicas-Aztecs inhabitants. It is evident in the early chronicles that Aztecs had mastered hydrologic technology: for instance, they constructed dikes and other water-control systems to maintain lake levels. Although the Spanish wanted to prevent the flooding of their churches, convents, and houses, the Mexican-Aztecs also were concerned about protecting great areas of chinampas, which are sometimes, today, referred to as "floating gardens."

Records of great floods—two during the prehispanic period and eleven in the viceroyalty times—were reported to the Spanish: in 1446 and 1499 when the city flooded in spite of the dikes; in 1555 when the viceroy Luis de Velasco ordered the construction of another dike named San Lázaro; the floods continued in 1579, 1604, 1607; and the worst was in 1629, the duration of which was four years; more were in 1714, 1747 and the last in 1951. All of these attest to the recurrence of floods in the main lake system in the Basin of Mexico.

An indigene pushing a floating chinampa with four "calli" or houses. Tlatelolco Codex (courtesy of the Biblioteca Nacional de Antropología e Historia, Mexico).

The complex urban center of Tenochtitlan, the Aztec capital, and the prosperity of Aztec society can be said to have been partly the result of the development and employment of chinampas. Chinampas were of two kinds as first describes Angel Palerm.[1] The first are lacustrine chinampas or "inland lagoon." These were artificial islands that were built in the lagoons and permanent marshes by using ditches and the gathering of grass, mud and earth in order to form a small barren island that was very narrow and long. These artificial islands only existed in the Basin of Mexico, where there was a combination of shallow lakes, fresh water springs, and abundant aquatic vegetation.

The second kind were the "inland chinampa," also known as "dry chinampas," located in marshy zones where drainage was insufficient. Ditches were excavated to delimit the parcel and earth and mud were piled to set the level of the planting surface above the water level. According to Teresa Ocejo Cázares,[2] this kind of chinampas existed in the environs of Cholula and Puebla state.

Any visitor to Mexico has heard of the chinampas, and knows them as the floating flower gardens of Mexico. I shall provide a

This codex reproduces one part of Mexico city, from "Garita de Peralvillo," next to Tlatelolco, to "Calzada de Guadalupe." It is possible to observe houses, main and secondary canals, water springs and chinampas. "Plano en papel de Maguey" (courtesy of the Biblioteca Nacional de Antropología e Historia, Mexico).

more complex view of the chinampas, explaining how they were made and cultivated. How the ancient Mexicas-Aztecs developed them, what they cultivated, how they were used after the European conquest, and what is left of their former glory.

The Chinampa Lands in the Basin of Mexico

The need for chinampas arose from the geography and hydrology of the Basin of Mexico and the history of the Aztecs. One of the terms by which they called their home was Anahuac, a name[3] that was, at first, only given to the Valley of Mexico because its main cities were founded in the small isles along the margins of two lagoons. Over the years, the name was used to refer to the land that during the viceroyship was known as New Spain.

The Valley of Mexico is surrounded by beautiful green mountains, whose circumference measured from the inferior mounts is greater than 120 miles.[4] The larger lagoon within this Basin contained fresh water while the minor one held brackish water. The two lagoons were connected through a large channel. The water that flowed from the mountains into the brackish lagoon was located in the lowest part of the Valley. Because of this, the Great Tenochtitlan was constantly flooded, not only under native rule but also during Spanish domination. It was in this very lagoon where the Great Tenochtitlan was founded in the year II *Calli*, or A.D. 1325.

The circumference of these lagoons was somewhat more than ninety miles, Francisco Javier Clavijero said that the form of these two lagoons were "in the shape of a camel, whose head and neck was the fresh water lagoon, that means the Chalco lagoon. The camel's body was represented by the brackish water lagoon, the Texcoco lagoon; the legs and the feet were formed by small rivers and rivers that flowed from the mountains into the lagoon. Between both lagoons is the little Iztapalapan Peninsula."[5] Apart from the courts of Mexico, Acolhuacan and Tlalpan, there were in this Valley four other cities, which were: Xochimilco,[6] Chalco, Ixtapalapan and Quauhtitlan which currently have hardly one-twenth of what they had in those days.[7]

One of the main problems the Mexicans faced was lack of land on which to build their houses, for the Tenochtitlán island didn't have enough room for all the inhabitants. They fixed this situation by extending those parts of the island where the water is shallow by reclaiming land with alternate beds of reeds and mud taken from the bottom of the lagoon. This, of course, was the same technique used to create the chinampas. While the inhabited islands could be enlarged, other lands could be created to support wildlife for hunting and gathering, while other chinampas served as extensive gardens.

At the end of sixteenth century, Clavijero mentions, "With the commerce of this hunting with the places settled on the

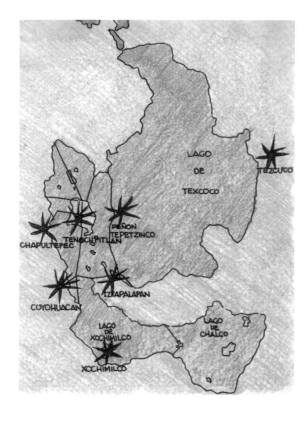

Prehispanic gardens in Ancient Mexico (drawing: author).

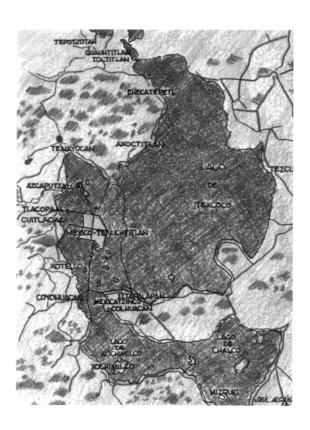

Reconstruction map of Tenochtitlan, 16th century, showing the location of chinampaneca's towns that pay tribute to the Texcoco region (drawing: author). Ground and section plan (drawing: author).

margin of the lagoon, they could acquire all what they needed. But where their efforts went further, was in the construction of floating orchards, or chinampas[8] which were made of embankment of rocks and mud which came from the lagoon. The chinampas were used to cultivate corn, capsicum, beans and squashes."[9]

Fernando de Alva Ixtlixochitl gets very close to the description of the districts that were devoted to the chinampa agriculture in the Basin of Mexico:

> The king Izcohuatzin and other important people, asked Nezahualcoyotzin grace of their lives which he granted. And he ordered that from then on they would offered him some recognizing or tribute which was known as the royal census of Tezcuco, "*chinampanacatla callacuilli*" which means the tribute of the chinampanecas that are the following cities, towns and places: Mexico-Tenuchtitlan, Xolteco, Tlacopan which are the heads of their kingdoms; Azcaputzalco, Tenayocan, Tepotzotan, Quauhtitlan, Toltitlan, Ecatépec, Axoctitlan, Coyohuacan, Xochimilco, Iquexomatitlan each of which gave tributes of jewelry and pieces of gold and all the vegetables, flowers, fish, and birds that grow in these parts of the lagoon.. . . . [10]

So the *tlatoani* Motecuhzoma came back to his city and, "ordered that the cities and villages of the Chinampa which used to give some recognition to the kings of Tetzcuco, wouldn't give more tribute . . ."[11]   In fact, Nezahualcóyotl exacted infamous tributes from the Aztecs. In 1521, the chinampanecas villages of Xochimilco, Cuitlahuac, Mizquic, Ixtapalapan, Mexicatzinco, etc; helped the Mexicans and "tlatelolcas" to fight against the Spaniards, but all of these cultures were witnessing the transformation of a water civilization into one that sees the ideal city without floods and the position of the religious ideals over the pagan indigenous temples.

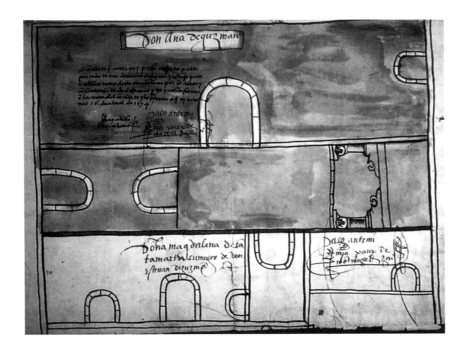

Plan of a house in Xochimilco, Fray Alonso de Escalona, 1574, Archivo General de la Nación, México (photo: author).

The Chinampa Techniques

Torquemada wrote on the fragility of the chinampa-based agriculture. While floods destroyed these cultivations, droughts rarely affected them.[12] Numerous authors have written on the chinampa technique, from those written by Sahagún, Durán, Bernal Díaz del Castillo, West and Armillas to the last publications by Teresa Rojas Rabiela. Nevertheless, Javier Clavijero's descriptions are the most comprehensive about the construction of the chinampas.

Torquemada also mentions that the natives ". . . with not so much effort cultivate and collect their corn, because all these are "*camellones*" that they call chinampas which are furrows that are on the water surrounded by ditches, there aren't any risks . . .".[13]

The zone that was regarded as the core of the chinampa horticulture was the Xochimilco-Chalco district. Additionally, in the bay of Texcoco lake, there were other chinampa zones, around island settlements, and from dry land, in "Huizilopochco," which nowadays is Churubusco, and Iztapalapa.

The form and procedures used to construct the chinampas have been described as follows:

> . . . they make their farming . . . fetching grass in canoes from dry land, throwing them into the water until it is about four to five feet thick, and rises about one and a half foot above the water being between eight and twelve feet wide. And the natives make a farm out of a number of them, go in their canoes, growing up herbs and profiting in a way that had not been seen before.[14]

Additionally, Father Alonso described the chinampas saying,

> [they] are built in the water, stacking many alternated layers of reed and mud from the same lagoon, and making them at times very narrow . . . . leaving one channel between two chinampas rising two feet or less above the water and carrying ears of corn that grow very big because they get water directly from the lagoon without need of rain water falling from the sky and they also make seedbeds of corn in their chinampas producing young plants to be later transplanted according to a very common practice in that land.[15]

The plant nurseries described by Father Alonso were made in movable sections that measured between twenty and thirty feet in length and the width that they desired, founded on the water over aquatic-grass, rush and reed, "where they made the nursery of their vegetables . . . to be transplanted later elsewhere; they tie them with ropes to take them from one place to another in the lagoon."[16] Father Alonso Ponce himself saw the corn growing in the chinampas of Xochimilco in 1585.

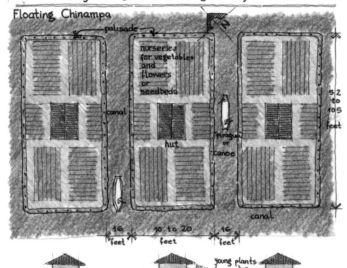

CHINAMPA CULTIVATION SYSTEM OF THE VALLEY OF MEXICO

CHINAMPA derives from the Náhuatl "CHINAMITL", which means reed boundary or hedge, or fence with sticks or interwined reeds. A chinampa if, generally speaking, a floating garden formed artificially either by dumping earth in a designated area near the shore of a lake until a small islet if formed, or by making a sort of raft gradually and touches bottom

Floating Chinampa

- palisade
- nurseries for vegetables and flowers or seedbeds
- canal
- piragua or canoe
- hut
- 52 to 105 feet
- canal
- 16 feet
- 10 to 20 feet
- 16 feet
- young plants
- hut

Saúl Alcántara Onofre 2005

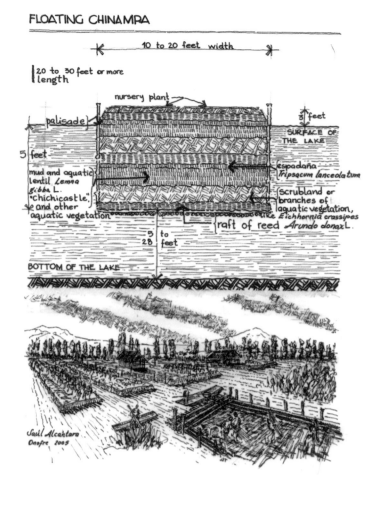

FLOATING CHINAMPA

- 10 to 20 feet width
- 20 to 30 feet or more length
- nursery plant
- palisade
- 5 feet
- SURFACE OF THE LAKE
- 5 feet
- espadaña *Tripsacum lanceolatum*
- mud and aquatic lentil *Lemna gibba* L. "chichicastle", and other aquatic vegetation
- scrubland or branches of aquatic vegetation, *Eichhornia crassipes*
- raft of reed *Arundo donax* L.
- 5 to 28 feet
- BOTTOM OF THE LAKE

Saúl Alcántara Onofre 2005

A chinampa formation is, generally speaking, a floating garden formed artificially either by dumping earth in a designated area near the shore of a lake until a small island is formed, or by making a sort of raft gradually loaded with herbs and earth until it reaches bottom. (drawing: author)

Section and sketch showing the construction of the floating chinampa (drawing: author).

The Aztecs, during their quest for a final place in which to settle down in the Basin of Mexico, cultivated vegetables in the chinampas of Xaltocan, and once they settled down in Tenochtitlan, they kept on doing so in the numerous chinampas that they built. The same thing happened with the inhabitants of the nearby kingdoms during the biggest expansion stage in their efforts to claim these plots in the Basin swamps.[17]

Javier Clavijero has described the chinampas as mobile isles or floating gardens where the plants were grown, and later transplanted to the fixed chinampas. When the Spaniards arrived, the natives could not be controlled and the chinampas zones weren't possible to have as registered land, so the Spaniards ordered all the floating gardens with "Ahuejotes" trees *Salix bonplandiana var. fastigiata André* to be fixed to the bottom of the lagoon.

According to the laws published by the famous *tlatoani* Nezahualcóyotl, . . . It also condemned to the death penalty the thieves of the "sementeras" or chinampas, declaring that it was enough to steal seven ears of corn for this rule to be applied and

Cultivation in the seedbeds. Florentine Codex, book IX. (courtesy of the Biblioteca Nacional de Antropología e Historia, Mexico).

One can see the seedbeds cut into half-foot cubes with a wood knife (coa) before the chinampero sows the seeds in the nursery and the cover of the nursery plant: in this manner the humidity is always in the seedbed (photo: author).

that the owner of the field had the right to enslave whosoever stole a certain number of ears of corn or took away from another's field a certain number of useful plants.[18]

The importance of chinampas is further in evidence by the great care taken in developing shelters for garden workers: "On the fields of corn in the chinampas they used to build a kind of stick turrets, where a man protected, from the sun and the rain, watched and hunted with the sling the birds that went there to damage the 'sementeras' or chinampas. Even now there are similar turrets in the Spanish fields, due to the abundance of birds."[19] They also built barns, some of them were so big that they could contain five or six thousand or more "fanegas" or big sacks of corn. At the end of the sixteenth century there were "similar barns in some distant places from the capital, and some of them are so ancient they seem to have been built before the conquest, and as a very intelligent person told me, the seeds are preserved better than in the barns made following the European style."[20]

Plants Cultivated in the Chimampas

Let us turn now to the plants that were cultivated in prehispanic times and then in the colonial period to show how a traditional horticulture was maintained for both Mexican and European plants.

First, I should stress that the numerous prehispanic plants cultivated in the Basin of Mexico were produced in the chinampas. Maize or corn *Zea mays* L., called by the Mexicans "tlaolli," was grown in many varieties of different weights, flavors, size, and colors. They were large or small, and white, yellow, blue, purple, red or black. Among other things the Mexicans baked their bread out of corn. The corn was sent from America to Spain and then to other countries in Europe.[21]

In the prehispanic epoch, corn was typically cultivated in the chinampas, from plants grown in mud nurseries. It is inaccurate that the chinampa had been inadequate for the production of corn and that it was only suitable for that of vegetables, as some authors have asserted.[22] The "chalqueño" chinampero, a kind of corn, is now found in all those lands that had or still have chinampas or in dried up lands of the old lacustrine territories, like Iztapalapa, Chalco, Tlalpan, Tláhuac, Mizquic, Xochimilco, San Gregorio, Santa Cruz, and so on.--

Due to small plot size in the Great Tenochtitlan,[23] the corn was only cultivated for obtaining very small ears of corn for eating

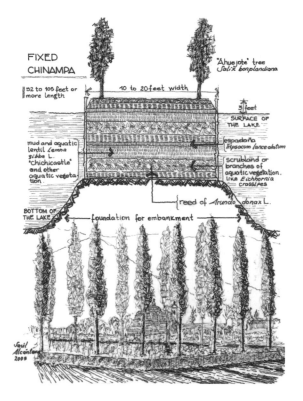

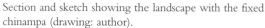

FIXED
CHINAMPA

"Ahuejote" tree
Salix bonplandiana

32 to 105 feet or
more length

10 to 20 feet width

3 feet

SURFACE OF
THE LAKE

mud and aquatic
lentil Lemna
gibba L.
"chichicastle"
and other
aquatic vegeta-
tion.

espadaña
Tripsacum lanceolatum

scrubland or
branches of
aquatic vegetation.
like Eichhornia
crassipes

reed of Arundo donax L.

BOTTOM OF
THE LAKE

foundation for embankment

Saúl
Alcántara
2005

Section and sketch showing the landscape with the fixed chinampa (drawing: author).

Panoramic view of the chinampas zone and the channels of the San Gregorio lake Tlapulco, Xochimilco, 1907, col. Casasola Bazar de Fotografía, from Memoria de la Ciudad de México, Cien Años (photo: Hugo Brehme).

on the cob. However, in the rural areas and in the districts of Chalco-Xochimilco much larger cobs were produced and used to fill sacks of grain for food or tribute.

In the sixteenth century, according to Alzate description, the natives of Culhuacán and Xochimilco used the corn nursery to produce early crops, especially corn that was regularly sown.[24] Zucchini, for example, was also cultivated in the same way. The plant nursery was covered with a roof for sun protection. And the leaves of the young plants were cut off for transplantation, in the "way that the farmers used with onions, garlic and other plants, and the cultivation of corn in the chinampas prosper very much: at first sight they look like forest . . ."[25]

The corn yield of the chinampas at the time is not known, but it can be inferred from a tributary imposition of the farmers of the county of Xochimilco in 1558, working either in neighboring mountain fields or in lacustrine chinampas. They had to produce the same amount of grain, and to that effect it took five times as much land in the mountains as in the chinampas.

During the viceroyship, and until the beginning of the twentieth century, the green or dry parts of the corn served to feed the horses, mules, oxen and cows of the new Spaniard city of Mexico, and its surroundings were maintained with the remains of the corn and with the spontaneous vegetation that grew in the marshes and in the hills.

The Mexicas-Aztecs cultivated six species of tomatoes of different sizes, colors, and flavors. The biggest was the "xictomatel" or "xictómat" *Lycopersicum esculentum* Mill. or "xitomate," as the Spaniards from Mexico called it. As Clavijero said towards the end of the sixteenth century and the beginning of the seventeenth, it was widely known in Europe, in Spain and in France under the name "tomato" or "pomodoro"[26] in Italy.

Huautli Amaranthus leucocarpus, this flower had great religious importance, Florentine Codex, book XI (courtesy of the Biblioteca Nacional de Antropología e Historia, Mexico).

"Tlaxochimanco" party, parties and offering of flowers "… at the party's day, the idols 'sátrapas' arranged with many flowers to Huitzilouchtli, and… statues of others gods were arranged with garlands and strings and collars of flowers… everybody arranged with flowers the statues they had at home…" (Sahagún, 1981, p.183) Florentine Codex, book II (courtesy of the Biblioteca Nacional de Antropología e Historia, Mexico).

The "tómatl" *Physalis angulata* L. was an important crop of the chinampas, of which all the information seems to indicate that it was produced in more than one cycle a year.

Beans chinampero *Phaseolus vulgaris* L. were also an extremely important vegetable staple for the ancient Mexicans, of which there were more species and varieties than corn, the largest species was the "ayacotli," the size of a lima bean *Phaseolus coccineus* L., produced from a red flowered plant. Yet the most appreciated was the small, black, and heavy bean. Clavijero wrote, "This vegetable was not well-known in Italy because it was considered a bad plant, on the contrary, in Mexico it was appreciated, not only as the major staple for the poor, but also as a delicacy for the Spanish nobility."[27]

Green bean *Phaseolus acutifolius* Gray., "alegría" or "uauhtli" *Amaranthus leucocarpus* S. Wats., "chayote" *Sechium edule* Sw., "chilacayote" *Cucurbita ficifolia* Couche and "quelites" *Amaranthus spp.*, complete the list of the Mesoamerican crops in the chinampas.

Chia *Salvia hispanica* L. or *Hyptis suaveolens* Pot. is the small seed of a plant whose shaft is straight and square-shaped, the branches extended toward four points and with opposed symmetry and blue flowers. There are two species, a small black one, out of which an excellent oil is extracted for painting, and the other one, white and bigger, from which a refreshing drink is made. The Chia is mentioned in some early viceregal documents but it is not known (currently) in the area, and the older chinamperos do not mention it. Nevertheless, sixteenth century documents refer to other areas in the Basin of Mexico, where cultivation of Chia was frequent.

Chilies[28] *Capsicum annum* L., and *Capsicum frutescens L.* were widely used among the Mexicans like salt among the Europeans. There are at least eleven species that differ in size, form, and texture. It was constantly cultivated in the chinampas, and even under freezing temperatures, its production was possible thanks to special care the chinamperos provided during the winter. The rain does not affect it but hail damages its fruit.

The sources only distinguish some of the varieties and state that chilies were sown and identified by Humboldt as Capsicum in the early nineteenth century. The common documented names suggest more variety: "chílchotl" or green chili, creole or "chicostle," "tornachile," "chilito," "carricillo," "poblano," "pasilla," "mulato" and "cascabel."

Let us say a few words about fruit production. The avocado, ahuacatl, is the fruit of *Persea Americana* Mill. The "guava," "xalxócot," is the fruit of *Psidium guajava* L. Mamey *Calocarpum sapota* Merr. is also an islander name. *Annona squamosa* L. in Aztec language it is known as "te-tzápotl" or "te-tzontzapotl." There are many varieties of "zapote" or "tzápotl" *Casimiroa edulis* Llave et Lex.; the white, yellow, blackish; among them one counts the "chicozapote" or "zapotilla" *Achras zapota* L. The "camote" or in Aztec language "camotli" *Ipomea batatas* L., is nothing else than the European sweet potato. "Xicama" or "xicamatl" *Pachyrrhizus erosus* Urb. is a tuber similar to the turnip, of sweet, almost insipid flavor. "Cacomite" is a Mexican name, "cacomitl" *Tigrida pavonia* Kerr. for a plant that

In ancient Mexico, the name for a garden was "Xochitla" (place of flowers) and "Xoxochitla", for a place of many flowers. A walled garden was called "Xochitepanyo" and "Tlatoani's" pleasure gardens were called "Xochiteipancalli", or flower palaces. The humble people gardens were called "Xochichinancali", which means a place of flowers, vegetables and fruit trees, surrounded by canes and "magueyes", like is shown in the illustration. These classification of ancient gardens refers to a long relation with horticulture. "...Mexicans are surrounded by countries with every kind of climate, they enjoy all their different fruits, vegetables and flowers. Mexican market is the centre of every nature gift". (Clavijero, 1853, p. 342). Florentine Codex, Book XI. (courtesy of the Biblioteca Nacional de Antropología e Historia, México). #17 ("Xochichinancali", was wrong instead it is "Xochichinancali")

God Tlaloc. He is also known as "Tlalocateuctli", master of paradise and chinampas', he is the god of water. He is the master of Talocan where "it is fresh all time, plants have sprouts all time, it is always springtime; where flowers and trees full of flowers are always standing up", "it is the place where water runs under and fog floats up. Tlaloc is observed as a farmer in a chinampa where, with his "coa" (a very old tool used to cultivate land) he takes care of corncobs. On the other side there is a "tlaloque", an assistant who, obeying the god's orders, destroys the corncobs with hail and lightning axes. In the Borgia Codex, Ediciones Tecolote, México, 2002 (photo: author)

A rare lithograph from the 19th century, that shows the floating chinampas, Anonymous (courtesy of the Biblioteca Nacional de Antropología e Historia, Mexico).

produces a beautiful flower and a root or tuber that is eaten. The name of the "mesquite" is also Mexican, "mitzquitl" *Prosopis juliflora* (Sw.) DC.; this tree produces some edible berries, and rubber that can substitute gum in Arabic medicine and the arts . . . [29]

Last, we shall turn quickly to flowers because of their great religious importance. The happiness plant "alegría" *Amaranthus leucocarpus* S. Wats. and, the flower of the dead or "cempazúchil" *Tagetes erecta* L., are germinated and grown first in seedbeds "almácigos" or nurseries and are later transplanted, in "chapines" or chopine, to the chinampas. It is useful to remember that the prehispanic production of flowers consisted basically of cut flowers for parties and to adorn religious altars.

The "cempazúchil" or "cempoalxóchitl," *Tagetes*, like the Spaniards said,[30] is very common in Mexico where it is known as the flower of the dead. There are different kinds varying in size and number of petals and leaves. Bernardino de Sahagún mentions that these flowers are yellow and fragrant, wide and beautiful, that they are born mainly in the chinampas, as well as others later cultivated in the orchards. They come in two ways, some called female "cempoalxóchitl" are large and beautiful, and others called male "cempoalxóchitl" are neither so beautiful nor so large. The flowers in old Mexico had an enormous meaning for their religious altars and gods, for example, in the ninth month, before the "Tlaxochimanco" party, the whole population went out to the fields, cornfields, and chinampas to look for flowers, wild and cultivated.

Some were called, in Aztec language: "acocotli" *Bidens pilosa* L., "huitzilxóchitl" *Loeselia mexicana* (Lam.), "tepesquisúchil" *Arctostaphylos pungens* H.B.K., "nextamalxóchitl" *Ranunculus petiolaris* H.B.K., "tlacoxóchitl" *Bouvardia ternifolia* (Cav.) Schl.;

others were named "oceloxóchitl" *Tigrida pavonia* Kerr., "cacaloxóchitl" *Plumeria acutifolia* Poir: P., "ocoxóchitl" *Didymaea mexicana* Hook., "cuauhtla-huitz-quílitl" *Cirsium* sp., "xiloxóchitl" *Caesalpinia pulcherrima* Sw., "yoloxóchitl" *Talauma mexicana* (DC.), and having many of these flowers, they joined them in the house of the temple where this "Tlaxochimanco" party took place. There they were stored all night long, and then at dawn they threaded them in their threads. They then made thick ropes of them, bent and long, and spread them in the yard of that temple, presenting them to that god whose party they made.

> Another day very early when the party of the Sun god took place, the "sátrapas" offered to this idol flowers, incense and food and they adorned him with garlands; having composed this statue of the Sun god with flowers and having presented him many flowers, very skillfully made and very fragrant, they made the same thing to all the statues of all the other gods, for all the cúes; and then, in all the houses of the gentlemen and main people they decorated with flowers the idols that each one had, and they presented them other flowers, putting them before, and the whole other popular people made the same thing in their houses.[31]

Flowers resonate in many ways with the Mesoamerican cultural life, like León Portilla mentions, "songs are a florid language. Flowers also are present as offerings and metaphors in ritual ceremonies, dances and in the human and the gods apparels. In the sacred calendar the last of the twenty signs of the days was "xóchitls," the flower. They were also present in the eighteen months of the solar calendar that ruled their liturgy. The first months were dedicated to the offering of the tender ear of corn, "xilomanaliztli"; the ninth month was "Tlaxochimaco," delivery of flowers, and the last month was denominated by "Xochílhuitl," party of the flowers and also "Izcalli," return of the life. [32]

The parties of the ninth month dedicated to the god "Huitzilopochtli," the solar god par excellence, consisted exclusively in floral offerings. A great quantity of multicolored flowers and garlands are represented in the parties and rituals in the codices: the flower of the cocoa *Theobroma angustifolium* DC., the climbing orchid, the vanilla *Vanilla fragans* Ames., the field dahlia *Dahlia coccinea* Cav., and many more. All the Mesoamerican flowers were cultivated in the "xochimilla" or chinampas.

In the indigenous languages, flowers evoke the community of friends, the power to pronounce true words, the memory that man leaves in his songs after death, love, sex, the search for the "giver of life," the yearning for happiness, and also death, as although flowers can be pleasant, some can intoxicate, and all wither in the long term.

The god "Tlaloc," with another name "Tlalocateuctli" (senior of the paradise and of the chinampas), was the rain god, the fertilizer of the earth and protector of the temporary goods. Ancient Mexicas-Aztecs believed "Tlaloc" resided in the towering mounts where the clouds are usually formed, like those of "Tlaloc," Tlaxcala and Toluca.

The Chinampas During the Colonial Period

The period after the conquest saw much continuity, and also many changes. For instance, the ancient type of chinampa could be moved around the lagoon, as described by Javier Clavijero. The mobile island held the nurseries where the plants were grown in "chapín" later to be transplanted to the inland chinampas. However, under early Spanish domination, natives could not easily be controlled and it was not possible to tax mobile lands. Authorities imposed planting "Ahuejotes" trees[33] *Salix bonplandiana* var. *fastigiata André,* around all the chinampas, which would grow roots into the bottom of the lake, thereby giving permanence to all the mobile chinampas or lacustrine orchards. The variety of corn grown was also modified. At present, the cultivated variety in the chinampas and wetlands of the ancient lacustrine zones is the "chalqueño," named after the Chalco district.

According to Teresa Rojas and other specialists, this corn type began to be conical races or "corn of the hill" as it is known in the area of Xochimilco and the tuxpeño.[34]

Some plant cultivation, like chile, for instance, withered away with the years. In their legendary migration to Tenochtitlan, the Mexicas-Aztecs had sown the "chilchotl," probably the green chile *C. annum*, in the chinampas of Xaltocan. It could also be the tornachile, which Molina translates as the chile of irrigable lands, with a green lemon color, that was still cultivated in the chinampas of Iztacalco in 1826.

European culture introduced a new economy, new issues of political control, and new religious perspectives. All of these contributed to large changes. Two major series of events should be highlighted to understand the history of the chinampas: the drying up of the lakes, and the introduction of European plants.

The draining of the lakes that lasted nearly four hundred years started at the beginning of the seventeenth century with the construction of the Huehuetoca tunnel, otherwise known as the Nochistongo channel. In order to prevent the growing viceregal city from flooding, the Cuautitlán river which flowed into the Zumpango river was deviated through a cut that measured eight kilometers and a channel of seven kilometers with a depth of fifty meters.

Dating from this time, the lacustrine environment became a "valley" connected to the Gulf of Mexico by the Tula, Moctezuma, and Panuco rivers. By the end of the eighteenth century another channel was constructed, the Guadalupe or Castera channel, which connected the last works to the lakes of Xaltocan and San Cristóbal.

In addition, some major transformations were a result of the introduction of plants from Europe after the sixteenth century. The great majority of vegetables, flowers, and spices introduced by Spaniards, immediately after the conquest and during the viceroyalty, were incorporated into the chinampas and cultivated according to its technique. For example, cucumber *Cucumis sativus* L., cauliflower *Brassica oleracea* var. *botrytis* L., cabbage *Brassica oleracea* var. L., Brussels sprouts *Brassica oleracea* var. *gemmifera* L., artichoke *Cynara scolymus* L., garlic *Allium sativum* L., onion *Allium cepa* L., spinach *Spinacia oleracea* L., beet *Beta vulgaris* L., cilantro *Coriandrum sativum* L., parsley *Petroselinum crispum* Hoffm., "betabel" *Beta vulgaris* var. crassa Alef., celery *Apium graveolens L.,* turnip *Brassica napus L.,* radish *Raphanus sativus L.,* pore *Allium porrum L.,* broccoli *Brassica oleracea* var. cauliflora, carrot *Daucus carota L., salsify Tragopogon porrifolius,* etcetera, were grown. Cereals and leguminous as wheat *Triticumsativum* Lam., barley *Hordeum vulgare L.,* tare (pea) *Lathyrus odoratus L. and bean Phaseolus vulgaris* L. have also been present, at some time, but their cultivation, does not seem to have prospered due, probably, to the excess of humidity in the plots. Among the plants of Andean origin, brought after 1519 to New Spain, the potato was, mainly sown as a "curiosity." [35]

Olive trees, introduced in Mexico from Europe in the sixteenth century, were also widely disseminated in the lacustrine plains of the Valley.[36]

However, it should be stressed that in spite of the introduction of so many species in other chinampas since the sixteenth century, two chinampa districts, Xochimilco and Chalco, have mainly stood out for flower production, and in that respect they have been instrumental in the efforts to uproot the ancient religion. *Cempoalxóchitl* flowers are the most significant flowers of the prehispanic world either as a gift for aristocrats or for religious altars. After the conquest the Spaniards brought the Castilla rose to compete with and displace these sacred flowers. This rose is associated with the cult of the Virgin of Guadalupe who appeared in the same place as the old Tonantzin, our mother for the natives, who had a temple on a mount three miles from Mexico to the north, according to Clavijero, where the sanctuary of the Virgin of Guadalupe is located at present.

934. A fiesta on the Viga · WAITE. Photo.

A party in the "Canal de la Viga," 1920, Archivo General de la Nación, México

With the arrival of European culture, with all their techniques and botanical knowledge, the floating gardens were enriched with the cultivation of the new imported flowers from distant lands. Flowers or "xóchitls" grew together with their relatives from Asia and Europe, the dahlias *Dahalia coccinea* Cav., different kinds of roses, the irises *Iris spuria*, the gladiolas *Gladiolus spp.*, the poppies *Papaver rhoeas* L. and the water-lilies *Nymphaea spp.* the current emblem of Xochimilco; white lily *Eicchornia crassipes* Solms., jasmines *Philadelphus mexicanus* Schl., carnations *Dianthus caryophyllus* L. and many more, that compete with the American flowers in the gardens of Mexico.

Poppy was another important cultivation in the chinampas areas. These beautiful flowers were sought during many years to help communicate with the divine world, until its consumption was prohibited for its high content of noxious alkaloids. By contrast, at the present time the cultivation of the traditional wallflowers *Matthiola incana* L., cloud, "cempoalxóchitl" *Tagetes erecta* L., paintbrush *Centaurea cyanus* L. and *Psium sativum* L. still goes on.

Reminders of the Chinampa Tradition into the Present

The draining process that started after the conquest continues to the present with extraction of water through clandestine wells. This systematic drainage and drying of the lacustrine zone has modified the traditional chinampa techniques, as well as the uses of the cultural landscape. Ancient corn is still cultivated in the areas of Chalco, Xochimilco, or in small agricultural remainders that can be found in Mexicaltzingo, Iztapalapa, and Tláhuac. Yet it is no longer cultivated in chinampas but in the fields with a plow or a tractor. These are alien techniques that were introduced as a consequence of the evolution of the chinampa-based agriculture in Mexico in a drying environment. They carry as a consequence the development of salty and impoverished soils, as well as the necessity to use fertilizers, mainly cow manure, but also chemicals, and the proliferation of plagues unknown or not very frequent until the present.

This is the result of a slow process of change. In the first years of the twentieth century, corn was grown together with beans and chilies. The chinamperos from San Luis and Mizquic remember growing corn with beans or with the "uauhzontle" in the same chinampa. Today, chili is cultivated on a small scale in the chinampas and in the hills, it is transplanted in the proper season from the mud nurseries in the chinampas to the fields.

Beans with chili and corn-made tortillas, "tlaxcalli," still form the main food of the peasants, and are served as the main dish at all tables. "The bean chinampero is a climber with a big seed. Nowadays it is cultivated only sporadically in the chinampas, but only one or two generations ago, it was very common to find it from June until October, climbing ahuejotes trees surrounding the chinampas and entangling them."[37]

SAÚL ALCÁNTARA ONOFRE

171

Transport by "piragua" of the final production of flowers (photo: author).

Chinamperos working the mud or "chichicastle" in the "piragua" or canoe to transport to the chinampas (photo: author).

During the nineteenth century, tomatoes, chilies, and pumpkins *Cucurbita foetidissima* H.B.K. remained important in spite of drought and changes in water supply for a growing Mexico City. At present, the "xitomate" is no longer cultivated in the chinampas; chilies and pumpkins can only be found in a few small plots. Pumpkins systematically appear in the historical registrations of every century from the sixteenth century. However, "the first detailed description of their production process was made in 1791 by the Ozumba wise man, José Antonio de Alzate. The data indicate that these were the most important commercial products for the chinampa economy, at least since the end of the 18th century until they were completely replaced by Old World vegetables introduced between 1930 and 1950. The chinamperos combined them with native plants."[38]

Tomato was produced on a large scale in the area of Xochimilco, and on a small scale in Iztacalco-Mexicaltzingo, according to early twentieth-century reports. Currently, it is cultivated in the neighboring hills, between Xochimilco and Mizquic, from seeds developed in nurseries in the chinampas. Some chinamperos of Xochimilco have mentioned that the "xitomate" was sown there up to 1940. The drought that had already prevailed in those days put and end to their cultivation.

The European plants that were adapted by the chinamperos represented the incorporation of species that had different values: botanical, religious, social, and commercial. These plants offered to the producers more combinations and options because of the capacity for adaptation of these plants to prevailing environmental conditions, means of production, and regional demands, especially for Mexico City market places. Some European vegetables that were very popular in the sixteenth century, especially the cucumber *Cucumis sativa* L., lettuce *Lactuca sativa* L., and cabbage *Brassica oleracea* var. L., were totally incorporated into the chinampas production.[39] The "commercial" native plants and the species that came from Europe (fast-growing plants and ornamental plants) were also gradually integrated into the agricultural system from the sixteenth century.

We may conclude that the chinampa tradition managed to survive, in spite of the Spaniards draining the Basin of Mexico. Yet its significance changed completely.

Originally it had been the source of all cereal, fruit, vegetable, and flower riches for the Mexicas–Aztecs, and it allowed the expansion of their cities around the lakes of the Basin.

The chinampero spreading mud and aquatic vegetation for making the "almácigo" or nursery plant or seedbed (photo: author).

The plants are ready to transplant (photo: author).

Mesoamerican vegetables were still very important after corn, dominating the chinampa–agrarian landscape during the viceroyship in the nineteenth century and the beginning of the twentieth century. These species, *uauhtli* and *chía*, which were mainly used to elaborate flour and "masa," were widely used in the prehispanic world and for a while during the viceroyship. The native vegetables did not lose their importance until the twentieth century. In the 1950s, as it has already been stated, the "xitomate" and the chili were rotated with the corn in the same chinampa; one year corn, another "xitomate" and chili.[40]

Yet some chinampas, in Xochimilco and Chalco, were devoted to flowers, and the introduction of foreign flowers may have contributed to obfuscating cultural memories linked to native flowers. The recent period of growth of Mexico City has engaged a new phase of development, with the abandonment of all but a few scattered plots of traditional vegetables, and a further specialization of Xochimilco and Chalco.

However, this may have contributed to the survival of the chinampas as a native agricultural system. It is remarkable that lately Xochimilco stands out for two kinds of cultivation: the production of bedding plants for later transplant to the mainland that fulfills a significant economic role; and the production of flowers important both for economic and symbolic reasons.

NOTES

1. Ángel Palerm, *Obras hidráulicas prehispánicas en el sistema lacustre del Valle de México* (México: SEP-INAH, 1973), 10.

2. María Teresa Ocejo Cázares, *Naturaleza y Paisaje en Cacaxtla-Xochitécatl* (México: Tesis de Maestría en Diseño, Universidad Autónoma Metropolitana, inédita, 2004).

3. The word *Anáhuac* means "by the water," and it seems that from this name the words *Anahuatlaca* or *Nahuatlaca* were formed and used to refer to the cultivated nations settled along the margins of the lagoon of Mexico.

4. Francisco Javier Clavijero, *Historia Antigua de México* (México: Editorial del Valle de México, S. A., 1853), 2.

5. Ibid., 2.

6. Ibid., 281. Xochimilco. This beautiful city, the largest after the main courts of the Valley of Mexico, was founded in the margin of the Chalco Lagoon, more than twelve miles from the main city. Its population was numerous, its temples so many, their buildings were magnificent, and particularly in the lagoon its beautiful floating gardens (chinampas). This is why the word Xochimilco means: Gardens and flower fields

7. Other important cities of the Valley of Mexico were: Mixquic, Cuitlahuac, Azcapotzalco, Tenayocan, Otompan, Colhuacan, Mexicaltzingo, Huitzilopochco, Coyohuacan, Atenco, Coatlichan, Huexotla, Chiautla, Acolman, Teotihuacan, Iztapaloccan, Tepetlaoztoc, Tepecolco, Tizayoccan, Citlaltepec, Coyotepec, Tzonpanco, Toltitlan, Xalttoccan, Tetepanco, Ehecatepec, Tequizquiac, Huipochtlan, Tepozotlan, Tehuilloyoccan, Huehuetoca, Atlacuihuallan, etc.

8. Chinampas, or *camellones,* or *sementeras* or *tajones.* The word chinampa has a *náhuatl* origin, it comes form *chinamitl* which literally means "hedge or enclosure of canes."

9. Francisco Javier Clavijero, *Historia Antigua de México* (México: Editorial del Valle de México, S. A., 1853), 60.

10. Fernando De Alva Ixtlixochitl, *Obras Históricas* (México: Biblioteca Nezahualcóyotl, I y II, 1997), 187.

11. Ibid., 187.

12. Fray Bernardino De Sahagún, *Historia General de las Cosas de Nueva España* (México: Editorial Porrúa, Tomo IV, 1981), 63.

13. Fray Juan De Torquemada, *Monarquía Indiana* (Mexico: Editorial Porrúa, 1995).

14. B. De Vargas Machuca, *Milicia y Descripción de las Indias* (Madrid, España, 1599).

15. Padre Alonso Ponce, *Relación breve y verdadera de algunas cosas de las muchas que sucedieron al Padre Alonso Ponce en las provincias de la Nueva España* (Madrid, España, Libro 13, Cáp.32, 1723)

16. Fray Fernando Ojea, *Libro tercero dela historia religiosa de la provincia de México de la orden de Santo Domingo* (México: Museo Nacional, 1897), 3.

17. Armillas locates this from the fourteenth century to the sixteenth century; Pedro Armillas, *Notas sobre el área lacustre. Apéndice a la ponencia Tecnológica y paisaje agrario azteca* (México: Ponencia XXXV Congreso Internacional de Americanistas, 1974), 658.

18. Francisco Javier Clavijero, *Historia Antigua de MÉXICO* (México: Editorial del Valle de México, S. A., 1853), 161.

19. Ibid., 168.

20. Ibid., 168.

21. See Bomare's words in his Natural History Dictionary; De Bomare, *Diction. univ. V. Histoire natur. V. Plante* (Francia, ca. 1750). The word wheat of Turkey, corn (bled of Turquoise). It was given to this curious and useful plant the name Indian wheat, due to its origins from it was sent to Turkey, and then to Europe, Africa, and America. The name Indian wheat, known by the Italians, was the name that Bomare erroneously used to refer to this plant, conversely to the testimony of the American writers and to the nations universal opinion. The Spaniards from Europe and America use the word maize, which came from the isle La Española.

22. Teresa Rojas Rabiela, et. al. *La agricultura Chinampera, compilación histórica* (México: Universidad de Chapingo, 1983), 186.

23. Edward E. Calnek, "Settlement pattern and chinampa agriculture at Tenochtitlan," *American Antiquity*, Vol. 37, no.1 (1972):104–115, see 111.

24. José Antonio Alzate y Ramírez, *Gacetas de Literatura de México* (México, Puebla, 4 tomos reimpresa en la oficina del Hospital de San Pedro, 1831), 11:392.

25. Ibid., 11:392.

26. Francisco Javier Clavijero, *Historia Antigua de México* (México: Editorial del Valle de México, S. A., 1853), 14.

27. Ibid., 14.

28. In other American countries, the chili is known as "axí." In Spain, it is known as "pimiento." In France "poivre de Guinée." In some parts of Italy, it is known as "peperone."

29. Francisco Cervantes de Salazar, *México en 1554 y Túmulo Imperial* (México: Editorial Porrúa, Sepan Cuantos, 1978), 117.

30. There are various and different species that are cultivated with this name, or with the common name *cempazúchil* in the gardens of Mexico, and all of them are described in the unregistered Mexican vegetation.

31. Fray Bernardino De Sahagún, *Historia General de las Cosas de Nueva España* (México: Editorial Porrúa, Tomo I, 1983), 183.

32. Miguel León-Portilla, *Universo de Flores la Palabra de Mesoamérica* (México: Artes de México, Número 47, CONACULTA-INBA, 2000), 11.

33. Quetzalcóatl and Tezcatlipoca decided to descend from their place among the stars so as to become trees. Quetzalcóatl chose to be a willow decked out in emerald-colored feathers, and took the name "quetzalhuéxotl," the lovely tree, which, descending like an emerald bird, sunk itself into the black soil, clutching the depths of the earth with its powerful roots: the "ahuejote xochimilca" which, in mythical times, multiplied in the marshland of the area. Tezcatlipoca, in turn, chose to become a mirror-tree, the tree whose image is reflected in the transparent waters of the channels and lagoons: the

"tezcáhuitl" or "ahuejote" which, by means of an optical illusion, seems to sink into the depths of the water and hold up the sky reflected on its surface.

[34] E. J. Wellhausen, L. M. Roberts and E. Hernández X, *Razas de maíz en México, su origen, características y distribución* (México: Secretaría de Agricultura y Ganadería, 1951), 172.

[35] Teresa Rojas Rabiela, et. al., *La Agricultura Chinampera, compilación histórica* (México: Universidad de Chapingo, 1983), 184.

[36] Charles Gibson, *Los aztecas bajo el dominio español (1519–1810)* (México: Siglo Veintiuno Editores, S. A., 1967), 328

[37] Teresa Rojas Rabiela, et. al., *La Agricultura Chinampera, compilación histórica* (México: Universidad de Chapingo, 1983), 202.

[38] Ibid., 194.

[39] Ibid., 205.

[40] William T. Sanders, *Tierra y agua. A study of the ecological factors in the development of Mesoamerican civilizations* (Ph. D. Dissertation, Department of Anthropology, Harvard University, 1957), 89.

*Botanical Progress,
Horticultural Innovations
and Cultural Changes*

*Botanical Progress,
Horticultural Innovations
and Cultural Changes*

*Botanical Progress,
Horticultural Innovations
and Cultural Changes*

*Botanical Progress,
Horticultural Innovations
and Cultural Changes*

Conservation and Diffusion of Species Diversity in Northern Italy: Peasant Gardeners of the Renaissance and After

Mauro Ambrosoli

As Emilio Sereni has suggested, the Mediterranean garden made an important contribution to the Italian agricultural landscape. Local peasants who performed labours and contractual obligations played an important role in the organization of Italian territory. Although Sereni made extensive use of sixteenth- and seventeenth-century cartography, he dedicated less space to peasant gardens and to commercial gardens of Northern Italy.[1] Nevertheless, next to field cultivation, gardens and orchards, large or small, became a regular feature of every dwelling, whether a gentleman's seat or a cotter's hut. Furthermore, commercial market gardening developed next to highly populated centers: every major city had its range of orchards. Everyday food consumption relied heavily upon vegetables and religious institutions praised a vegetarian diet as part of general well-being.

Numerous studies on garden history in Italy have approached a variety of subjects, such as Roman gardens, the gardens of religious orders, formal gardens of the Renaissance, and garden architecture. Yet, they generally have not approached the main topic of this symposium. The aim of this chapter is to take in due consideration the role of Italian peasant gardeners in horticulture and in botanical progress. In order to do that, we have to make few initial remarks: who is a peasant and what is a peasant garden.

In comparison to the vast literature that deals with the history of formal gardens, printed sources about peasant gardens are very limited. In order to answer the major topics of this symposium, we had to make a conflation of new data and contemporary printed sources of a vast area, the Po plain, embracing a triangle whose angles are the cities of Turin, Udine, and Ravenna. In this region, Italian cities have developed since the eleventh century and a mixture of agriculture and architecture became a feature of this densely populated area, where cities and towns grew numerous and helped to transform poor alluvium soils into a rich countryside. Toward the end of the sixteenth century, the urban population was particularly high in Milan (120,000 inhabitants) and Venice (200,000); larger cities had between 20,000 and 40,000 inhabitants (Bergamo, Brescia, Verona, Vicenza, Padua, Mantua, Bologna, Ferrara) and smaller towns between 10,000 and 15,000 (Treviso, Cremona, Novara, Turin) were extremely numerous by the standards of European population.[2] Many of these cities were also the seat of princely courts (Turin, Milan, Mantua, Venice, and Ferrara), the residence of wealthy nobility, and all were the seats of bishops and prosperous religious orders. In the cities lived numerous families of the regional elite: their conspicuous consumptions marked a sharp contrast with poor urban diets and food supply was a common political problem throughout the sixteenth century.

The definition of peasant gardener requires some attention. "Peasant" is an easily understood sociological term, although it varies sensibly in relationship to age and areas. The main characteristic of ancient, modern, and present peasantries is the allocation of resources through the family: the husbandman manages his possessions and organizes production and consumption in a businesslike manner, taking into consideration gender and age of his family members. Access to economic resources, either natural (woods, pastures, marshes) or man-made (mills, irrigation canals, mines), substantially contributed to family income, which varied from place to place according to the institutional powers over the land exerted by the ruling social elite, civil or religious, local or national. The Italian peasantry of late medieval and early modern centuries underwent a continuous redefinition of its economic status. Its share of the land, either private or common, was variable and, by the mid-sixteenth century, owner-occupiers had lost a good part of their landed possessions because of debts and evictions. Nevertheless, peasant families did not disappear from the countryside; they simply went down the social scale and generally became the tenants on the same farms, eventually ending as landless labourers. Thus, we should keep in mind that Italian peasantry was far from being a static social group, that some peasant families were more successful than others. Furthermore, the term "peasant" applies better to the culture than to the level of family incomes. On the one hand, market gardening gave some gardeners the chance to improve their monetary income; on the other hand, the technical and botanical know-how was shared by a very large social body. For the purpose of this chapter, we could well define peasant gardens as the result of the labors exerted by various people employed in agriculture, who, regardless of their social status, all earned their living through manual work. Therefore, peasant gardeners could practice their art as laborers, owner-occupiers, or market gardeners in gardens of very variable sizes, petty cottagers' gardens, family gardens within a larger peasant farm, and larger commercial gardens near cities.

This brief introduction is necessary in order to make comparable the body of archival information, which I have been collecting in recent years. This information is extremely variable in quality and quantity and helps to shed fresh light on the present topic. Private records on peasant gardens are usually scarce, even on a local scale: Italian peasants, generally illiterate, kept written documents to a minimum. Thus historical documentation on gardens turns up irregularly and only in connection with other institutional bodies, such as the city, the Church, and larger estates. Furthermore, when we are lucky enough to find some documents on gardens, information on horticultural practices are disappointing. Common names described cultivated plants (such as celery, turnip, vine, olive tree, and the like), without any reference to the botanical names employed by Renaissance botanists (Linnaean classification became useful only after 1750 and it remained confined to the landed classes). It will be necessary to compare this archival information with the books on agriculture and gardening printed for the use of the landed classes.

The Middle-sized Garden: Fruits and Vegetables for the Table

At the very beginning of 1300, Pier de Crescenzi, in his well-known agricultural *summa*, described three types of gardens, according to the income of the landowner.[3] Pier de Crescenzi, a citizen and civil servant from Bologna, wrote the textbook that founded the villa system in the Italian countryside for future generations. First came the verger (*viridiarium*, *vergerie*), which was a square meadow surrounded by ornamental, kitchen herbs, and flowers; the second, properly called the garden, was defended by a ditch or moat, a hedge of prune trees and roses, and a few rows of vines, trees, and pergolas, which forced nature into a formal order. Finally came the gardens of kings and wealthy lords, places of social emulation, surrounded by a high wall to provide intimacy for the royals and their kinfolk. In this orderly environment, fruit trees grew next to forest trees and wild animals ran free for the pleasure of the very wealthy.[4] In light of these definitions, it is clear that the division between verger and garden was very thin and only formal gardens were the result of projects carried on by specialized laborers and architects. Instead, vergers and gardens were the direct extension of the house and kept under the supervision of the landlord or the tenant with his family labor.

The quality of plants growing qualified a site as garden or verger according to Pier de Crescenzi: the attentive landowner had to choose trees carefully to shade a grass lawn. The gardener avoided planting walnut and similar trees to prevent damage to the lawn and the growing herbs. The mulberry, the cherry, the prune, the fig, the hazelnut, the almond, the quince, the pomegranate, and similar "noble trees" grew, ordered in rows according to size and variety: different sorts of vines grew in alternate rows with the above trees. Pergolas, built as a pavilion if possible, gave the final touch to the garden of the average gentleman. Though he did not personally work in his garden, he enjoyed direct supervision of the garden area. The frescoes of Schifanoia Palace in Ferrara are the best illustration of the landowner's supervision of specialized peasants building pergolas and arranging the cultivation of vines within an architectural context.[5] The duke Borso d'Este greatly improved his estates around Ferrara, transferring to the countryside the concept of the well-managed garden. Borso turned his estate at Belfiore into a building closely connected to the vegetal and animal world outside; in 1451, he had wild animals, such as gazelle and deer, brought into the garden. Outside, large fields produced grains for raising these animals: laborers carefully mixed dung collected around the pigeon tower with the garden soil. In 1472, Ercole I d'Este extended the walls to enclose a much larger area. Although few landowners were willing to waste grains to afford the pleasing companionship of a flock of deer, such gardens became the ideal of many. The Dukes d'Este carried on an extensive plan of land reclamation in their states: they turned marshes and wetlands into tree-lined ploughing fields. New farmhouses with vineyards, gardens, and orchards became the main feature among the reclaimed lands.[6] For a longer period, perhaps 1450-1550, the division between vergers and gardens (*brolo* and *giardino*) was very narrow.[7] The main feature that turns up in land registers of the fifteenth and early sixteenth century all over the Po plain is the diffusion of brick buildings both in the countryside and in the outskirts of towns and cities. Brick farm buildings, with tiles and attics (*'caxe solerate et coppate'* as one reads in contemporary registers) took the place of thatched roofs, along with gardens defended by walls and hedges.[8] This happened both to tenant farms and to owner-occupiers' dwellings. Three outbuildings qualified the Renaissance estate and were located next to the farmhouse: the brick oven, the pigeon-house and the enclosed garden (Figs. 6, 12, 13.1, 14.2). Walking through a live arch, appropriately made with box or rose bush, one could enter the fruit orchard and the vegetable patch. Large or small, brick or cane, this spatial organization proved to be long-lasting, possibly because the new class of city landowners, who took possession of peasant holdings from the late 1400s, enforced a strict utilitarian approach to the newly acquired gentle style of life. They followed the suggestion given by Cato the Elder that the landowner had to be a seller and not a buyer.[9]

How large was a house garden? Could it provide for family needs? Two, three, four acres, roughly one hectare was the space required for this type of garden: documents will show this figure to be realistic. This dimension proves that utilitarian aspects of such pleasure grounds were a primary concern. The garden had to provide the owner's table with fresh fruit. The illustrious humanist Bartolomeo Platina (1421-1481), himself the owner of a small garden (*vigna*), gave numerous recipes for every fruit mentioned earlier.[10] Large or small, the garden provided continuity between the landowner and his peasants: the private documents of the literate owner recorded the manual labors performed by illiterate peasants.

Between 1505 and 1509, Leonardo Bevilacqua (1450-1524), the head of a prosperous merchant family who kept a large house in Verona, wrote numerous entries about fruits, vegetables, and the garden in his account book (Fig. 10, 10.1).[11] Although wealthy, when it came to the garden or to the table, Leonardo was attentive to the last detail. In February 1505 he bought green salads and leeks (£0.1.6), in March lemons (£0.0.9), in April white cabbage (£0.0.6), on 22 May he spent £0.2.6 for cherries, another £0.2.9 for salads, and bought more cherries for £0.1.6. He certainly had acquired a taste for them: on 3 June, on 8 May 1506, in May 1509 (a total of £0.7.6), in June 1509 (£0.6.0) he bought cherries and sour cherries. Leonardo had more fresh fruits bought in Verona: melons on 20 June (£0.3.6) and again in July 1505 (a total of £1.0.8 together with some cheese and headed cabbage). On other occasions, Leonardo bought pears, figs, and apricots for similar sums.[12] In one instance, he mentioned Polo the gardener (*l'hortolano*), to whom he was indebted of £3.10.0. He was ready to pay good money for fresh fruit: to give a comparison, such a luxury as a pair of reading glasses [*sic*] cost him £0.3.0 in 1509. Without taking into consideration the cereal crops, in 1505 eight tenants paid in 18 *quarte* (i.e., 162 litres 162) as part of his bean crop; in 1507 family servants got 20 *quarte* (i.e., 180 litres) of linseed out of the old crop. On St. John's Day 1506 he had the grass in the orchard mown (£0.6.0). Leonardo's accounts reveal how profitable the cultivation of trees was: he owned sixty-three mulberry trees, which brought him a total rent of £166.9.0. Such trees were growing in groups or alone: two, six, one, eight, eleven were standing in one orchard (*brolo*). One tenant rented thirty-three of them. Leonardo rented them for cash. Leaves were necessary to feed silkworms and very much sought by landowners and peasants. It was another connection between the city and the country: every space available in the neighborhood proved to be useful.[13]

Lying south, a few miles away from Verona, the borough of Villafranca reveals more about the diffusion and the cultivation of the mulberry tree in the mid-sixteenth century. Land registers (*estimi*) and records collected for fiscal purposes show a crescendo of public interest into private affairs: the 1534 and 1549 census described the borders and the quality of each landholding. In 1569 every possession, such as vineyard, mulberry, olive trees, grains, wine, hay, cattle, and so forth were registered.[14] Villafranca near Verona was a prosperous agricultural borough, where numerous husbandmen thrived. Yet, wealth was unequally distributed: out of 507 heads of family, sixty-seven declared property valued at less than one ducat. Seventy-three husbandmen, or fourteen percent, declared mulberry trees: the dispersion of trees throughout the village and the village lands is extremely revealing. Mulberry trees could grow either alone or to provide a support for a growing vine. Husbandmen owning between one and five trees were the more numerous group (fifty), another small group of fifteen declared between six and nine plants and only five had ten trees or more. In most of the cases (fifty again), trees were growing separately; in another twenty-five declarations trees served to hold vines: in total, there were 288 trees standing alone, whereas only thirty-nine were part of a row of vines (Fig. 13.2). In the early twelfth century, the freemen of Villafranca came together to build irrigation canals and to keep the custom of common field farming: in 1549, freemen held an auction to rent out the arable common fields for five years.[15] Nevertheless, trees were private property; either they belonged to the landowners who gave written instructions in the rent agreements about their cultivation and replacement of fallen or dead trees, or they belonged to the small peasantry and this seems the case of the trees growing in courtyards. Taking now into consideration only the trees of more substantial tree owners, we find Bon and Zuan Andrea Trivisan with twenty-six trees, all coming up in their courtyard. One Francesco in society with Delegio owned twenty-five mulberry trees, all growing in courtyards; a Domenego Corticello declared two lots of seven trees each, one lot in the courtyard and another lot in the field. It seems plausible that all these trees growing in courtyards belonged to people busy in the silk trade. Such is the opposite case of Pietro and Francesco Baldorio who declared seventy heads of sheep and eleven cattle, who owned only two mulberries with eight plants of vines, as did Vincenzo Cagliero with twenty sheep and one mulberry tree with fifteen plants of vines.[16]

The diffusion of the *piantata*, the tree-lined arable fields, with rows of vines planted along the selion to give multiple cropping, became the main agricultural feature all over the Po plain (Fig. 15.2).[17] The demand of silk for Venetian workshops pushed to increase the planting of mulberry trees: in the central decades of the sixteenth century, landowners requested more attention to the trees growing on their property and forced tenants to plant mulberries throughout the Venetian states. In the district of Treviso, new irrigation canals made the best use of the water of rivers and streams, and from the 1450s, it became a common opinion that the landscape turned to look like a garden. The general picture was that the whole Treviso district acted as the garden for one large town house, the city of Venice (Figs. 13.1, 13.3). This movement, founded on local and private initiative at the beginning, got new momentum after the 1540s when the Italian wars had ended, debts paid off by the Venetian government, and new magistrates (*Provveditori ai beni inculti*) received orders by the Senate to take care of land reclamation. The canal system, known as the Brentella, that had begun in 1436 to bring water around the Montello Woods, at the end of the sixteenth century reached fifty-nine villages. In 1572, irrigation watered twenty-five hundred hectares of arable land and pastures, not counting the extensive orchards, gardens, vergers, fountains, and fisheries, which surrounded private villas.[18] The general replanting of the country began around those newly built estates, which became the feature of the rising merchant classes and urban nobility during the fifteenth century, taking advantage of the improving economic situation. In 1526, in the garden of the estate of San Parisio at Zermana (a nun monastery near Treviso) fruit trees such as the peach and the walnut grew next to willows; along fields and pastures, few newly planted mulberry trees turned up next to poplars and willows (Fig. 13.2, 13.3). Also in this district, the mulberry remained confined mainly into gardens and orchards: when lucky, the tenant could halve the income from the valuable leaves with the landowner. In other cases the income was all for the owner. The care of these newly planted trees became a new obligation for local tenants and peasants. These included sorting out chopped woods according to size, hoeing twice a year mulberries, young walnut, vines and fruit trees, up-keep of the pergolas, cleaning yearly the third part of every ditch, spreading the collected mud, and forbade the felling of any tree. These were the most common obligations for the care of trees such as those ordered from the same Monastery to Tonio Conte and sons in 1612.[19]

The extensive investigations carried out for the *Campagne trevigiane in età moderna*, a research program sponsored by the Benetton Foundation, are full of valuable data. Trees such as walnut, apple, pear, quince, cherry, sour-cherry, fig, chestnut, rowan, pomegranate, prune, almond, peach, and apricot, turned up in agreements written in 1563, 1564, and 1567. Some of these were growing alone in fields, others as almond trees, sour cherry, prune, grafted pear, in a specialized orchard (*brolo*) at Cendron (Zosagna di sotto).[20] Similar plants grew in the orchard of Girolamo Federici at Chiarano, and Nicolò Salmon rented out his estate that included 1,932 grown trees with vines, 311 trees of various quality, 136 vines, with or without grown trees, for the tenant house, and finally fifty-one cherry and walnut trees growing in the small and large garden (*horto grande et piccolo*) (Fig. 14.3).[21] Trees of this quantity overcame family needs: we have shown that, when necessary, fresh fruits turned up in the market.

Gardening for a Living: An Integration of the Laborers Incomes

Far from being part of the projection of the dwelling house into the fields, small gardens and orchards were a necessity for the living of the smallest peasantry. Open space was at a premium in built areas, and Italian towns retained similar characters to those they had inherited from Roman cities, such as Pompei, where three-story blocks existed next to larger town houses built around a garden, or medium-sized suburban dwellings with an orchard attached. When medieval towns lost the square reticulum of the Roman city planning, small gardens appeared in the empty spaces left by collapsed buildings, such as in thirteenth-century Turin. Even in Venice, a city that had no Roman foundation, small gardens and vegetable patches were cultivated in early medieval times within the built area. Small gardens were more numerous along the main roads, which developed outside the cities, when medieval ramparts, containing larger built areas, replaced roman walls. Small hamlets with their gardens happened to be included within the new town walls, as in fifteenth-century Verona. In the sixteenth century, new ramparts encircled even larger areas of open space, where small orchards and greater gardens developed freely. Cities such as Mantua and Florence displayed this feature that, at any rate, reached its maximum in Rome: in the sixteenth and seventeenth centuries, larger areas of Rome remained un-built within the long circle of walls. Patrician gardens and vegetable orchards were competing for the same spaces: the Tribunal of Agriculture discussed numerous cases that took long sessions for what seemed very trivial matters (such as goats trespassing in what is nowadays via Del Corso, artichokes grown and sold without a licence from the guild and so on).[22]

If space was at a premium in city areas for leisure or for money, it was a necessity of life in smaller villages and hamlets. Small gardens and orchards were part of cottages and remained attached to them in different areas. In the hills overlooking Turin, a thriving city such as Chieri, which founded its fortunes on cotton weaving, made the best use of fruit orchards and vineyards to reclaim the wooded areas during the fourteenth century and free them from Church tithes and common rights (Fig. 3).[23] Nearby at the same time, the villagers of Revigliasco built sparse dwellings to turn local woods into vineyards and fruit orchards with cherry, apple, pear, chestnut, walnut, and olive trees, all enclosed to defend them from wild animals and common rights. Down the hill, bordering the Po River, just before entering Turin, the landscape looked little different with vineyards, tree-lined arable fields, and fruit orchards. In 1643 in the small hamlet of Cavoretto, fourteen heads of family, out of 106, declared gardens for thirty-six *giornate* (twelve ha). To be more precise: eight gardens were smaller than one *giornata* (that is roughly one acre), six between one and fifteen acres. However, seven owners out of eight declared gardens, without a solution of continuity with woods, vineyards, tree-lined arable fields, meadows, and vegetable gardens. The borderline between a garden, orchard, and vineyard was very fine: gardens did not offer pure aesthetic values but had to produce an income. Thirty-five husbandmen declared a vegetable garden, all below one acre in size, more often one-half, one-third of an acre and even less. Every dwelling had its own patch, either in the easy reach of irrigation waters and even in the very center of the village, on a small hill, where water, running from above, was in no shortage at all.[24] Even in the village of Sermide, on the left bank of the Po in the estates of the duke of Mantua, the situation was similar. In 1460, the lesser inhabitants of the borough had to make do with just a house attached to a vegetable or kitchen garden. Lorenzo di Berto, fisherman, Pedrobon Moruzo and Pietro Anton Zuan Moruzo, the village chemist, Pietro di Domenico della Stazza, and others, all paid rent for just one small vegetable garden or, if lucky, managed to draw together a few garden patches. Yet, because of the economic trend and social institutions, there was often a shortage of labor because of runaway peasants in debt. Immigrant workers often replaced them: the house and the vegetable patch provided the shelter for incoming families, who came from a wide region, reaching Milan, Verona, and Bologna. However, opportunities existed: for example, between 1483 and 1486, an immigrant from Mantua, Feretto di Ruboli, managed to get possession of various small buildings and vegetable patches for a total of about two acres (2. 18 *biolche*). Eventually, he turned these patches into a beautiful garden (*brolo*) that measured a third of an acre for a yearly rent of £1.9.8 and £8.15.4 for the whole lot. Nevertheless, the petty landholders could not improve their social position over the centuries. In the cadastre of the 1780s, their individual possessions hardly reached one acre of land: only horticulture could provide a living from such small possessions.[25] In the states of Mantua, the situation was very complex, with the Duke being the largest feudal owner. In broad terms, he managed his large estates as a medieval *curtis* with customary peasants paying various rents, heriots, and fines for the right to dwell on the estate. Since late medieval times, two major socioeconomic institutions organized the landscape around Mantua: the large estate and the cottage with a vegetable patch. Throughout the sixteenth century, immigrant workers, who paid rents for houses rented on a short let base, replaced customary tenants. The house and vegetable patch provided a good income to the landlord, or his tenant-in-chief, and a living for the laborer's family.

Gardens and orchards of various dimensions and quality dotted the countryside: lay and Church estates were similar in structure and size. The bishop of Mantua owned thirty-eight estates, which were regrouped into five larger estates (*corti*) for a total of almost six thousand acres (5,989 *biolche* equals 2,000 ha): whereas tree lined arable fields took the largest area (fifty-seven percent), buildings, gardens, and orchards took only 3.3 percent of the total. Yet, each estate had a variable garden area, varying from the Corte di Nuovolato with 100 *biolche* to the average estate, where only between one and five *biolche* of gardens were kept (the largest group with twenty-seven cases and five estates had gardens between seven and eleven *biolche*). In ten estates, there were no gardens at all. The land register (*cabreo*), drawn in 1690, gives us a vivid picture of the situation. The most frequent house and garden complex was around two acres with house building, garden, and orchard, eventually with other outbuildings, vineyard, and meadows. Yet a number of small estates, made out of a thatch cottage, a brick dairy, and large meadows, had no garden or orchard at all. A certain division of labor applied: the shepherd and the dairy farmer did not require a vegetable patch. Given the custom of transhumant pastoralism, they were absent from the farm from June to September.[26]

It is important to point out that it was the size that qualified the garden, peasant, or gentleman, not the owner. Population growth did not alter considerably the existing agricultural structure: farms changed dimensions rarely and the *podere* system took over the organization of both hills and plain country. *Poderi* changed hands frequently throughout early modern centuries, the status of peasant owner-occupiers lowered during the sixteenth century, new landowners reclaimed new lands roughly between 1660 and 1760, new owners bought monastic and private lands during the Napoleonic period, yet farms were rarely split. In this context, landowners who had the status of a gentleman often owned small farms let to peasant tenants. Such was the case in Friuli: in 1641, Count Leandro Antonini and his brothers exchanged a small estate with the local church of Santa Margherita. In 1639, the estate, which had been valued at 109 ducats, contained a house, courtyard, vegetable garden, a small arable field, meadow with five mulberry trees, thirteen walnut, two pear, three chestnut, two cornel-trees, three fig, five cherry trees, and four trees lined with vines.[27] In the outer part of the Venetian state and thinly populated Friuli, the rebuilding trend of the Po plain appeared later. Still, in years such as 1677, farm houses had two storeys, brick walls and thatched roofs, and had individual small gardens and small orchards, to comply with local individualism. Brick tiles substituted thatch in 1722, 1726 and new gardens appeared (Fig. 14.1).[28] Other documents point out that local churches and confraternities had invested their money and put together a variable patrimony of small strips to rent out either to the needy or to the members. Venetian bylaws obligated a correct financial management of these religious and lay institutions: hence there was a wealth of information about their patrimony. In 1677, the confraternity of the SS Rosario at Santa Margherita carefully illustrated its possessions with simple yet very significant drawings. Vegetable gardens usually grew next to every house or building, yet one could find vegetables growing in gardens enclosed within tree-lined fields (Fig. 14.2). One still finds them today in old-fashioned mountain areas.

Growing Vegetables for the Market: Turin and Venice

Commercial market gardening became an early feature of Italian cities. In the 1840s, von Thünen explained that the city[29] was able to organize the agricultural exploitation of its territory in circles, in order to maximize space and that market gardening was best located in the inner circle. This was not particularly true of the cities of early modern centuries, when transport necessities required that lands neighboring the cities produced fodder. Therefore, market gardening and meadows competed for access to water and the limited space to grow vegetables for the market kept the prices high. By contrast, horse dung and industrial waste, such as textiles, leather, and tallow, were more easily accessible to the market gardeners. The cases of Turin and Venice will provide examples to understand the role of market gardeners to sustain botanical progress and horticultural innovations.

In 1559, Emmanuel Philibert, duke of Savoy, recovered his Italian states from the French king, and in 1566 moved the capital from Chambery in Savoy to Turin in Piedmont. On that occasion, he declared his subjects living in Turin territory exempt from the land tax and in 1601 Charles Emmanuel I undertook works to improve the new capital. Thanks to immigration, population rose steadily from a few thousand people to twenty thousand people in the early seventeenth century, and reached seventy-five thousand people toward the end of the eighteenth century. Opportunities opened for market gardening: the vast pastures and meadows offered scope for improvement. The Po River crossed the city's territory and received waters of three major tributaries: customs and bylaws regulated the numerous canals and the use of water. Yet trespassers were numerous, such as the tenants of the aristocracy, who opened illegal canal gates to water their meadows, thus depriving already established gardens of their share. The fathers of Saint Augustin complained in 1608.[30] Meadows around the northern side of the town walls became vegetable gardens during the seventeenth century. In 1573, a nobleman, Aleramo Beccuti, left his possessions to the newly founded Jesuit convent: among these, the order inherited over thirteen acres of meadow. Actually, this piece of land was a small island between the Dora River and the important canal that kept the city mills working, just outside the city ramparts and the hamlet where folks played traditional ballgames. Eventually, around the year 1600, because of some unknown transactions, three owners shared this meadow: the Jesuits themselves, the Municipality of Turin, and Count d'Harcourt. In January 1623, the Jesuits spent 360 *écus* to improve their meadow into a vegetable garden: they opened a new gate into the canal, for irrigation purposes, and a small building to store the vegetables. The Jesuits happily reported in their accounts that the garden was finally let for 132 *écus* instead of the 45 *écus*, which were paid in before as a meadow. In the same year, the municipality held an auction to sell another two acres of meadow and pasture, in the same region watered by the Dora and its canals. A Tommaso Ferrero, dealer in cheeses, won over the offer made by Antonio Garino, a market gardener: both had recently immigrated into the city from the valleys of Lanzo, a specialized area for cheese making and gardening (Fig. 5). The Jesuit fathers rapidly resorted to land improvements as a solution to the financial crisis of 1622 that caused the collapse of the entire Italian financial system. The connection between the growth of vegetable gardens and the main economic cycle remained even after the worst decades of the century (1629-1659) had gone by. Local practice reserved hay and pastures to the landlord and, in consequence, the cost of hay on the Turin market rose, making a team of oxen very expensive for the average tenant or sharecropper. Hence, market gardening became an option for those peasants, who were relatively free in the choice of their crops. In the light of Turin urban development, which had to mediate between military defense and the necessities of the capital city, very few buildings could afford an inner garden and orchard within the ramparts.[31] Extensive vegetable plots established just outside the city gates of Porta Nuova and Porta Susa, thus completing the green belt that circled the whole city (Figs. 1.1, 1.2), established around the complex of sixteenth- and seventeenth-century Savoy chateaux, which made the city famous.[32] In 1706, the vegetable gardens around the city provided the camp for the besieging French army and the battleground for the final Savoy victory. The steady urban growth of Turin during the eighteenth century enforced a functional division of the territory: woods for gentlemens' sports, fields and meadows for high farming, gardens, orchards, and vineyards for peasants' toil and labour. The Napoleonic cadastre of 1806 witnessed the final point of a series of transformations, which pictured an ordered countryside around villas and farmhouses that belonged to the city elite. Since late medieval times, religous institutions had granted possession of the woods and moors in the hill part of Turin's territory to the most important families. The labor of tenants and sharecroppers, as in Revigliasco, turned these barren lands into vineyards, gardens, and orchards, grounds for the villas where their masters spent a pleasurable country life in view of the city.

The market gardeners of Venice took advantage of the easy access to natural resources that favored horticulture, such as sandy soil, plenty of humidity, silt, and city waste to substitute for dung.[33] Early settlers practiced horticulture in the islands of the lagoon and vegetables had a relevant place in the diet of early modern Venetians. Furthermore, the city map drawn by Jacopo de' Barbari (early 1500) pictures numerous gardens, large and small, which were among the wonders of the city: in the gardens of the nobility and the small gardens of the lesser folks grew a high number of rare and foreign plants.[34] Throughout the 1500s and the early 1600s, Venice was the largest Italian city, housing between 200,000 and 220,000 inhabitants, who lived in the major islands of the lagoon. Most of them were skilled workers employed in the arsenal and those many industries, which made the city famous. The city's magistrates had a difficult task in providing the markets with staple food for this industrious population: a variety of merchants, stall keepers, and peddlers bought fruit and vegetables from the growers and sold them around the city. The nobility and wealthy merchants had bought estates in the mainland to invest their funds and to feed their large households. By contract, tenants had to deliver rents paid in kind at their masters' doorstep, and tenants of the numerous religious houses of the city did the same, with holdings in the nearby countryside. Large households provided for themselves in this manner, whereas authorities and private merchants imported a great quantity of foods, grain, cattle, cheese, wine, and oil, from a variety of distant places. Yet, in one respect, one thing favored the Venetians: within the city's boundaries, meadows were not necessary to provide hay for horses. In any other European city, market gardeners competed for land and water with local farmers and manufactures, as we have just seen in seventeenth-century Turin. In Venice, anyone with access to land, void of buildings, could profit and turn it into a vegetable patch, taking advantage of consumers who paid cash for the daily foodstuffs.

Horticulture, providing continuous cropping, gave the best return for money. The great demand for everyday foodstuffs expanded in the neighboring areas of the mainland. Since the mid-sixteenth century, Mestre and Zosagna provided vegetables for the Venetian markets. Yet, even if large estates were growing vegetables for the market, the most common management of commercial gardening was peasant cultivation: private or institutional landowners chose to split and rent to individual gardeners their largest ground laid to vegetable growing.[35] Tough, even in the glorious days of St. Mark's Republic, taxmen were not welcome to enquire into the private affairs of the landed class: only the 1661 Census gives us some information of the practical aspects of vegetable growing in the shores of the lagoon. The 1661 Census valued incomes and gave a description of possessions: a comparison with the 1582 Census is revealing. In 1582, landowners declared only the value of their incomes, without any description of the size and quality of their estate. The great plague of 1629 and 1630 took away one third of the city's population and a great many changes occurred afterward, as by 1661, because of immigration, food demand was high again. Between 1582 and 1661, over two thirds of the gardens owned by the nobility changed hands; only major families such as Contarini, Gritti, and Pisani dal Banco did not lose their gardens. New families who moved into the island after the plague built three hundred new homes in the Lido.[36] This situation reflected the recovery of market gardening and some 169 hectares of land were under vegetable cultivation in the islands of Lido, Sant Erasmo, and Torcello. The same situation was seen in Vignole and Mazzorbo: however, the noblemen who owned every parcel of garden managed to keep the taxmen at bay, as their predecessors usually did, and did not declare the dimension of the lands owned. At any rate, it seems that the Venetian nobility had a keen interest in market gardening for its financial returns.

The largest single owner was the famous Alberto Gozzi, whose family immigrated into Venice from Bergamo, and who practiced the silk trade and became so rich that he easily bought his accession into the Golden Book of Venetian nobility in 1646. He declared fifty-three *campi* (20.5 ha) of gardens in the island of Sant'Erasmo, where nobles and religious houses owned seventy-seven percent of the available gardens (81 ha). He received 1,360 ducats of rents, having invested carefully in the real estate market between 1620 and 1650. The religious houses did the same. The Benedictine monastery of San Nicolò at the Lido was the second largest owner of market gardens: the fathers owned forty-three *campi* (16.6 ha), which gave rents for 1,188 ducats in 1661. Once again, large landowners chose to rent out individual plots of gardens to landless laborers rather than risking direct management. At Vignole, few noble families owned every garden (that was never measured, by the way) and received an average rent of thirdty-five ducats per hectare; in Torcello the aristocracy and religious houses shared 14.6 hectares of gardens.[37]

Noble landowners, peasant gardeners, and landless laborers shared the profits that derived from the city's demand for vegetables. When daily toil and labor of the gardeners and their family were not enough to keep up with cultivation requirements, then immigrant workers from Udine's countryside turned up in the islands of the lagoon. The vegetable patches fetched high rents because their money returns were significant, especially after costly investments. The most popular were artichoke or asparagus beds; then followed vineyards, rose gardens, and flower gardens. One official stated the case of a gardener who made the neat income of 200 ducats in 1559 from the sale of pinks. To evaluate this figure, we should keep in mind that an average civil servant earned about fifty ducats every year. In his memoirs, Agostinetti (see later) quoted a gardener who made 150 ducats from the sale of rose water that he produced out of his rose garden in 1670s. Yet the continuous demand for asparagus and artichokes[38] caused real problems when the landowners had to pay compensations for such improvements. For a ten-year lease, compensation often equaled the rent paid in the same period. For this reason, at the beginning of a lease, the landowner and the tenant subscribed an inventory, listing poles, hedges, gates, palisades, canes, and crops left in the plot. The tenant had to leave everything listed or pay for what was missing. Among the fifty-three tenants who signed agreements with the Benedictine fathers of San Nicolò between 1599 and 1704, eight contracted debts, valued up to twice the yearly rent.

Everlasting Structures: City and Country

So far, we have pointed out three major features which regulated peasant gardening in Renaissance Italy: the middle-sized garden that produced for the landowner's table; the vegetable patch valued as part of the peasant salary; and the market gardens that cared for city consumers.[39] These structures proved to last well into the eighteenth century and beyond, possibly until the railway age, when specialized horticultural areas began to produce for the national market. Thus, at first sight, it may seem that horticultural innovations were very scarce throughout the centuries. Yet, in different ways, these structures provided the frame for diffusion of botanical progress and certainly cultural change sustained horticultural innovations.

Pier de' Crescenzi's *Agricultura* made available to generations of landed gentlemen information on a small, but consistent, patrimony of cultivated plants and cereals, which had been partly inherited from the classical antiquity and partly were medieval imports. He introduced his readers to 217 wild and cultivated plants, cereals, fruit trees, trees grown for timber, and medicinal herbs. In the following century, the manuscript tradition added Asian plants such as rice, pepper, papyrus, and the palm.[40] Onto this body of knowledge and not from the wider body of plants described in the Greek and Roman herbals, Renaissance landowners grafted their curiosity and the practice of Italian peasants. On this inheritance, the Italian peasant gardeners built their own patrimony: they distinguished and produced local varieties that fitted best under different climate and soil. Yet, to be precise, a limited number of cultivated plants were available to the average farmer, who selected the best-growing cultivars.[41] Furthermore, the list of plant names quoted in the documents shows that botanical information printed in books was not available to everybody at the same time. Writers and speakers alike employed generic names and rarely described varieties when they compiled public or private acts. The printed body of horticultural information was available in such books as Crescenzi's *Agricultura*: they may serve as guidelines to trace new plants or new cultivars into the gardens. The artichoke was a relevant novelty of the second half of the fifteenth century. Though the wild artichoke (*Cynara cardunculus*) was spontaneous in Sicily in ancient times and grew in Roman gardens, the common artichoke (*Cynara scolymus*) was first introduced in Naples in the same century. Its name, *carciofo francese*, betrays its country of origin: the Angevin kings ruled Naples in the first part of the century. Only in 1466, a young scion of the Strozzi family brought it back to Florence. Furthermore, Ermolao Barbaro, the humanist and commentator of the Elder Pliny's *Naturalis Historia*, declared the novelty of having seen one plant of artichokes growing in a Venetian garden. He died in 1493 and might well have made this observation in the last quarter of the century. There was a similar fate for the eggplant: its use remained confined to Genoa's Riviera, Rome, and one social group, the Hebrews.[42] In the early decades of the sixteenth century, various small new books, translations from Latin and Greek authors and inexpensive editions of Crescenzi's agricultural textbook offered Italian landowners a substantial, although dated, body of gardening technique.

In the mid-1500s, the agricultural dialogues of Agostino Gallo appeared on the book market. He was a rich wool merchant, who lived in Brescia and belonged to the local elite: here he published his dialogues in 1564. In 1566 and 1567, he put forth new editions with numerous additions in Venice: the book then began its fortunate career to become the basic instruction on farming in Northern Italy for a very long time. Thanks to the many editions, both landowners and estate stewards read the book and Gallo's agricultural ideals stimulated a sensible land management. His words and instructions describe carefully two types of gardens: the gentleman's garden and the common garden. Both had a square shape, cut in four by regular paths and enclosed with pergolas and hedges made by lemon trees, climbing plants, rosebush, jasmine,[43] bay tree, and myrtle. In the beds, a number of herbs were regularly sown to keep the supply throughout the year: lettuce, endive, estragon, balm, garden savory, lady's mantle, rocket, sorrel, borage, parsley, radish, spring onions, betony, peppermint, penny-royal, spinach, bugloss, basil, bog-myrtle, violet, and pinks.[44] Gallo carefully described the content of the common gardens, grown for food and let to a tenant either for money or for a share of the crop. Four varieties of cabbage (best in spring and autumn), Savoy cabbage (particularly appreciated in Venice),[45] turnips (best at Carnival and Lent), broad beans,[46] leeks, garlic, echalotte, onions, cucumbers, melons, pumpkins, carrot, parsnips, bitter radish, radish, fennel, rue, wormwood, dill, sage, rosemary, asparagus, artichokes, strawberries,[47] barberry, lavender, rose (four wild varieties and white, wild and cultivated), and lily.[48] Gallo insisted on paying great attention to the choicest seed and he freely mixed vegetables, herbs, and flowers in the gardens. If the taste for eating was the main object of cultivation, it came always with an eye for decoration. Yet flowers brought an income to the household: roses were for rose waters and flowers for decorative use in festivities.[49]

Gallo's precepts became common lore in the following century and other writers quoted him directly. Giovanni Maria Bonardo (1584, 1586, 1608, 1619), Marco Bussato (1573, 1592, 1593, 1599, 1612), Giuseppe Falcone (1577, 1599, 1601, 1612, 1602, 1619, 1691), and Africo Clementi (1572, 1623, 1677, 1692, 1696) all wrote books, which were partly compilations and partly original, and aimed to a reader of a lower social status. Yet, it appears clearly that new plants, either flowers or vegetables, did not come into common use rapidly and that they remained safe in the largest gardens of the great houses. Peasant gardeners were somehow more conservative in their choices. They had to produce for their own pot and mistrusted plants from the New World, which were foreign to their gastronomic tradition. Market gardeners produced for consumers, who shared this conservative approach to food. Scholars should keep in mind that since late medieval times and the Renaissance the medical sciences warmly suggested caring of the daily diet as the leading way to keep healthy. Attention to diet and preparation of food certainly helped the population to recover after the 1450s. Hence, the extreme cautiousness of city consumers toward untested foodstuffs. Yet, peasants of particular areas were the first large group of consumers of new American crops. Most likely, an unknown Venetian landowner tried maize in 1550s in the Polesine area: nevertheless, the Po delta did not turn into a center of diffusion of the new crop. Other regions did in the mountain parts of the Venetian state, such as those bordering Belluno, Treviso, Verona, and Bergamo: they eventually took the leadership after major cereal crop failures in the 1590s, 1620, and 1630. Yet one had to wait for the 1660s to find maize growing in the fields of the plain (around Milan in

1677 and in Piedmont in the 1660s).[50] Vegetables such as tomatoes, peppers, Chile peppers, beans (*Phaseolus sp.*), pumpkin, and courgettes did grow in gardens between the 1520s and 1540s.

Yet they remained confined to private gardens for a long time and it does not seem that they were much in favor with the consumers.[51] Contracts between landowners and gardeners insisted on traditional cultivations usually described as greens (*herbazi*) or veggies (*ortaglia*). The historian has to resort to printed works to find better descriptions.

In the mid-seventeenth century, when the secular growth of the city of Bologna was fading away, Vincenzo Tanara published his *Economia del cittadino in villa*.[52] He gave a whole body of husbandry of every farm product for the benefit of the city dweller. The garden became very central to the family economy: it raised five crops in a year. Furthermore, the gardener could profitably sell every product that his family did not consume. The square shape was the best layout for the garden: vines decorated the central access to the plant beds, each with its specialized cultivations (Fig. 15.1). Different sorts of bushes and climbing trees enclosed the garden and improved its exterior: bay tree, sage, rosemary, jasmine, juniper, myrtle, and so on. In a second bed grew lavender, rose, thyme, marjoram, roman mugwort, and wormwood, all cut short. Next came four, six, or eight beds for vegetables according to family need. Flowers, trees, and bushes provided "beauty" and "utility."[53] Even more plots could be useful: to nurse plant and vine cuttings, to grow flowers, simple medicines, and, if possible, asparagus and strawberries. Pergolas with vines holding to the stalks provided shade and a final touch of beauty (Fig. 12, 13.1). Tanara was very attentive to varieties: he quoted three artichokes (white, dark, thorny).[54] He suggested changing the seeds of cabbages every year: the best seeds for the *Brassica oleracea capitata* came from Chioggia and Romagna, from Bologna those of the *B. o. viridis*. He described a third cabbage, the cauliflower (*B. o. botrytis*), as a novelty in contemporary gardens: seeds came in great quantity from Crete, Cyprus, or Istanbul, via Venice, Genoa, or Florence. He quoted three varieties especially sought after in Lombardy, Tuscany, and Rome and the use of sauerkraut in Germany. Finally, he suggested making good culinary use of broccoli, the seedlings of the cauliflower grown in August.[55] Tanara's gluttony took him to test carefully every variety that he mentioned. Such as three varieties of onions: "Caetane" in Rome, "Beneventane" and "di Romagna" in his own city, "Maligi" in Florence. To these he added echalotte (*Allium ascalonicum*) and similar flower bulbs of the *Liliacea* family, such as the tulips, which he enjoyed eating.

Unfortunately, the German soldiers who besieged and sacked Mantua in 1630 had a taste for the same luxury: they dug up and devoured every tulip bulb found in the Gonzagas' gardens, without any regard for their extreme value.[56] Tanara keeps us informed about other typical Italian vegetables telling us the local names of lettuce, the fame of fennel from Bologna, and his curiosity for melons. He suggested to steep melon seeds in sugar to improve its sweetness, water and dust them with cinders when growing, select the seed choosing among the many present in the market, cut and hang them from ceilings to ripen (as it is still practiced around Naples). He spoke of the famous melon seeds from Romagna, developed from Turkish and Black Sea varieties, watermelons, two varieties of melons from Verona, long and green melons from Genoa, round and yellow melons from Tuscany. He acknowledged a Marquis Bianchini, who gave him a gift of Spanish melon seeds: climbing and extremely large, some were oblong, green skin, red inside with black seeds, others were round, green skin and red inside, others with white skin and purple inside, others round and sweet, all extremely tasty.[57] Along with traditional vegetables, Tanara found some space in his garden to grow tobacco, not to smoke its leaves but to boil them into a juice to destroy lice.[58] He took in some consideration also other American products, beans and green beans, the red chili pepper (an inexpensive alternative to the Indian one), very briefly the tomato (a vegetable more decorative than useful) and maize, eaten by peasants in time of dearth, and not personally tested.[59] The pages he dedicated to flowers, fruit, and ornamental trees are too numerous to be taken into consideration here.[60] It is just worth remembering that he quoted nineteen varieties of lemons and twenty-eight of oranges, plus twenty-eight cultivars of various citrus fruits.[61]

Tanara's handbook became very popular among land stewards as landowners were losing their interest in direct management of their estates. In the latter part of the seventeenth century, Giacomo Agostinetti, from Treviso, was a particularly capable steward (*fattore*) and wrote a small tract on farming after thirty years of practice.[62] It is worth our while to point out that

Agostinetti called for a more efficient type of farming than the baroque cultivations suggested by Tanara. Talking of gardens and orchards during a long period of monetary deflation, Agostinetti praised the daily cash flow from vegetable sales: although they were small sums of money, this income became relevant at the end of the year. In view of this maximization of resources, he changed the shape of the garden, opting for a rectangular shape, with alternate rows of cultivations that required similar soil and sun conditions. Therefore, he suggested planting the peach trees in the vegetable garden, between rows of strawberries, as the shade of the peach tree was harmless to other crops. Furthermore, he reduced the number of cash crops to peach, strawberries, asparagus, and melons. All were arranged in long strips that allowed the gardener to work freely without damaging the growing plants (Fig. 14.2). Agostinetti, on the basis of daily experience, suggested a relevant revolution: the substitution of the square shape of the Renaissance garden, as recommended by Gallo, enriched by its wealth of baroque cultivations as described by Tanara, with row cultivation. Agostinetti was a farmer in a landscape where rows of trees supporting vines alternated strips with cereal crops and meadows to maximize production. Square or rectangular fields were still competing in the Po plain to provide the best soil organization (Figs. 7, 11, 12, 15.1,2). Yet this silent but consistent revolution happened some fifty years before Tull in his *Horse-hoeing husbandry* (1731) put forth to farmers the advantages of row cultivation. Once again, the garden provided the experimental ground for changes, which were going to take place in the following century.

Conclusion

At this point, it is necessary to draw some conclusions. In the Italian situation, peasants alone performed manual labor; upper classes never (or very, very rarely) extended their interests in gardens beyond planning and plant collecting. Even if, especially after 1750, landowners signed agreements with gardeners, which forbade pulling up trees and taking away fruits and veggies, they could never prevent cuttings from leaving the enclosed gardens of the nobility to become items of cultivation in the neighboring countryside. Cosimo I de Medici invested a lot of money in foreign plants for his garden. He was particularly proud of being the only Italian owner of an Asian jasmine (*J. sambac*) and forbade his gardeners to take away any cutting, even its leaves. However, the rumor is that a young gardener in love stole a cutting for his maiden girl, who planted it in her garden: one year later, the new plant produced so many beautiful flowers, which they sold very well in the Florence markets and multiplied the original plant.[63] Something similar happened to the *Robinia pseudoacacia* at the end of the eighteenth century: imported as a garden plant from France, it went into the wild and became invasive in the poor soils of Northern Italy. In the early days of Napoleonic rule in Turin, the *préfet* complained that walkers by the main public gardens pulled up the newly planted trees and tore branches off to grow them in their home orchards.[64] Therefore, it was peasants' manual skill that kept gardens and orchards going on both sides of the gates of Italian villas. In this way, plant varieties, grown in hardier conditions than the botanic gardens, improved greatly: eventually, some foreign species naturalized in the Italian countryside. The Italian flora grew richer in the seventeenth and eighteenth centuries, thanks to the toil and labor of peasant gardeners.

There were many agents in the conservation and diffusion of species: lords with connections with distant countries,[65] landowners who exchanged cultivars within a tight social network, gardeners more curious of others, women who took care of the vegetable patch next to the family dwelling, consumers' tastes and the general economic cycle. Yet, we perceive that some plants got around more easily than others. Following the studies of Saccardo, important plants, such as the sugar cane, white mulberry, varieties of garlic, turnips, and spinach found their way into the Italian peninsula during the fourteenth century. The eggplant and the artichoke found their way in the peninsula in the first part of the fifteenth century; around 1450, rice cultivation began near Mantua and Verona. Fruit trees, such as the apricot, sweet and bitter almond, and citrus (the lemon and the bitter orange) spread around, thanks to the new taste for sweets of the European upper classes. These plants came out of the enclosed gardens of the Middle Ages to grow in specialized tree orchards, that is, the Mediterranean gardens that Sereni had mentioned. Saccardo counted 127 exotic plants and varieties, which came this way during the sixteenth century. Among these were buckwheat, sesame, four varieties of cabbage (*Brassica campestris, B. sabauda, B. botrytis, B. acephala*), four varieties of wheat (*Triticum aestivum n dicoccum, T. monococcum, T. aestivum b hibernum, T. ae. Compositum*), two varieties of orange (*Citrus aurantia grandis, C. a. bigardia*), the plum, and the nectarine (*Prunus avium duracina, P. a. Juliana, P. persica nucepersica*).[66] The main American plants remained as curiosities unless they were a variety of a plant that was already familiar to the Italian peasant. Such was the case of the bean (*Phaseolus vulgaris*): Emperor Charles V sent a sample as a gift to Pope Clemens VII and a secretary of the papal Curia took a few to Belluno, his hometown, in the Venetian Alps. There, the beans found a favorable climate and soil and Belluno rapidly became a center of diffusion of this species on both sides of the Alps. Other well-known American plants (such as the potato, maize, tobacco, pepper, chili pepper, and pumpkin) found their way into the major botanical collections, herbaria, and printed herbals. Yet they came in use gradually and much later: maize from the 1620s, the potato from the 1790s, and tomatoes from the 1850s. The main current of plant imports remained the Asian route and foreign plants acclimatized and naturalized in Italy came thence in the greatest number.

Italian peasant gardeners provided shelter for a larger number of plants than any botanic garden, public or private, could provide: since the sixteenth century, local varieties became more esteemed for their qualities and sought after as seed (for example, wheat from Treviso and from Sicily). Italian peasant gardeners grew exotic plants under a great variety of conditions, which became a good test for the acclimatization of these species. Before commercial nurseries were established in the early 1800s, local market gardeners provided the best seed and plant varieties: that is a point that is still worth researching. It would not have been possible to grow new crops, when suddenly necessary, such as maize in the 1620s and later, or to replant lemon and olive trees after the 1811–1812 terrible frost, without using all the vegetable patrimony built up throughout centuries of peasant gardeners' work.

I would like to thank friends and colleagues who have generously helped me: Furio Bianco, Gigi Corazzol, and Danilo Gasparini. The thesis of Paolo Candio, Roberta Padovano, and Michela Giorgiutti provided relevant information. Dr Joan Thirsk's comments have been valuable and are always welcome. However, the writer is solely responsible for the views expressed here.

Prices are given in account money (*lire venete, di Piemonte, di Milano, abbrev. £ s d*) if not otherwise stated. Note on Italian land measures: pre-metric agrarian measure in Italy depended closely on the Roman *Jugerum*, 0.25 ha as much as the *Statute acre* did. For a rapid comprehension of the actual size, the reader could easily substitute the names of the Italian measures quoted in this study with the more familiar *acre*, if he or she wishes so.

Botanical names in Latin follow the Linnean system, unless otherwise stated. List of abbreviations: BSBS: Bollettino Storico Bilbiografico Subalpino. ASVR: Archivio di Stato di Verona. ASCT: Archivio Storico del Comune di Torino. ASMN: Archivio di Stato di Mantova. ASDMN: Archivio Storico Diocesano di Mantova. ASU: Archivio di Stato di Udine. ASR: Archivio di Stato di Roma.

NOTES

1. See *History of Italian agricultural landscape* (English translation with an introduction by Raymond Burr Litchfield, Princeton: Princeton University Press, 1997).

2. Paul Bairoch, *La population des villes européennes: banque de données et analyse sommaire des resultats 800-1850* (Geneve: Librairie Droz, 1988), and David Herlihy, *Cities and Society in Medieval Italy* (London: Variorum reprints, 1980); AA. VV, *Paesaggi urbani dell'Italia padana nei secoli VIII-XIV* (Bologna: Cappelli, 1988).

3. Pier de' Crescenzi wrote in Latin his agricultural text book that had many translations in manuscript and numerous printed editions in different European countries: for the Latin text see now Reinhlielt Richter-Bergmeier and Will Richter (eds.), *Ruralia Commoda Petri de Crescentii: das Wissen des Wollkommenen Landwirts um 1300* (Heidelberg: C. Winter, 1995-1998, vols.1-3); for the Italian text see *Il Trattato dell'agricoltura di Pier de' Crescenzi traslato nella favella fiorentina rivisto dallo 'Nferigno, accademico della Crusca* (Bologna, 1784 reprint, Bologna: G. Forni, 1987); for the French translation see Crescenzi, Pietro de, *Le Livre des profits champêtres, de Pierre des Crescens* (Paris: Jean Bon Homme, 1486), and Jean Louis Gaulin, *Recherches sur Pier de Crescenzi* (Thèse, Paris: Sorbonne, 1980). See also *L'ambiente vegetale nell'Alto Medioevo* (Spoleto: Atti della XXXVII Settimana di studio del Centro Italiano di Studi sull'Alto Medioevo, 1990).

4. See especially books V, *Degli alberi,* VI *Degli orti,* VIII *De' giardini.* Of the vast bibliography about the theory and practice of Italian gardens see the standard works by Alessandro Tagliolini, *Storia del giardino italiano* (Florence: Casa Husher 1988); Claudia Lazzaro, *The Italian Renaissance garden: from the conventions of . . . planting, design, and ornament to the grand gardens of the sixteenth-century central Italy* (New Haven and London: Yale University Press 1990), chapter 2 *passim* and Appendix I; David R. Coffin, *Gardens and Gardening in Papal Rome* (Princeton: Princeton University Press, 1991) and Margherita Azzi Vicentini, ed., *L'arte dei giardini: scritti teorici e pratici dal 14. al 19. secolo* (Milan: Il Polifilo, 2000).

5. The frescoes of Schifanoia are well known. For their agricultural context see now Franco Cazzola, *La città, il principe, i contadini. Ricerche sull'economia ferrarese nel Rinascimento, 1450-1630* (Ferrara, Ed. Corbo: 2003), 75-97.

6. Cazzola, *La città, il principe, i contadini,* 53-67, especially 56-8, 88-9.

7. See Margherita Azzi Vicentini, *L'Orto Botanico di Padova e il giardino del Rinascimento* (Milan: Il Polifilo, 1984), 61. See also note 4, especially Lazzaro, *The Italian Renaissance Garden,* 34.

8. See ASCT, Coll. V. 1126, 1558 *Registrum forensium* and 1127, 1559 *Registrum Portae Susinae.* For the evolution of building materials, see also Maria Teresa Todesco, *Oderzo e Motta. Paesaggio agrario, proprietà e conduzione di due podesterie nella prima metà del secolo XVI* (Treviso: Fondazione Benetton, 1995), 147-57; Giampier Nicoletti, *Le Campagne. Un'area rurale tra Sile e Montello nei secoli XV e XVI* (Treviso: Fondazione Benetton, 1999), vol. 1, 291-323, especially 320 and Gianmaria Varanini, *Le campagne veronesi del'400 fra tradizione ed innovazione* in Giorgio Borelli, ed., *Uomini e civiltà agraria in territorio veronese. I secoli IX-XVII* (Verona: Banca Popolare di Verona, 1987), 185-262.

9. Marcus Porcio Cato, *De agricoltura,* I.1.

10. Bartolomeo Platina, *Il piacere onesto e la buona salute,* (1st ed. Rome?, 1474?) (Emilio Faccioli Ed.) (Turin: Einaudi, 1985); David R. Coffin, *The Villa in the Life of Renaissance Rome* (Princeton: Princeton University Press, 1979), 182.

11. ASVR, Famiglia Bevilacqua, xxvi.340, 1505-1509. In general see Giorgio Borelli, *Patriziato e nobiltà: il caso veronese* (Milan: Vita e Pensiero, 1975).

12. Leonardo's eating habits did not differ from those of the religious houses, such as the Jesuits, who ate lots of fresh fruits. Fruits were sold when abundant or bought when crops failed. See later.

13. ASVR, Famiglia Bevilacqua, xxvi.340, 1505-1509, ff. 11, 17, 22, 29, 38, 41, 44, 46, 58, 68,71, 72, 143, 197, 200.

14. ASVR, Villafranca, *Estimi,* nn. 42, 49. In general see Gian Maria Varanini, *Per una storia di Villafranca Veronese,* in Sante Bortolami, ed., *Città murate nel Veneto* (Cinisello Balsamo: Silvana, 1988).

15. ASVR, Villafranca, *Estimo,* n. 41.

16. ASVR, Villafranca, *Estimo,* n. 42, ff. 35, 43, 62, 74, 81.

17. See Andrea Castagnetti, *La pianura veronese nel medioevo. La conquista del suolo e la regolamentazione delle acque,* in Giorgio Borelli, ed., *Una città e il suo fiume: Verona e l'Adige* (Verona: Banca Popolare di Verona, 1977), vol. 1, 62-76.

18. Danilo Gasparini, *La città e la campagna: contadini, patrizi e fattori tra Piave e Sile,* in Aldino Bondesan (ed.), *Il Sile* (Verona: Cierre, 1998).

19. Ibid., 173.

20. Nicoletti, *Le Campagne,* 48-49.

21. Todesco, *Oderzo e Motta,* 47-48.

22. Both works by Coffin, *Gardens and Gardening in Papal Rome* and *The Villa in the Life of Renaissance Rome* discuss this situation. For one example, see ASR, Tribunale dell'agricoltura, vol. 1, 1616, fos. 26 ff.

23. Maria Clotilde Daviso di Charvensod, *I piu antichi catasti del Comune di Chieri* (1253), in BSBS, v. 38, a. 1936, n. 1-2, 66-102 and Luciano Allegra, *La citta verticale: usurai, mercanti e tessitori nella Chieri del Cinquecento* (Milan: F. Angeli, 1987).

24. Mauro Ambrosoli, *Orti, vigne, giardini,* in Rinaldo Comba and Stefano Benedetto, eds., *Torino, le sue montagne, le sue campagne* (1350-1840) (Turin: Archivio Storico della Città, 2002), 149-74.

25. Mauro Ambrosoli " . . . *laudato ingentia rura / exiguum colito . . .": grande e piccola proprietà nella formazione del paesaggio agrario mantovano di '500 e '600,* in Eugenio Camerlenghi and Mario Vaini (eds.), *Il Secondo Convegno sul paesaggio mantovano* (Mantua: Accademia Virgiliana, 2006, forthcoming) small holdings dedicated to vegetable gardening were extremely numerous throughout the whole area considered here (Figs. 8.1,2,3)

26. See ASDMN, Registro Vialardi, 1690; see also Giovanni Vareschi, *Le proprietà vescovili nel 1690,* in Eugenio Camerlenghi and Mario Vaini (eds.), *Il Secondo Convegno sul paesaggio mantovano.*

27. ASU, Congregazioni Religiose Soppresse (C.R.S.): busta 247, bound fascicule, 105v.

28. ASU, C.R.S.: busta 247, fasc. 1, 3.

29. Johann Heinrich Von Thünen (see *Von Thunens isolated state* translated by Carla M. Wartenberg; edited with an introduction by Peter Hall (Oxford: Pergamon Press, 1966) more precisely spoke of the "isolated state": the Italian city-states of the Renaissance, which enforced strict control over the surrounding countryside provide many good cases to discuss the model.

30. ASCT, Carte sciolte n. 1966, year 1608.

31. See Martha D. Pollak, *Turin, 1564–1680: urban design, military culture, and the creation of the absolutist capital* (Chicago: Chicago University Press, 1991).

32. The *chateaux* of Miraflores, Valentino, Villa della Regina, Regio Parco, Venaria Reale, to which were added Rivoli and Stupinigi in the eighteenth century.

33. Paolo Candio, *Note sull'orticoltura lagunare nel secolo XVII*, Unpublished Thesis at the Faculty of Letters, University of Venice, 1986.

34. See Tagliolini, *Storia del giardino italiano*, 143-44, and Francesco Sansovino, *Venetia citta nobilissima et singolare*, (Venice: Iacomo Sansouino, 1581), 174 a,b.

35. Candio, *Note sull'orticoltura*, 2-16.

36. Richard John Goy, *Chioggia and the Villages of the Venetian Lagoon: studies in urban history* (Cambridge: Cambridge University Press, 1985).

37. Candio, *Note sull'orticoltura*, 20-33, 46-65.

38. In 1604 magistrates had to intervene against the "artichoke dearth" caused by the frost and the high prices that followed.

39. Serlio in the mid-sixteenth century provided plans for the houses of landowners and tenants, see Sebastiano Serlio, *Serlio on domestic architecture*, text by Myra Nan Rosenfeld; foreword by Adolf K. Placzek; introduction by James S. Ackerman (New York: Architectural History Foundation, ca. 1978). So far, only Coffin, *The Villa in the Life of Renaissance Rome*, 190 ff. and 233 ff. and more recently the volume by Lucia Bonelli Conenna and Ettore Pacini, eds., *Vita in villa nel Senese: dimore, giardini e fattorie*, (Siena: Monte dei Paschi, 2000) have devoted attention to the question of middle-sized gardens.

40. Mauro Ambrosoli, *The Wild and the Sown: Botany and Agriculture in Western Europe, 1350-1850* (Cambridge: Cambridge University Press: 1997), 97.

41. See Pietro A. Saccardo, *Cronologia della flora italiana* (Padua: Tipografia del Seminario, 1909); Bruno Andreolli, *Il ruolo dell'orticoltura e della frutticoltura nelle campagne dell'Alto Medioevo, in L'ambiente vegetale*, 175-219; Ambrosoli, *The Wild and the Sown*, 97 ff.; Ada Segre, *Le metamorfosi e il giardino italiano*, in Maria Adriana Giusti and Alessandro Tagliolini, eds., *Il giardino delle muse: arti e artifici nel barocco europeo: atti del Quarto Colloquio internazionale, Pietrasanta, 8-10 settembre 1993* (Florence: Edifir, 1995), 110; Lazzaro, *The Italian Renaissance garden*, Appendix I.

42. See note 49.

43. The jasmine was a recent introduction in to the Italian gardens: it was quoted as a novelty by the humanist Ermolao Barbaro who died in 1492, see Ippolito Pizzetti (ed.), *Enciclopedia dei Fiori e del Giardino*, (Milan: Garzanti, 1998), 462.

44. Ibid., 124-28.

45. Ibid., 130.

46. Gallo mentioned other important legumes, such as lentils, chickpeas, beans among the field crops (quotations are from the Venice 1569 edition by Giovanni Percaccino, 49-51). Gallo spoke of the Old World bean (Dolichos sp.).

47. Ibid., 140, Gallo clearly spoke of the European strawberry.

48. Ibid., 129-42.

49. Jack Goody, *The Culture of Flowers* (Cambridge: Cambridge University Press, 1993), 166 ff.

50. Marco Fassina, *L'introduzione del mais nelle campagne venete*, in "Società e storia," 15, 1982, 31-59; Danilo Gasparini, *Una provvida gloria regionale: il mais nel Veneto*, in "Venetica," 1999, 11-42.

51. So far, I found only one landowner who declared growing in 1793 green beans in his Turin garden among other more traditional crops, onions and spinach. See ASCT, Collezione V, 1159 (1793).

52. The book had many editions in Bologna 1644, 1658, Rome 1651, Venice 1665, 1674, 1670, 1680, 1687, 1700, 1745.

53. *Economia del cittadino in villa*, (Venice: Prodotti, 1680), Libro IV, *L'orto*, 203-64 and Libro V, *Il giardino*, 266-367, see especially 207-10.

54. Ibid., 222.

55. Ibid., 224-27, 229.

56. Ibid., 234–35. See also N. W. Posthumus, *The Tulip Mania in Holland in the years 1636 and 1637, i*n "Journal of Economic and Business History," I, 1929, 434–55; Sam Segal, *Tulips portrayed. The tulip trade in Holland in the 17th century*, (Hillegom: Museum voor de Bloembollenstreek, 1992); Goody, *Culture of Flowers*, 188-90.

57. Tanara, *Economia*, 245-46.

58. Ibid., 237.

59. Ibid., 238, 248-49, 406.

60. Ibid., 166-205.

61. Ibid., 286-87.

62. Filippo Re, *Dizionario georgico*, (Milan: Silvestri, 1812), *sub nomine*. Giacomo Agostinetti, *Cento, e dieci ricordi, che formano il buon fattor di villa,* (Venice: Stefano Curti, 1679) see also *Dizionario Biografico degli Italiani* (Rome: Istituto dell'Enciclopedia Italiana, 1960 vol.1, *sub nomine*.)

63. Pizzetti, ed., *Enciclopedia dei fiori e del giardino*, 462.

64. Mauro Ambrosoli, *Alberate imperiali per le strade d'Italia: la politica dei vegetali di Napoleone Bonaparte*, in "Quaderni storici," vol. 99, 1998, 707-738.

65. Osvaldo Raggio, *Storia di una passione: cultura aristocratica e collezionismo alla fine dell'ancien regime* (Venice: Marsilio, 2000).

66. Saccardo, *Cronologia della flora italiana*; see also note 42.

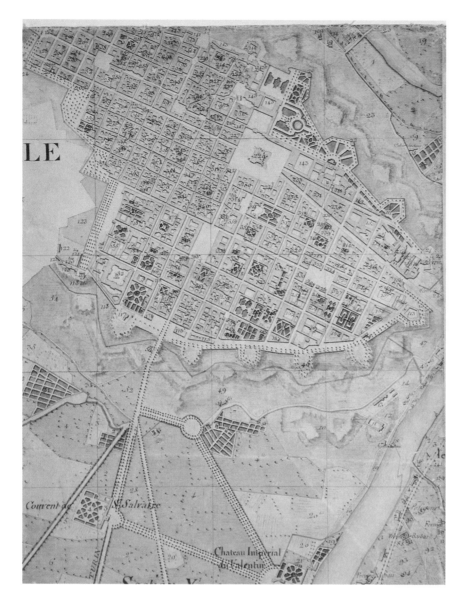

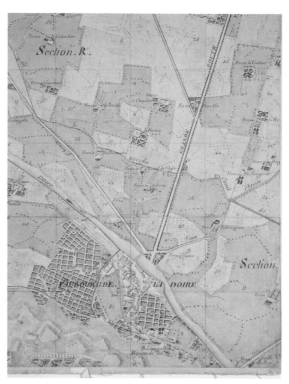

Figure 1.1. Turin: city gardens and commercial orchards outside Porta Nuova, West of the city walls. Source: Napoleonic Cadastre (1810), sheet 8, Archivio di Stato di Torino, Sezioni Riunite: Catasti.

Fig. 1.2. Turin: commercial vegetable gardens outside Porta Susa, northeast of the city walls. Source: Napoleonic Cadastre (1810), sheet 13, as earlier.

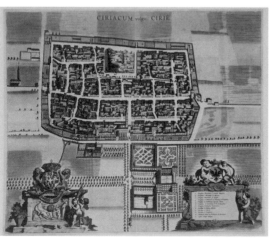

Fig. 2. Cirié, Piedmont: vegetable gardens and orchards around the town walls of this small township. Source: *Theatrum statuum . . . Sabaudiae ducis . . . ,* (Amstelodami: Blaeu, 1682).

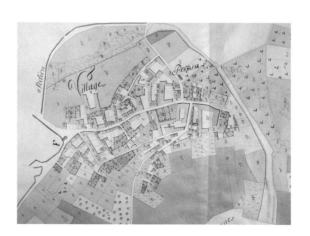

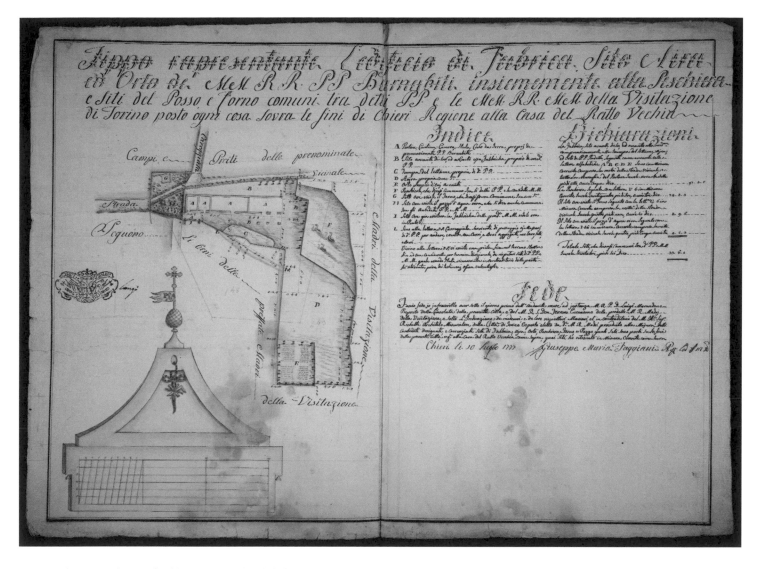

Fig. 3. Chieri, Piedmont: buildings and orchards belonging to the Barnabiti Fathers. Ink and watercolor map drawn by G. M. Faggiani, 1777. Source: Archivio di Stato di Torino, Sezione Corte, Carte topografiche e disegni, III, n. 8.

Fig. 4. (left) Village of Perosa in Val Chisone, Piedmont: orchards, trees, and village buildings. Source: Archivio di Stato di Torino, Sezioni Riunite, Catasti, Perosa, Allegato B, 68 Sezione H, about 1810.

Fig. 5. (right) Quagliuzzo in Val Chiusella, Piedmont: vegetable gardens and village buildings. Source: Archivio di Stato di Torino, Sezioni Riunite, Catasti, Quagliuzzo, Allegato B, 73, Sezione B; about 1810.

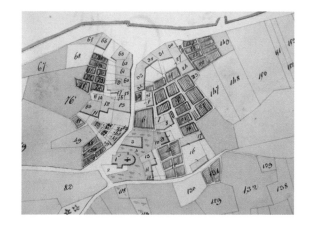

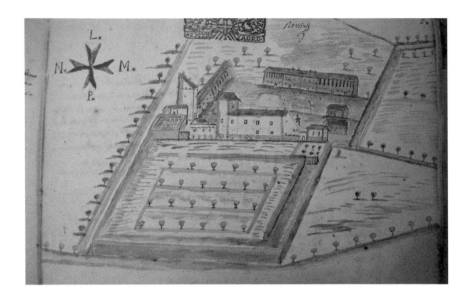

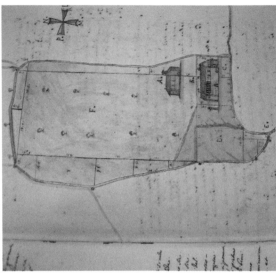

Fig. 6. The enclosed orchard of the Castle and Church at Candiolo (Turin).
Source: Archivio di Stato di Torino, Archivio Ordine di Malta, Cartella 64, 1728.

Fig. 7. A farm at Cavallermaggiore, Piedmont, about 1750: the vegetable garden is the plot marked with the letter D. Source: as above, Cartella 92.

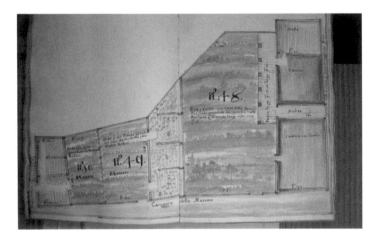

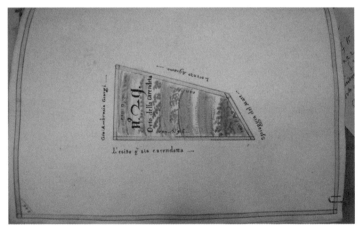

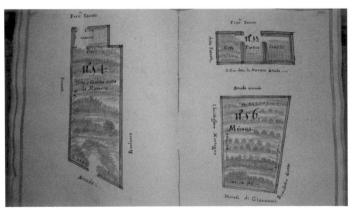

Fig. 8.1, 2, 3. In the hills of the Riviera growing vegetables was a local specialization. Source: as above, Cartella 222, Savona, fos.317, 320, 321, 446.

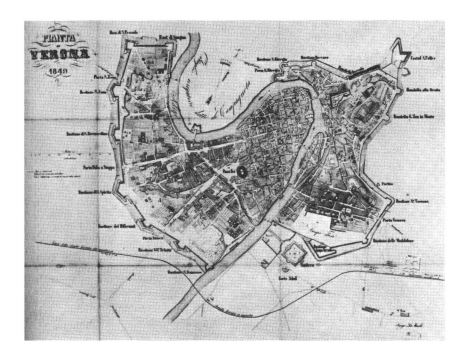

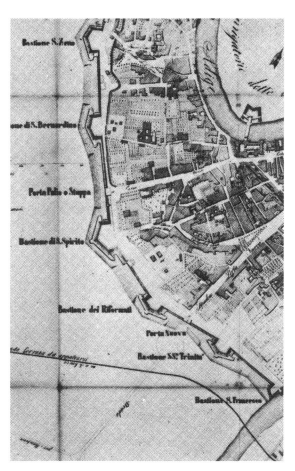

Fig. 10. Verona, 1840: note the gardens between the town walls and the city.

Fig. 10.1. Enlargement of the west side of the above map.

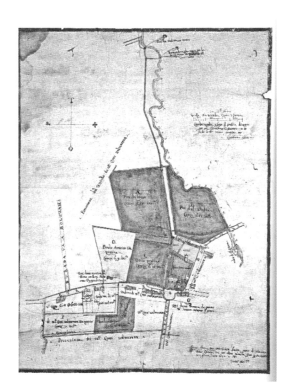

Fig. 11. Valmarana estate, Lisiero near Vicenza: the small orchards and vegetable gardens along the road were given to customary tenants and laborers in exchange of their work. Source: Vicenza, Biblioteca Bartoliana, Cabrei.

Fig. 12. Mocenigo Soranzo estate, original drawing 1696, copy made in 1823. The tenant farmer rented the buildings, courts, outbuildings (marked I), and the orchard (K), the customary tenants and laborers received the small vegetable plots in exchange of their work (C,E). Source: Vicenza, Biblioteca Bartoliana, Cabrei.

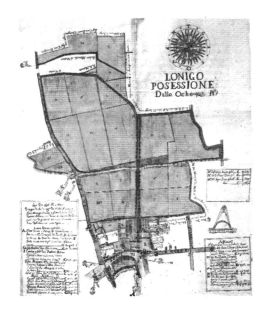

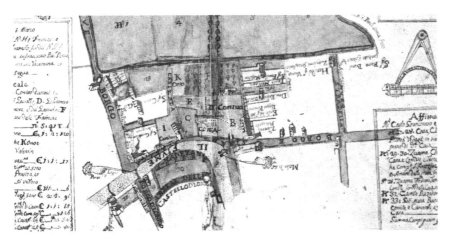

Fig. 12.1. Particular of the estate in Fig. 12.

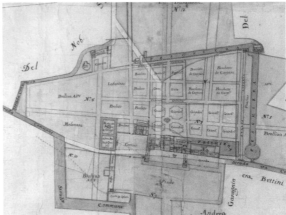

Fig. 13. 1. Posmon at Montebelluna, Treviso: villa and gardens of Antonio Giustinian. Note the enclosed orchards, *brolo* and *broleto,* and vegetable plots, *orto,* as part of the formal gardens. Source: Archivio di Stato di Treviso, Giustinian Archives.

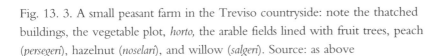

Fig. 13. 2. The *piantata* system in the Venetian countryside: vines supported by maple (*opio*) and willow (*salgaro).* Source: as above.

Fig. 13. 3. A small peasant farm in the Treviso countryside: note the thatched buildings, the vegetable plot, *horto,* the arable fields lined with fruit trees, peach (*persegeri*), hazelnut (*noselari*), and willow (*salgeri*). Source: as above

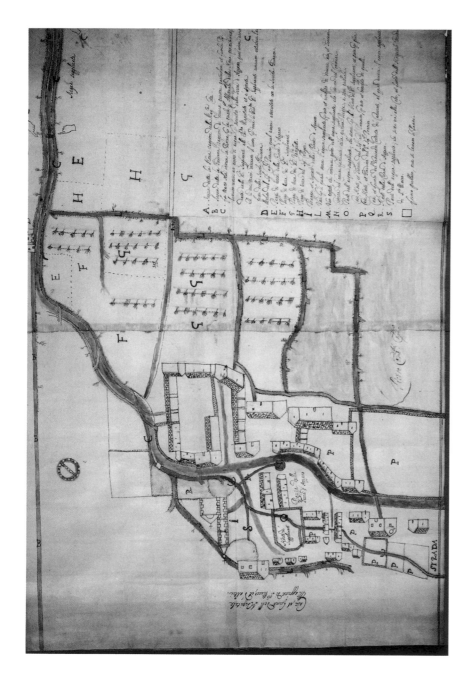

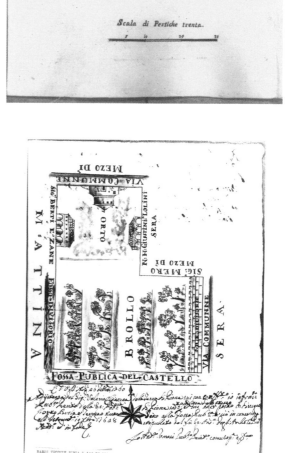

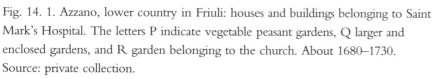

Fig. 14. 1. Azzano, lower country in Friuli: houses and buildings belonging to Saint Mark's Hospital. The letters P indicate vegetable peasant gardens, Q larger and enclosed gardens, and R garden belonging to the church. About 1680–1730. Source: private collection.

Fig. 14. 2 Villa Claviano, near Udine, estate owned by the Confraternity of the Calegari (Shoemakers) of Udine: the three farms displayed a well-planned vegetable garden contemporary to that described by Agostinetti, see text p. 192. 1709. Source: private collection.

Fig. 14. 3. Friuli, A well-established walled fruit tree orchard, *brolo,* attached to a farmhouse with outbuildings. Source: private collection.

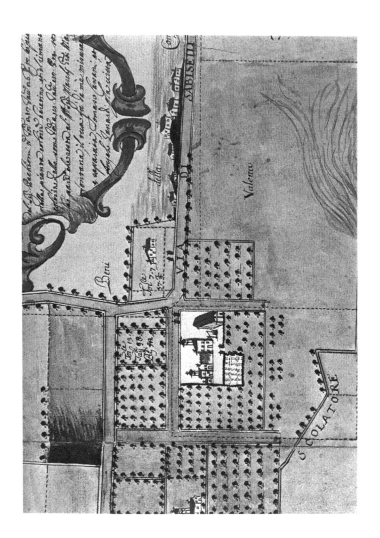

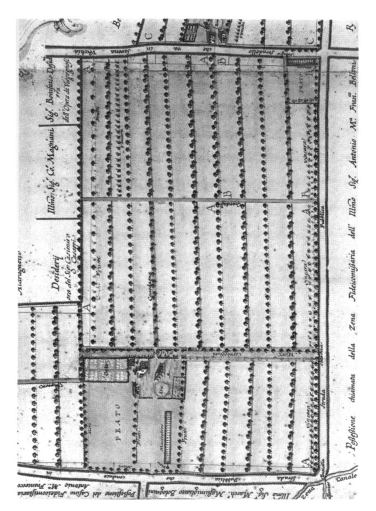

Fig. 15.1. La Mazzara estate, Medicina near Bologna, 1682.
Source: F. Varignana (ed.), *Le collezioni d'arte della Cassa di
Risparmio di Bologna, I: Disegni, II: Mappe agricole e urbane del territorio
bolognese,* (Bologna: Cassa di Risparmio di Bologna, 1974).

Fig. 15.2. Farm belonging to the Great Estate of Santa Maria
Maddalena at Cazzano, Bologna, 1743. Source: as above.

Botanical Progress,
Horticultural Innovations
and Cultural Changes

Botanical Progress,
Horticultural Innovations
and Cultural Changes

Botanical Progress,
Horticultural Innovations
and Cultural Changes

Botanical Progress,
Horticultural Innovations
and Cultural Changes

Horticultural Utopianism in France in the Late Eighteenth Century

Michel Conan

The horticultural society of Paris, the first one in France, was created in 1827. In 1835, its first general secretary, Etienne Soulange Bodin,[1] the inventor of the *Magnolia soulangiana*, presented a "Historical Coup d'Oeil upon the Progress of French Horticulture Since 1789."[2] His introduction leaves no doubt that this is going to be a glowing account of a success story. He compares his task to the felicitous moment when a traveler who has crossed the rich landscapes of a beautiful valley starts climbing the mountainside and turns around to embrace the whole spectacle in his gaze before leaving it to pursue his future advances. First, Soulange Bodin describes the course of French gardening—please note that he refrains from using the word horticulture—in four parts. First, the beginning with the "Instructions on Gardens and the plants to be cultivated there" in an eighth-century A.D. capitulary by Carolus Magnus, whom Soulange Bodin credits with obtaining the best kinds of vegetables, melons, peaches, figs, and other oriental fruits from Harun al Raschid. Bodin notes that gardens at the time were very poor and mostly devoted to the cultivation of medicinal plants of dubious value for health. Second, the introduction of flower gardens in France eight centuries later by Francis I who borrowed this art from Pope Leo X. This also heralds the introduction in France of physics, botany, and other branches of natural history. Third, he highlights the reign of Louis XIV that

Fig. 1. Hubert Robert: Méréville, Le Grand Lac et le Chateau, in François d' Ormesson & Pierre Wittmer, Aux Jardins de Méréville, Paris: Editions du Labyrinthe, 1999 [45]

introduced the "royal munificence of design, and the absolutist reign of geometry."[3] Yet such rules were untenable and the "freedom necessary to the development of natural beauties was introduced long before the time when the monarchy was

Fig. 2. New gardens at Ris Orangis in Jardins en Ile-de-France : dessins d'Oudry à Carmontelle : collection du cabinet des dessins du Musée de l'Ile-de-France : Orangerie du Château de Sceaux, 15 octobre-15 décembre [1990] / [catalogue par Isabelle de Lannoy, Jean Georges Lavit, Gérard Rousset-Charny] (26).

precipitated towards its demise under Louis XVI."[4] Soulange Bodin quotes in the same breath great baroque gardens such as Vaux and Chantilly, the famous landscape gardens of Méréville (Fig. 1), and the orchards of the fruit groves in Montreuil, a small village near Paris. He deliberately sees the progress of cultivation and design as a whole and chooses to praise the developments of horticulture for both pleasure and utility before 1789. This date marks the beginning of the fourth period he hails as the true origin of horticulture. I quote " . . . it is in those times of universal emancipation resulting from the Revolution, as the first outpouring of generous feelings led to a sense of fraternity between all the sciences more noteworthy than the sense of political fraternity, that not only plant physiology and botany, but the science of trees and flowers as well as many other branches of science, all of them growing more popular, lent a hand to what later came to be known as horticulture"[5] (Fig. 2). Then Soulange Bodin continues, insisting that the need was felt for a new one to replace the old name of gardening in the 1770s, when men such as Morel were attempting to account for the new practices of landscaping. I quote: "This name has brought together the common husbandman and the philosopher or the artist who endeavors by his studies, as the other does with his own hands, to achieve a new way of improving and embellishing the earth."[6]

This is a significant text for the history of French horticulture that contributes to the self-serving myth of the Revolution as a moment of illumination when the darkness that surrounded all human minds subsided. Soulange Bodin, however, was well aware of the continuities between the developments of botany, plant physiology, soil chemistry, and the cultivation of trees or the naturalization of exotic plants since the mid-eighteenth century, when Duhamel du Monceau started publishing his Treatise of Land Cultivation." He was a very well informed horticulturist. Thus, his claim that the name horticulture was introduced to name a new practice that concerned husbandmen, philosophers (whom we would call men of science today) and artists at the same time, and to which each of these groups contributed with different means, demands our attention. This implies that people whose interests and activities were quite different and who had little reason to cooperate before the Revolution had been brought together by the general sense of fraternity that united enlightened men in a quest for the common good after it. This claim is sustained by the whole description of the progress of horticulture by Soulange Bodin. He also argues that horticulture contributed to the development of agriculture itself: "The larger cultivation received from horticulture a large number of useful plants introduced and first cultivated in gardens with a mandate to propagate and disseminate them. Such was the case of the potato and all its improved varieties; the sugar beetroot; the sweet potato; chicory, which became an important substitute for coffee; oil rich seeds,

Fig. 3. André Thouin in 1780, in *Le Jardin des plantes à la croisée des chemins avec André Thouin, 1747–1824* / [édité par Yvonne Letouzey], Paris : Editions du muséum, 1989 (Pl 1)

plants for cattle fodder; vegetable plants for artificial meadows; several varieties of mulberry trees, and in particular '*morus multicaulis*'—the most beautiful gift in a long time that horticulture has made to agriculture, and that lent such a timely hand to recent efforts of agriculture in favor of silk production." He goes on for five pages to describe a number of trees and grain crops that were first naturalized, cultivated and improved in gardens before they were introduced into the rural economy.[7]

This claim rests on a large number of observations and was not challenged by Soulange Bodin's contemporaries. They shared a sense of the novelty of horticulture as a philanthropic activity that rested upon the conjunction of practical husbandry, science and art. The examples he quotes to sustain his point of view are well ascertained. I must say that I was persuaded at first. In fact, an earlier study of the activities of André Thouin[8] (Fig. 3) (the head gardener since 1766 at the *Jardin du Roi*, the royal botanical garden in Paris, and the first professor of agriculture to give public lectures in Paris beginning in 1799), had enabled me to study first hand in the archives the traces of the wonderful practice of philanthropy by a European network of botanists, naturalists, and devoted husbandmen during the French Revolution and the Napoleonic era.

All of this indicated the development of a new type of activity after the Revolution that was later called horticulture and contributed new directions of development for agriculture. This reversal of the expected course of history between horticulture and agriculture seemed worthy of some interest.

Fig. 4. Peach tree pruning schema in the eighteenth century at Montreuil, in Marie-Rose Simoni-Arembou, *Parlers et Jardins de la Banlieue Parisienne*, Paris: Klinkcsieck, 1982 (pl.V)

There were, however, a few slight difficulties. For instance, Thouin himself explicitly intended his lectures for nine types of husbandmen in France: ploughmen, vine-dressers, foresters, fruit tree pruners and grafters, *fleuriste* gardeners, pleasure gardeners, and botanist gardeners. Thus, he sees the horticulturalist, whom he calls the *fleuriste* gardener, as only one of the different husbandmen whom he addresses, as if his own broad view of rural economy and practical cultivation could be applied to different domains of practice such as plowing, vine growing, and flower cultivation. It should also be noted that the Royal Society of Horticulture was created in 1827 by Héricart de Thury who was then also the president of the Royal Society of Agriculture.[9] This does not seem to imply any precedence of horticulture over agriculture and the fact that Thouin's experimental fields of cultivation practice were in the botanical garden does not prove that they were gardens. By contrast, one could not ignore that the development of a scientific knowledge and practice of agriculture originated in the royal botanical garden and was spurred by the activities of the European network of botanists, naturalists and devoted gardeners who engaged in earnest exchanges and communication with André Thouin. This invited further study of the origins of cooperation between naturalists, artists, gardeners, and husbandmen before 1789.

The *Jardinistes* and the Beginnings of Experimentation

There is a continuous history of cultural innovations introduced by gardeners after the publication of his book on gardening by La Quintinie in 1690. It was pursued by the *jardinistes*, as André Bourde calls fervent gardeners. Yet only a few of them committed their experiences to writing.[10]

The development of peach culture in Montreuil and Bagnolet near Paris provides an interesting example. At some point in the seventeenth century, gardeners in Montreuil understood that peach production could be greatly increased when the places where they were cultivated were divided into small areas by walls exposed to the south on which trees could be trained.[11] It encouraged them into developing a system of cultivation of espaliered peach trees that was already successful at the time that La Quintinie was appointed head gardener at Versailles. Pépin, the son of a Montreuil gardener, went to work in the kitchen gardens at Versailles and attempted in vain to convince La Quintinie of the superior value of the Montreuil system for management and pruning of trees (Fig. 4). La Quintinie fired him and he walked away in anger. He went back to Montreuil and contributed, according to Schabol, to the development of the most excellent method of peach cultivation in the Paris region. Roger Schabol (1690-1768) was a *jardiniste* convinced that gardeners should develop a better understanding of the

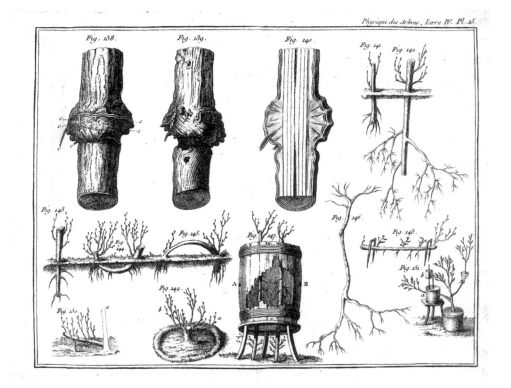

Fig. 5. Traité de la physique des arbres in Henri Louis Duhamel du Monceau, La physique des arbres, où, Il est traité de l'anatomie des plantes et de l'économie végétale : pour servir d'introduction au traité complet des bois & des forests : avec une dissertation sur l'utilité des méthodes de botanique : & une explication des termes propres à cette science, & qui sont en usage pour l'exploitation des bois & des forêts . . . Paris : chez H.L. Guerin & L.F. Delatour, 1758. (Book IV, title and plate 15)

anatomy and physiology of plants through direct observation and experiments. He had already collaborated with a well-known anatomist and a surgeon for many years while he learned about the practices of the gardeners cultivating peaches at Montreuil. He learned from Pépin and Girardot at Montreuil and succesfully tested their methods in his own gardens for several years. He noted that gardeners are usually considered poorly because they allow themselves to be guided by routine and traditions, whereas the gardeners of Montreuil had developed a sense of experimental practice from which they developed their knowledge. After he adopted their methods, he conducted many experiments that enabled him to propose solutions to many common defects in the Montreuil gardeners' practice.[12]

The New Agriculture of Duhamel du Monceau and Anglomania

The major discussion of agricultural methods, however, had a completely different origin. Between 1730 and 1750 some of the rich landowners who were taking some pride in the improvement of their lands published their ideas and were even able to catch the attention of regional and national authorities. Buffon, the great naturalist, as early as 1742 reported some of his ideas and experiments in forestry at the Academy of Sciences.[13] The Marquis of Turbilly started a large rural rehabilitation program around his castle in 1737 and drew inspiration from German methods he had observed in Bavaria. And, most important, Henri Louis Duhamel du Monceau (1700–1782) at Denainvilliers in Gâtinais, after many publications on tree growth, cultivation, and uses, published in 1750 the first volume of his "Traité de la Culture des Terres." (Fig. 5) It was inspired by the theoretical writings of Jethro Tull and by agricultural developments in England. This book proved extremely influential because it proposed an answer to the great concern of philanthropists and public authorities: how to produce more wheat grains with a lesser quantity of seeds. Jethro Tull developed a method of wheat cultivation inspired by his observations of the cultivation of vineyards in rows that were constantly hoed. He proposed a theory of plant physiology that encouraged a similar practice for any type of culture including wheat. He also claimed that land could be replanted with wheat every year and that manure was detrimental and should be replaced by ploughing and hoeing. The French admiration for Bacon, Newton, and English agriculture insured the success of his physical interpretation of plant development.

After 1715, a few French travelers visited England and accounted for their wonder at the richness of the countryside that they discovered, in stark contrast to the wretchedness of the French countryside.[14] To the contrary, French visitors such as l'abbé Le Blanc[15] compared England, in 1749, to a garden where all rural inhabitants were industrious and thrived in a fertile countryside, and noble landowners, living on their lands, were keen on agricultural improvement. He also hailed the action of the Royal Society. "It is this scholarly society that has made agricultural improvement fashionable. Its care, works, experiments have enabled Englishmen to learn what riches can be obtained from plants . . ."[16]

A few years later, the first volume of the treatise by Duhamel du Monceau, that presented the ideas of Tull as a contribution to the quest for the public good, triggered a large number of independent experiments by agricultural improvers. He reported their work in the subsequent volumes. He changed his points of view as a consequence of the diversity of communication and critiques that he received. Moreover, the first description, in France, of agriculture practices in Norfolk was published in the *Encyclopédie* in 1754. In 1758 Duhamel du Monceau had achieved a personal synthesis of his knowledge of agricultural practice in Norfolk and of the system of Tull that came to be known as the "New Agriculture." In spite of the flurry of publication that followed, however, the new agriculture did not introduce major changes in the practice of rural people who could not read or understand it.

The Dubious Success of the Chemists

Henri Léonard Jean-Baptiste Bertin (1720–1792), a minister of agriculture in the governments of Louis XV and Louis XVI, between 1759 and 1780, adopted the ideas of the New Agriculture and engaged in largely unsuccessful reforms.[17] Following the advice of Turbilly, he also promoted the creation of the societies of agriculture in the various regions of France on the model of the assembly of Brittany that was created by the Parliament of this province in 1756.[18] They were meant to allow practical problems to be discussed by people who were tilling the land, but they did not meet the expectations of the minister. Instead, they provided an outlet to the activities of agricultural improvers who were testing theoretical ideas rather than discussing practical issues.

The disappointing results of cultivation according to the principles of the New Agriculture led to its near abandonment by landowners between the years 1765 and 1770. After the departure of Bertin in 1780, the different domains that he controlled were disputed by several administrations. It allowed the entry of a number of agricultural improvers such as Antoine Augustin Parmentier,[19] Pierre Marie Auguste Broussonnet,[20] Louis Jean Marie Daubenton,[21] and Antoine Laurent Lavoisier,[22] the celebrated chemist, into public office. As a result, the grip of the New Agriculture and of the ideas of the physicist and chemists on public policy of the royal administrations tightened even though it had achieved very little change in practice in the country as a whole.

Agricultural improvers were too easily persuaded of the validity of their theories by short-term results of their practices. Yet, by contrary, they were to a large extent engaged in pragmatic improvements and most of them were doubtful of the possibility of a systematic approach to agricultural issues. So they presented themselves as men of practical experience as opposed to rocking chair reformers: "We are not looking for speculative thinkers who claim that without any direct knowledge of agricultural practice they can outline the methods and dictate the laws of husbandry from the depth of their study. Their systems resulting from their imagination only, lead astray those who trust them."[23]

In fact, they shared many views of the rural world, such as common views attributing plant illnesses to blight or blast that seemed to result from a fire that would have consummated the plant, or to honeyfall, or honeydew, or mildew that looked like an oil covering the plant and choking it. So the agriculturist reformers, without questioning the categories of perception of plant illness, attempted to isolate the chemical substances that were presumed to be the source of these illnesses and proceeded from their interpretation to propose methods of treatment. This is quite fascinating, because as early as 1705 Tournefort had identified mold as a mushroom and in the 1720s Jussieu had recognized that parasitic mushrooms were not the same plant as their host. The Italian botanists J. G. Fontani and G. Targioni-Tozzetti were, however, the first to publish in 1767 that parasitic microplants are at the origin of plant illnesses. This was underlined again in 1770 by Spallanzani, but the agriculturist reformers in France ignored this research. Plant illnesses were, however, a great concern of the public administration. Tillet, the director of the Royal Society of Agriculture, a member of the Academy of Sciences, was commissioned to study them and find a remedy in 1779, and he adopted a chemical paradigm for his research searching for a product to treat the seeds before sowing.[24] Men such as Lavoisier, Parmentier, Tessier, Lapostolle,[25] who played an eminent role in the orientation of unsuccessful efforts at agricultural reform directed to the increase of wheat production in France, were swept aside after the Revolution. This created an opportunity for a new discourse on the contribution of botany to the public good that developed during the early years of the republic and during the period of the empire, and definitely gave a sense of a new scientific start for horticulture and agriculture.

The Impact of the American Revolution

The admiration for English agriculture that was shared by a number of large landowners led to the adoption of English landscape gardening by a number of agricultural improvers such as René Louis Gérardin, who owned more than a thousand hectares of land at Ermenonville, or the duc d'Harcourt at Betz, who commissioned Hubert Robert to design his gardens. Most of these landowners were rather critical of French courtly life and emulated the life of great aristocratic landowners in England by living on their domains. They also used their landscape gardens to advertise for the brave new world that they hoped to achieve in a reformed monarchy (Fig. 6). Joseph-Antoine-Joachim Cérutti[26] (1738-1792) visited the gardens of Betz in 1785, and wrote a long poem celebrating the charms and the romance of the gardens as well as the enlightenment of their owner. He told of his admiration for the transformation of a marshy vale surrounded by dark inhospitable forests into an idyllic place where the landowner's mansion and the prosperous farm at the opposite end of the domain engage visitors into some philosophic thoughts. The sight of a hill crowned by a monument attracted his attention, and led him to the top where he

Fig. 6. Betz, view of the house of Harcourt in Alexandre de Laborde, *Description des nouveaux jardins de la France et de ses anciens châteaux; mêlée d'observations sur la vie de la campagne et la composition des jardins, par Alexandre de Laborde. Les dessins par Ct. Bourgeois. Description of the modern gardens and ancient castles in France.*, Paris, Impr. de Delance, 1808 [pl. 101]

found a modern pyramid on which it was written in golden letters "The American Independence." This discovery inspired him a further encomium of Betz, as the first place to praise the Republican revolution and to call for freeing mankind from tyranny. Following a spring he observed the variety of exotic trees and noted: "The greatest merit of English gardens lies in my own eyes in having doubled in a matter of a few years the number of foreign trees, and having enriched the European soil with plants from Asia, Africa, and the New World"[27] (Fig. 7). In fact, the role of France during the American Revolution made it possible for many French naturalists to come to America, to explore its botanical riches and to introduce a large number of trees in France. At the same time, other naturalists were sending more plants from all around the world to the Royal Botanical Garden. This triggered a great interest in naturalizing and improving many of these plants, in particular under the direction of André Thouin, already several years before the French Revolution.

Perspective on the Public Good at the Botanical Garden in Paris

Buffon[28] was appointed director of the Royal Botanical Garden in 1739, a position that Duhamel du Monceau coveted. The botanists whom he directed were not directly interested in agricultural issues. André Thouin, the young head gardener whom he later appointed and helped become a member of the Academy of Sciences, went in a different direction. He was himself the son of the previous head gardener of the botanical garden and very much interested in contributing to the common good, and became the first professor of agriculture at the Museum of Natural History after its creation in replacement of the Royal Botanical Garden. He shared many views with Duhamel du Monceau and the agricultural reformers, but he was trained in botany and plant physiology and he was not obsessed with the development of wheat cultivation. He was more concerned with the possibility of improving natural species and naturalizing foreign plants that could be useful for the country. He heartily welcomed the Revolution and even played a minor role in the revolutionary administration of Paris. He succeeded in avoiding the destruction of the Royal Botanical Garden and in having this institution transformed into the Museum of Natural History, still in existence. He saw the garden as a national resource where all the plants that could be useful or pleasant should be cultivated. He also considered that it was his responsibility to encourage the cultivation of these plants throughout the country and beyond in the entire world in an effort to increase the general welfare of all populations. He shared with a number of other naturalists the belief that this would bring an end to the wars of

Fig. 7. Rhus Tiphinum by Redouté in Duhamel du Monceau, *Traité des arbres et arbustes que l'on cultive en France en pleine terre.* Paris: Didot ainé, Michel, Lamy, 1800–1819 (Vol 2, # 47)

the European nations, as in a prosperous country where all plants were available no one would covet the products of another country.

André Thouin was twenty-three years old in 1770 when he started naturalizing exotic plants from seeds received at the Royal Botanical Garden in Paris from all over the world.[29] In 1773 he took the initiative of preparing two thousand bags of seeds reaped in the Botanical Garden and to send them as a countergift to all the donors and heads of botanical gardens he knew, noting that "since we are much richer (in plants) than all others, we should give back three times more than we receive."[30] In 1783 he wrote a proposal for the creation of a national nursery for the naturalization of exotic trees potentially useful in France,[31] and in 1786 he proposed the creation of an institution dedicated to agricultural research and education, with a group of traveling students who would have to travel throughout Europe and in other parts of the world with a similar climate. Exchanges with overseas correspondents suggested to Thouin the proposal to establish in each colony a Royal Botanical Garden in which all spice producing and medicinal plants as well as plants that yield other amenities would be cultivated, as a way of encouraging colonists to diversify their production rather than concentrate on quick profit derived from only one type of crop.[32] It also seemed possible to naturalize a few useful exotic plants in the south of France.[33] The same concerns led in 1793 to a proposal sent to the national convention to create a botanical garden in each department of France in

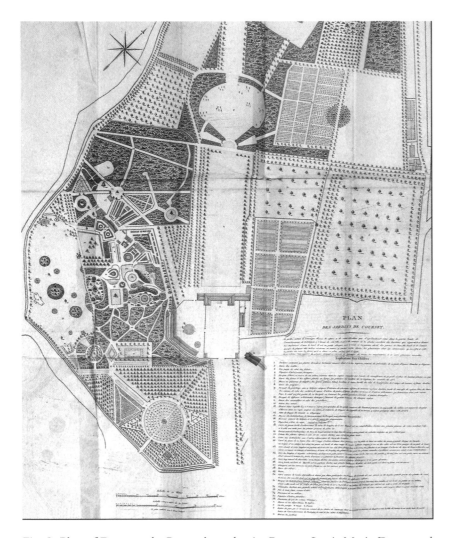

Fig. 8. Plan of Dumont de Courset's garden in Georges Louis Marie Dumont de Courset, *Le botaniste cultivateur; ou, Description, culture, et usages de la plus grande partie des plantes étrangères, naturalisées et indigènes, cultivées en France, en Autriche, en Italie, et en Angleterre, rangées suivant la méthode de Jussieu*: Paris, Deterville [etc.] 1811–14 (Vol 1, Pl. 1)

order to multiply and disseminate all the plants already cultivated in different parts of Europe and to start naturalizing useful plants from other parts of the world.[34]

Thus the interest for naturalization of exotic plants that had developed during the last decades of the monarchy among botanists kept growing during the revolutionary years. It was largely shared by devoted horticulturists like Georges Louis Marie Dumont de Courset, whom he encouraged by his policy of distribution of seeds.

Dumont de Courset and the Quest for Plant Variety

Du Mont de Courset was a dedicated philanthropist. He first published in 1784 a study of the plants indigenous to the regions between Boulogne and Calais, in order to guide local husbandmen toward a better agriculture and a more profitable economy.[35] He was thirty-eight years old.

In the preface to the second edition of his book, *The Botanist-Husbandman*, in 1811, he explains that he has added a plan of his own garden to satisfy a number of demands from readers of the first edition.[36] (Fig. 8) He then apologizes: "It is only the garden of a modest husbandman that does not compare, insofar as design or charm, with so many others created in France by skillful architects."[37] The gardens however offered a remarkable example of cultivation on a hilltop in the absence of a water source. They contained a very large number of plants that could be raised for their utility or their aesthetic qualities. They comprised no less than four thousand exotic plants, besides a large number of indigenous and already naturalized flowers to be found in many gardens[38] (Fig. 9).

(1) Parterre containing nine hundred foreign herbaceous perennial plants (comprising five to six hundred different species) surrounded by a ten-foot palisade of hawthorn and piceas;

(2) Maple trees;

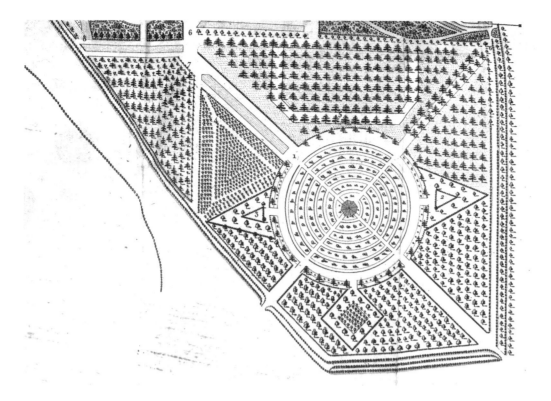

Fig. 9. Plan of the Parterre of foreign plants in Dumont de Courset, Le Botaniste Cultivateur . . . (Detail of Pl.1)
(1) Parterre of foreign plants with five to six hundred species; (2) Grove of maple trees; (3) Part of the genre of Ash Trees;
(4) Nursery of foreign trees; . . . (7) Quincunx of Tulip poplars . . .

(3) Part of the ash trees;

(4) Nursery of foreign trees;

(5) Winter grove made of two hundred resinous trees;

(6) Alignment of thirty species of ash trees and twenty-six species of medlar trees;

(7) Quincunx of tulip poplar, plane, linden, beech and alder trees with three rows of rose bushes on the nursery side.

His text begins with presentation of some ideas about the development of knowledge. Although he shuns the pursuit of hypothetical theories supposed to account for the origin of things because they cannot be verified, he suggests that knowledge of nature should strive to understand the processes of production of natural beings beyond their description. In that respect, he insists that plants cannot be understood unless one pays attention to the animals that help their reproduction or dwell on them, and that botanical knowledge should proceed from a classification based on the plant's organs of sexual reproduction that he borrows from Jussieu's Genera Plantarum. He gives a table of passage from the Linnean system to Jussieu's system of classification.[39]

He takes the *Encyclopédie's* authors to task for having separated their presentation of botany from the presentation of plant cultivation, and of the husbandry of shrubs and trees growing in the open, which forces readers to consult several volumes when they should have found all scientific and pertinent information about its culture and its uses under the name of each plant. This remark echoes similar complaints by André Thouin, that botany was given more attention than husbandry as if a part could be given more attention than the whole. He underlines that all the directors of public gardens are both botanists and husbandmen, and that their work calls on their understanding of the development, culture, and use of each plant.[40] He advocates the collaboration between the husbandman and a chemist in order to discover the chemical properties of the different parts of each plant in order to contribute to a scientific knowledge of their pharmaceutical use. He also demands close collaboration with a medical doctor, and he advises doctors to use the Linnean classificatory system in order to ascertain the plants that they prescribe from pharmacists, even though he notes that good doctors prescribe, with tongue in cheek, a number of innocuous plants to treat imaginary illnesses.

He develops several arguments in favor of the study of botany: first, it procures an infinite delight in the discovery of the variety of plants; second, it gives a sense of awe in front of the Creation that leads to the idea of God; third, the arts cannot compete with the organization of even the smallest living being, and with the beauty of the scenes created on a sunny day by plants in nature; fourth, looking for plants in nature keeps men healthy, and is conducive to simple, and virtuous mores; and, finally, it procures men with a satisfying self-image that is the very condition of true happiness.

His book comprises plants from thirteen hundred to fourteen hundred different genera, and eighty-six hundred species, and even more varieties. They include the largest number of known indigenous plants growing in France, as well as other plants cultivated in gardens at the time (1802), but he was limited in his presentation because his method demanded that he would cultivate himself any plant in order to study it before introducing it in his book.[41] He qualifies a plant as indigenous only when it grows between Amiens and Calais, as this work is a development of the first book that he produced in his *Mémoires sur l'Agriculture du Boulonnais* in which he wanted to study, first, the indigenous plants of the Boulogne region, and, second, only a few garden plants. He has, however, introduced in the second edition a copy of the catalogue of Kew gardens by Aiton (Hortus Kewensis), almost all the plants from Miller's dictionary, those described by Lamarck, those cultivated at the Museum of Natural History in Paris, and plants of different gardens (La Malmaison, Cels,' Hortus Schoenbruensis by Jacquin) or described by artists such as Redouté or Andrews, or botanists such as de Candolle.

His *Botanist Husbandman* pursued the same philanthropic goal as his first book on a larger scale. It was directed to people who could benefit the whole nation by their contribution to the development of useful plant cultivation. And he called forth the day when rhubarb, tea, and camphor trees would be naturalized and cultivated in open land in France. He insisted that truly devoted gardeners should not be led by a direct interest and short-term profit but, rather, by the hope that their efforts would benefit the nation and its future generations.[42]

He proposes to distinguish only two kinds of gardens: the kitchen garden and orchard that he conceives as a unit as they call for the same kind of cultivation, and the ornamental garden devoted to the cultivation of foreign plants that demands more experimenting because the specific demands of a foreign plant are rarely well known and the owner is bound to go through a lot of trials and errors. Thus, it is clear that the differences he makes between gardens do not result from considerations of formal appearance, and that he addresses an audience of gardeners. This is the most obvious in his treatment of the generalities about the cultivation of ornamental gardens devoted to the cultivation of foreign plants.[43] He advises whoever wants to gather a large number of newly arrived foreign plants in his garden to distrust the habits of gardeners he could hire, because they would blindly follows ideas that do not take into account the specificity of each foreign plant, and he recommends that the owner, having learned the major principles of botany, takes direct responsibility for planting, observing, and monitoring the growth of his plants.

He defers to the taste of a garden's owner in matters of spatial organization. In spite of his own appreciation of English landscape gardening, he did not take any stand on issues of garden design except practical ones that had a bearing on the growth of plants, arguing that everyone may take a personal view on the question of garden design. The gardening issues he sees fit to address concern the contribution of gardens to the public interest rather than to aesthetic disputes. He contrasts the disinterested attitude of the amateur gardener who develops a collection of foreign plants, to the mercantile attitude of most plant merchants.

However, the book is not written solely for the philanthropist who creates a garden with all his attention devoted to the improvement of the economy of his country. It caters also to the needs for better knowledge of a number of amateurs who are mostly interested in making their garden into a delightful sight. There is, however, such an insistence upon the understanding of the various conditions under which some plants may succeed or fail, that the reader is continuously invited to study the text in order to prepare himself to study the foreign plants he has selected for his garden, as in this passage: "One has to come back to trial and experiments, and to repeat them for oneself. The best theory is not by far as useful as practice, and even the general outline of its description must be subjected to the changes that the husbandman will provide according to the soil and the situation (of his garden.)"[44] Thus, he encourages his reader to adopt an experimental attitude, in the tacit hope that many of them will follow this advice, and meanwhile discover the pleasures of the pursuit of disinterested knowledge that he fully endorses: " . . . Maple trees, several ash and oak trees from America, the Tulip tree, should not only enter into the design of our gardens, but they should also embellish our forests with the diversity of their canopy, and give in some future day more material for the development of our arts."[45]

Dreams for a Better World at the Botanical Garden in Paris after the Revolution

This concern for naturalization of exotic plants was shared by many fervent botanist-gardeners who wrote to Thouin to let him know about their successes and publicize them. There were even polemical debates about the possibility of developing the cultivation of coffee, cotton, and sugar cane in France. The distribution of seeds organized by Thouin was meant as a deliberate effort to encourage the kind of local experiments of plant naturalization that people like Dumont de Courset performed. In 1810, for instance, Thouin sent 60,743 bags of seeds to botanical gardens devoted to public education, large landowners, nurserymen, botanists-husbandmen cultivating exotics, amateur gardens cultivating foreign decorative plants, small garden amateurs, and foreign botanists. In that same year, 350 bags of seeds were sent to a Mr. Black in New York, fifty to Thomas Jefferson landowner at Monticello, 253 to James Mease, the secretary of the agricultural society in Philadelphia, sixty-one to Martin Baker in Washington, and sixty to Bernard MacMahon, a botanist-husbandman in Philadelphia. These seeds were chosen among nine large groups of plants: cereals, vegetable roots, fodder, medicinal, oil-rich seed plants, tinctorial, picturesque and fragrant plants, trees, and assorted seeds for botanical education. As a whole, they belonged to 1,300 different varieties or species of plants. The plants for pleasure gardens accounted for only 125 species or varieties. This shows to what extent Thouin and the Botanical Garden in Paris were devoted to the development of useful cultivation, including the care for pleasure gardens.

This concern was reflected in a strange project by André's brother, Gabriel Thouin, for the enlargement of the Botanical Garden, published in 1820 in the *Plans Raisonnés de toutes les espèces de Jardins* (System of Plans for all Species of Gardens). It is presented in the book just after an ideal pharmaceutical garden, and just before an ideal garden for naturalization.[46] So it may be

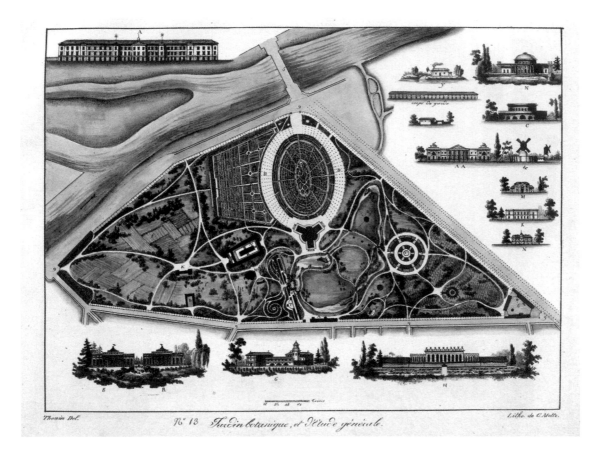

Fig. 10. Proposed enlargement of the Botanical Gardens in Paris in Gabriel Thouin, *Plans Raisonnés . . .* (Pl. 13)

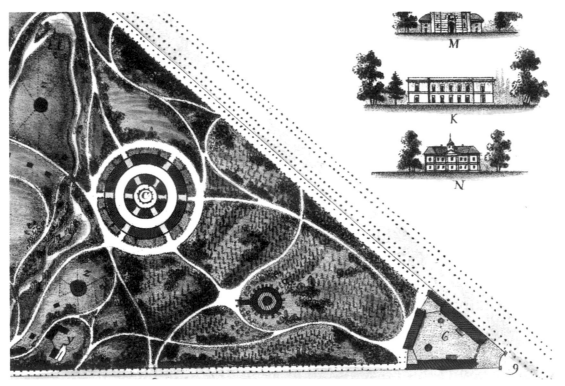

Fig. 11. Detail of the farm and vineyards, in Gabriel Thouin, *Plans Raisonnés . . .* (Pl 13)

Fig. 12. Experimental farm
in the tropics, in Gabriel
Thouin, *Plans Raisonnés . . .*
(Pl 51)

N.º 51 · Projet d'une ferme expérimentale de la Zone Torride

Thouin Del.

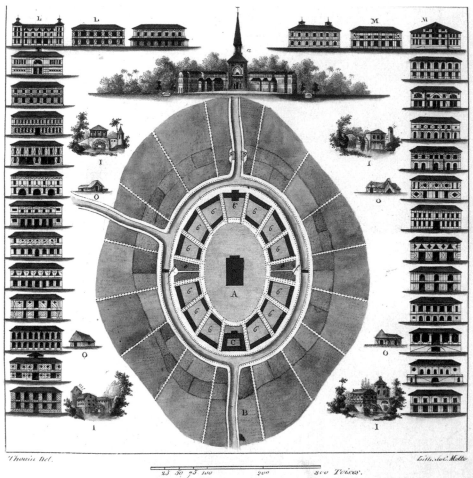

Thouin Del.   Lith. de C. Motte.

25 50 75 100   200   300 Toises.

N° 52. Île et Batimens de la Ferme Expérimentale.

Fig. 13. Detail of the Central Island, in Gabriel Thouin, *Plans Raisonnés . . .* (Pl 52)

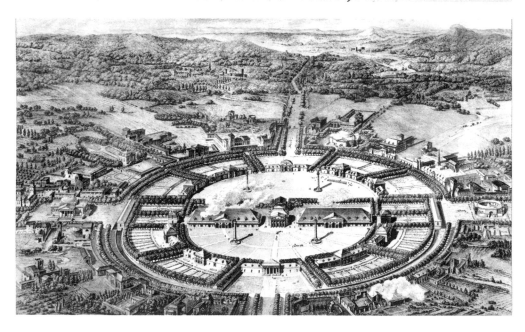

Fig. 14. Aerial view of the first plan for La Chaux de Fond by Claude Nicolas Ledoux in *Nicolas* Ledoux, *L'Architecture considérée sous le rapport de l'art, des mœurs et de la législation]* Paris, Lenoir, 1847 (Vol 2, 15)

Fig. 15. Aerial view of
the second plan for La
Chaux de Fond by
Claude Nicolas Ledoux
in *Architecture de C.N.
Ledoux. L'Architecture
considérée sous le rapport
de l'art, des mœurs et de la
législation]* Paris, Lenoir,
1847 (Vol 2, 16)

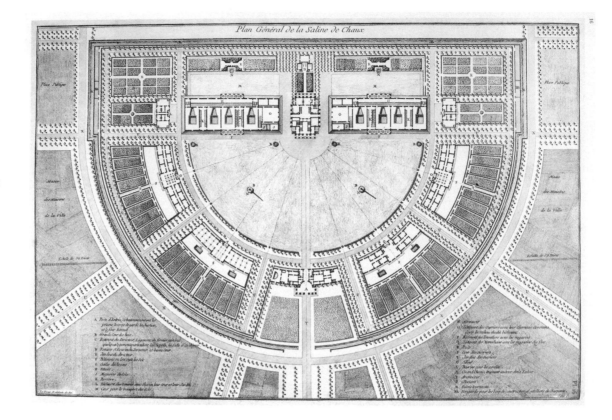

Fig. 16. Detail of the
mountainous farm
environment, in Gabriel
Thouin, *Plans
Raisonnés…* (Pl 52)

seen as an idealized view of the Botanical Garden itself (Fig. 10). The proposed enlargement comprised several existing buildings, and proposed a reorganization of the Botanical school, the nursery, the kitchen garden, and the seed beds as a well as of the collections of trees and flowers, in the center of the plan. At the westernmost corner of the 140-acre domain, he planned for a very large farm comprising a cowshed, a sheep, a barn, a dovecote, and a pigsty, close to a large place planted with all the kinds of vineyards existing in Europe (Fig. 11). It would be separated by a round building for fierce animals, from a very large meadow surrounded by an artificial river where all sorts of peaceful animals would be raised and could be seen by visitors. On the opposite side, to the east, he planned for large cultivated fields for the "grand cultivation of diverse kinds of grains," as well as fields for the usual medicinal plants distributed to the poor, as well as plants for dyeing and for spinning. Gabriel Thouin gives no explanation about the source of the intentions for this plan. By contrast, he claims that the program for an experimental farm for naturalization in a tropical country was given to him by his brother André (Fig. 12). This project located in an Arcadian valley seems to proceed from the same kind of will to create a botanical garden that would be a direct contribution to the improvement of the general welfare of the human race. Its panoptical plan, with farms located at various altitudes in the mountains around a central scientific establishment on an island in the center of the valley, constitutes the most achieved expression of the botanical utopia of the times (Fig. 13). It was obviously inspired by the plans proposed by Claude Nicolas Ledoux between 1773 and 1778 for the Salines at La Chaux de Fond (Fig. 14) an industrial complex with the house of the director at the center. A second project (1774-1778) shows only half the initial circular plan was conserved, but it looks even more like a panoptical device with lines of surveillance issuing from the director's house into the surrounding countryside (Fig. 15). Gabriel Thouin, who drew inspiration from but did not slavishly copy architects of his time, moved the house of the director to the border of the central place where the militia was supposed to exercise. Surveillance thus became a "public good" exercised by the citizens as a group. The project was, in fact, described at great length in the last chapter of the lectures on cultivation and naturalization by André Thouin, and the plan sketched by Gabriel follows very closely André's program. These lectures were posthumously published in 1827 by André's nephew. The idea, however, dated back at least to the end of the monarchy. It stemmed from a common observation that many botanists had made after Linneus had discovered that he could find, in the Alps, the same plants that he had discovered in Lapland, yet more numerous. In 1798, a professor of Natural History in Perpignan, close to the Mediterranean border with Spain, described the ease with which exotic plants could be naturalized in his region, and proposed to temporarily host North African plants on their way to the Museum. "I might be useful to the national botanic garden in Paris by receiving first the Spanish and North African plants that are sent there, and that die in this very long journey." This letter may have inspired André Thouin with the idea of the experimental farm that he presented in his "*Cours de Culture*" as located either in a tropical region or in a mountainous region in the south of France (Fig. 16). It also might have inspired his idea of a network of botanical gardens cooperating in an international project of universal naturalization of useful plants. He proposed that European governments cooperate in establishing, on their respective territories, naturalization grounds for the plants of both hemispheres. Plants from the arctic regions would first stay in a plot in Norway, then would move to Petersburg, from there to Vienna, and then to Paris, from where they would move to warmer regions. Plants from equatorial climates would first stay in the "Île de France," or Bourbon, and then in Madeira, and from there go to Nice before coming to Paris. In this way, governments would fulfill part of their imperative duty "to look after the needs and to augment as much as possible the welfare of their fellow-citizens."[48]

Thus, we can see that the utopia fostered by the developments of horticulture were directed not only to the improvement of the public good in France but also in the rest of the world. Thouin was extremely clear on this point. He expressed it very directly in a letter written to James Mease in Philadelphia on June 16, 1808:

> Do not be surprised by the package I sent last year to the botanical garden of your city in the absence of any demand from its management. Our institution aims at gathering, multiplying, and disseminating all the plants sparsely distributed over the different parts of the earth, that may be useful to the progress of sciences, arts and agriculture. And since the men who manage this institution belong to the great Republic of Letters whose members are all over the world, they hardly acknowledge the limits set by the governments for their own profit and consider as their brethren all men who devote themselves to education and who seek to improve the plight of other humans.[49]

Conclusion

I started with the assumption that the French Revolution had triggered a holy alliance of the experimental sciences that had transformed gardening into horticulture, and horticulture into a major contributor to agricultural development.

Instead we saw that, starting in the first half of the eighteenth century, men in very different social positions called on the experimental sciences to improve either horticulture or agriculture, depending on the attention they gave to the development of botanical knowledge itself. All of them shared the cultural optimism of the French enlightenment and cherished utopian dreams. However, they held quite different views of the contribution that scientific husbandry could make to the public good. Agricultural reformers mostly informed by physical and chemical theories hoped for covering France with wheat fields; horticultural reformers mostly informed by botany and theories of naturalization and grafting hoped for a broad diversification of the plant species living in France and saw gardens as their testing grounds. The French Revolution only shifted the control of official discourse from one group to the other. This was neither completely innocuous, however, nor impressive in terms of global impact. The monarchy attempted to push landowners to cultivate lands in fallow and failed. Napoleon's administration shifted its attention to botanical gardens, and created many in France and a few in other European countries. The dreams of the revolutionary botanists did not change, however, the poor overall development of kitchen gardening in France. It is only soon after 1823, when the occupying armies left the country and pleasure gardens in France quickly increased in number, that the horticultural society was created. These were different times. The number of landowners had increased as a consequence of the Revolution and, during the restoration, they indulged in the luxuries of vegetable and flower culture. Thus, they saw horticulture as a handmaid to the comfort of a rural domain. A new moralistic utopia then stimulated the growth of a conspicuous horticulture. "We strenuously hope for the day when each thatched cottage in France will be decorated with a bed of flowers. One should compare the moral state of populations among which the head of the family usually devotes to drinking his day of rest, with those where after attending the church he devotes the remainder of his Sunday to his flowers, his sweetest indulgence, a source of exchanges of neighborly support. This constitutes a powerful instrument for civilizing the countryside . . ."[50] The enduring theme of gardening as a handmaid to social control was set into place.

NOTES

1. Etienne Soulange Bodin (1774-1846) As a French cavalry officer during the Napoleonic wars, he is credited with a celebrated phrase after his return from Waterloo: "It would have been better for both parties that they stay at home to grow their cabbages." It is, nowadays, a common saying in France. He later devoted himself entirely to horticulture, created the Royal Institute of Horticulture at Fromont near Paris. He invented the Saucer Magnolia (*Magnolia soulangiana*) by crossing M. Yulan (now *M. denutata*) with M. purpurea (now *M. quinquipeta*.) This hybrid first flowered in 1827, was proven fertile, and has given rise to a large number of other hybrids.

2. Etienne Soulange Bodin, "Coup d'Oeil Historique sur les Progrès de l'Horticulture Française depuis 1789", in *Annales de la Société d'Horticulture de Paris* (Paris, 1835); 19-52.

3. The text follows: "They emulated the principles of the monarchy even in the symmetry of the bosquets . The gardens at Versailles, under the garb of a munificent ordering, were still revelatory of the Great King." Note the political interpretation of the XVIIth c. French garden art. Soulange Bodin, "Coup d'Oeil Historique" . . . (21).

4. Soulange Bodin, Coup d'Oeil Historique . . . (21).

5. Soulange Bodin, Coup d'Oeil Historique . . . (23).

6. Soulange Bodin, Coup d'Oeil Historique . . . (23).

7. Soulange Bodin, Coup d'Oeil Historique . . . (46,47).

8. André Thouin (1747-1824).

9. Vicomte Louis Etienne François Héricard de Thury (1776-1854), a mine engineer and an academician, who was known as an excellent agronomist. He presided until his death over the Royal Society of Agriculture, and was the founder of the Royal Horticultural Society of Paris in 1827.

10. See André J. Bourde, *Agronomie et Agronomes en France au XVIIIe siècle* (Paris: SEVPEN, 1967), 2, note 4. He quotes the Abbé François Rozier, *Cours complet d'agriculture théorique, pratique, économique, et de médecine rurale et vétérinaire, suivi d'une Méthode pour étudier l'agriculture par principes, ou Dictionnaire universel d'agriculture, par une société d'agriculteurs* . . . Paris, 1781- 1800 (T X).

11. See, for a more detailed account: Marie-Rose Simoni-Aurembou, *Parlers et Jardins de la Banlieue de Paris au XVIIIe siècle* Paris: Klincsieck, 1982.

12. Roger Schabol, *Dictionnaire pour la théorie et la pratique du jardinage et de l'agriculture* (Paris 1767).

13. Georges-Louis Leclerc Comte de Buffon (1707-1788) conducted important experiments in reforestation of his domain at Montbard. He was a highly regarded scientist, a member of the Academy of Sciences, the director of the Royal Botanical Gardens in Paris, and his fame reached its apex with the publication of his *Histoire Naturelle*, a forty-four-volume encyclopedia about the natural world.

14. All French travelers did not agree with the positive views of England that were so influential with the agricultural improvers as can be read in Josephine Grieder, *Anglomania in France 1740-1789: Fact, Fiction and Political Discourse* (Genève: Librairie Droz, 1985). Her conclusion proposes a survey of the changing impact of anglophilia and anglophobia in France during the period she studies (147-150). The anglophobia of many French travelers should not, however, lead us to ignore the importance of the admiration for England of noteworthy reformers, and prevent us from seeing the Enlightenment as a European cultural movement with its own contradictions.

15. L'abbé J.B. Le Blanc, *Lettres concernant le gouvernement, la politique et les moeurs des Anglais et des Français* (1749, Amsterdam) 3 Vol.

16. Le Blanc, *Lettres*, T II, (98-99)

17. Bourde, *Agronnomie et Agronomes* . . . (1080-1109) See also (1140)

18. Bourde, *Agronnomie et Agronomes* . . . (1106)

19. Antoine-Augustin Parmentier, an agriculturist, born at Montdidier, 17 August 1737; died in Paris, 13 December 1813. Left an orphan at an early age, he was compelled before taking a college course to become a pharmacist, in which capacity he joined the army of Hanover in 1757. Taken prisoner several times in the course of this service, he profited by his captivity in Prussia to gain knowledge, which he later put to valuable use. He resumed his studies, on his return to Paris in 1774, and was appointed pharmacist at the Hôtel-des-Invalides. At this time, he introduced the use of potatoes as food in France. He also promoted the improved cultivation of maize and chestnuts, and tried to reform the methods of baking. During the Revolution, he had charge of the preparation of salted provisions, and manufactured a sea biscuit. He wrote a number of books on horticultural and agricultural topics, which betray his lack of early education. André Parmentier (1780-1830), who attained distinction as a horticulturist in the United States, was a collateral relative.

20. Broussonnet (1761-1807) was a physician and natural scientist from Languedoc. In 1787, he created the Linnean Society of Paris with six colleagues and friends (Lavoisier, Redouté, Daubenton, Desfontaines, Fourcroy et Thouin). He is credited with introducing the first Ginkgo biloba in Montpellier and the first Broussonetia papyrifera as well as the Judea Tree in the Botanical Garden of Montpellier.

21. Louis Jean-Marie Daubenton was a natural scientist (Montbard, 1716 - Paris, 1800) He was a pioneer in studies of comparative anatomy. George Louis Leclerc Comte de Buffon enrolled him in the preparation of his momentous *Natural History* (1749-1788.)

22. Antoine Laurent Lavoisier (1743-1794), chemist, philosopher, economist; born in Paris, 26 August 1743; guillotined 8 May 1794.

23. Duhamel du Monceau, *Traité de la Culture*, . . . 1755 T IV, (135) quoted by Gilles Denis, "La Représentation de la Maladie des Plantes" in *Nature en Révolution*, edited by Andrée Corvol, 94–106. (Paris: Ed. L'Harmattan, 1993).

24. Mathieu Tillet, *Précis des expériences faites par ordre du Roi à Trianon sur la cause de la corruption des blés et sur les moyens de la prévenir ; à la suite, duquel est une instruction propre à guider les laboureurs dans la manière dont ils doivent préparer le grain avant de le semer* (Paris: imprimerie royale, 1787). Mathieu Tillet (1714-1791) was a physiologist. In 1755 he published an essay on what makes wheat grain turn black inside the spike and on ways of preventing it. He later became a member of the Academy of Sciences. He also published: *Expériences et observations sur la végétation du blé dans chacune des matières simples dont les terres labourables sont ordinairement composées, et dans différens mélanges de ces matières, par lesquels on s'est rapproché de ceux qui constituent ces mêmes terres à labour*. Tirées des Registres de l'Académie royale des sciences. (Paris: Impr. royale, 1774).

25. Alexandre-Ferdinand-Léonce, Lapostolle, *Sous les auspices de Mgr le Cte d'Agay, intendant de la province de Picardie. Plan d'un cours de chimie expérimentale, raisonnée et appliquée aux arts,* par M. Dhervillez, . . . et M. Lapostolle (Amiens: impr. de Vve Godard, 1777). Lapostolle, Alexandre-Ferdinand-Léonce, *Traité de la carie ou bled noir, dans lequel on prouve . . . que la chaux est le principal remède pour détruire cette maladie,* par M. Lapostolle (Amiens: impr. de J.-B. Caron l'aîné, 1787).

26. Joseph-Antoine-Joachim Cerutti, *Mémoire pour le peuple* (Paris, 1788), 63; Jean-Paul Rabaut Saint-Etienne, quoted in Mona Ozouf, "La Révolution française et la formation de l'homme nouveau," in *L'homme régénéré: Essais sur la Révolution française* (Paris: Gallinard, 1989), 116-157, at 125.

27. Joseph Antoine Joachim Cerutti, *Les Jardins de Betz, poème accompagné de notes instructives sur les travaux champêtres, sure les arts, les lois, les révolutions, la noblesse, le clergé, fait en 1785* (Paris, chez Desenne, 1792).

28. One hundred years before Darwin, Buffon, in his *Histoire Naturelle,* a forty-four-volume encyclopedia describing everything known about the natural world, wrestled with the similarities of humans and apes and even talked about common ancestry of man and apes. Although Buffon believed in organic change, he did not provide a coherent mechanism for such changes. He thought that the environment acted directly on organisms through what he called "organic particles." Buffon also published *Les Epoques de la Nature* (1788), in which he openly suggested that the planet was much older than the six thousand years proclaimed by the church, and discussed concepts very similar to Charles Lyell's "uniformitarianism," which were formulated forty years later.

29. He started corresponding with Linné in the same year. Yvonne Letouzey, *Le Jardin des plantes à la croisée des chemins avec André Thouin, 1747-1824* / [édité par] Yvonne Letouzey (Paris: Editions du muséum, 1989).

30. Letouzey, *Le Jardin des plantes . . .* (58).

31. Letouzey, *Le Jardin des plantes. . . .* (86–87).

32. Museum d'Histoire Naturelle, Manuscrit Ms 308, note of 12 January 1788.

33. Museum d'Histoire Naturelle, Manuscrit Ms 308, pièce 13.

34. There were at this date (an II) thirty-two botanical gardens in France, three in the colonies of the West Indies, and three in other European countries recently occupied by the French Republic. His proposal aimed both at protecting these gardens from being destroyed or ill maintained, and to make them into a systematic means for the improvement of agriculture. (Museum d'Histoire Naturelle, Manuscrit Ms 308, dated 1793.)

35. Georges-Louis-Marie Dumont de Courset, *Mémoires sur l'agriculture du Boulonnais et des cantons maritimes voisins par M.D.C (Dumont-Courset)* (Boulogne, F. Dolet : 1784).

36. Georges Louis Marie du Mont de Courset, *Le Botaniste Cultivateur, ou Description, Culture et Usages de la plus grande partie des Plantes étrangères, naturalisées et indigènes, cultivées en France, en Autriche, en Italie et en Angleterre, rangées suivant la méthode de Jussieu* (Paris: Chez Deterville et Goujon, 1811 première édition: 1802).

37. Dumont de Courset, *Le Botaniste Cultivateur,* . . . (page viii).

38. Héricart de Thury (186).

39. Dumont de Courset, *Le Botaniste Cultivateur,* . . . (503-52).

40. Dumont de Courset, *Le Botaniste Cultivateur,* . . . (18).

41. Dumont de Courset, *Le Botaniste Cultivateur,* . . . (29).

42. Dumont de Courset, *Le Botaniste Cultivateur,* . . . (83).

43. Dumont de Courset, *Le Botaniste Cultivateur,* . . . (80–108).

44. Dumont de Courset, *Le Botaniste Cultivateur,* . . . (149–50).

45. Dumont de Courset, *Le Botaniste Cultivateur,* . . . (153).

46. Gabriel Thouin, *Plans Raisonnés de toutes les espèces de Jardins* (Paris, Lebègue 1820). See a pharmaceutical garden, plate 12, and a garden for naturalization, plate 14.

47. I thank Richard Etlin for calling my attention to the parallel between Les Salines de Ledoux and Gabriel Thouin's design for the experimental farm. For a detailed presentation of the project and its utopian aspects, see Anthony Vidler, *Claude Nicolas Ledoux, Architecture and Social Reform at the End of the Ancien Regime* (MIT: MIT Press, 1990).

48. André Thouin, *Cours de Culture et de naturalization des végétaux, publié par Oscar Leclerc-Thouin* (Paris: Huzard, 1827) (459).

49. Museum d'Histoire Naturelle, Manuscript letter, dated 16 June 1808.

50. Ysabeau et Bixio, *La Maison Rustique du XIXe siècle* (Paris : Libraire Agricole, 1845), Tome V Horticulture, avant propos ij.

Botanical Progress, Horticultural Innovations and Cultural Changes

Botanical Progress, Horticultural Innovations and Cultural Changes

Botanical Progress, Horticultural Innovations and Cultural Changes

Botanical Progress, Horticultural Innovations and Cultural Changes

From Practice to Theory: The Emerging Profession of Landscape Gardening in Early Nineteenth-Century America

Therese O'Malley

In 1849, Sidney George Fisher, a Philadelphian gentleman, noted in his diary a complaint about the rural architect, A. J. Downing. He wrote: "Landscape gardening with him is a profession & not a liberal taste, and he talks with a professional air, I dislike artisanship, and the smell of the shop destroys my pleasure in any subject however interesting in itself."[1]

Downing was, by this time, the preeminent American practitioner and theorist of landscape and garden design (Fig. 1). The *Magazine of Horticulture* in 1841 claimed that "Indeed, we believe he is the only person at present in the country, who is consulted professionally as a landscape gardener."[2] He had elevated the field to the status of a fine art in America and disseminated the taste for rural architecture and ornament throughout the antebellum states.

By 1841, Downing himself wrote: "Landscape Gardening bids fair to become a profession in this country."[3] Yet Sidney Fisher spoke of such professionalism with disdain. Why was Downing so sharply criticized for his achievement by this gentleman who was a person of refined taste himself? This was from a man who just a few years earlier, when he still only knew Downing through his publications, had entered a note in his diary of quite a different tone:

> In the morning read some chapters of Downing's *Landscape Gardening,* a very well-written & most useful book. It exhibits taste and a true feeling for nature. . . . This book has had immense influence throughout the North in introducing a superior style of adorning country residences.[4]

After a trip to upstate New York in 1846 Fisher wrote:

> Nice lodges at the gates, graveled drive winding thro the park clumps, groves, masses of noble trees . . . in short, every appearance of wealth & good taste. The influence of Downing's books is seen everywhere in buildings &

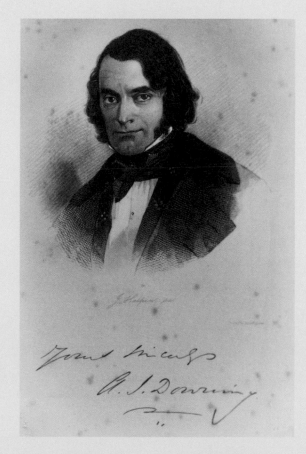

Figure 1. J. Halprin, Portrait of A.J. Downing in *Horticulturist* November 1852 (private collection)

grounds. He has done a vast deal of good in reforming the style of country residences and suggesting new & beautiful embellishments.[5]

Fisher finally came into direct contact with Downing in 1847 when his brother, Henry, decided to build a new house outside Philadelphia, in the Elizabethan mode, as promulgated by Downing's publications. Fisher wrote:

> He will have a beautiful place. He has had Downing, the author of books on landscape gardening & cottage architecture, to look at the place & advise him about the house & grounds. He liked Downing very much. It certainly is an indication of some advance in refinement that a "landscape gardener" can find employment, & constant, profitable employment, in this country.[6]

It was only when Fisher met Downing personally at a party in Philadelphia that his opinion cooled. He described the gathering:

> Downing, the writer on landscape gardening, there also. Had a good deal of talk with him on the subject . . . & I think gave him one or two new ideas. I like his books better than himself. He is a Yankee & not a thoroughbred.[7]

Later that year, Fisher noted in his diary, "His charge is $20 per diem."

It is curious that of all people, Downing, whose whole being was directed toward the refinement of American society, should be the target of condescension against his profession. A money-loving Yankee was a phrase used derisively by those who featured themselves above business concerns.[9] Fredrika Bremer, an ardent follower of Downing, however, wrote that discussions of money "would never defile the lips of . . . Mr. Downing," in spite of his dedication to his new profession.[10]

What had transpired in the first half of the nineteenth century in the development of landscape architecture in this country to generate this uneasiness with garden design as a profession? Was this a deep-seated resistance to the practice because of its ties to manual labor? Or was this occupation more appropriately considered a noble pursuit or "liberal taste," as Fisher called it, so refined that it should be considered a fine art and not regulated by a fee schedule? At this point in the nineteenth century, landscape gardening was caught somewhere in the middle between a vocation and an avocation, and remained stuck there for a long time.

For it was not until the end of the century that the American Society of Landscape Architects was founded (a full half-century I might add, after the American Institute of Architects); A. J. Downing was in an unstable, sometimes contradictory, position. Sidney George Fisher expressed a dislike shared by many who loathed the mixing of aesthetics and money-making.[11]

As his obituary stated, "Mr. Downing first

Figure 2. David Kennedy, McAran's Garden Arch at Seventeenth Street 1840, Historical Society of Pennsylvania, Phildelphia

claimed our attention as a practical Horticulturist and Nurseryman."[12] It seems useful on this occasion of a symposium on horticulture to trace the origins of the emerging profession of landscape design in this country, to the achievements of its predecessors, the nurserymen. I hope to show how the manner in which nurseries were promoted and business conducted paved the way for the profession of landscape gardening to take root. Struggling to assert a new identity in the face of increasing literacy and demand for "mental culture," the nurserymen, like so many contemporary trades and businesses, learned to exploit the

Taken on the Spot and lithographed by A Hoffy, Phil.ᵃ 1847.
South View of **THE OLD LANDRETH NURSERIES**, Philadelphia.

Figure 3. A. Hoffy, South View of Old Landreth Nurseries, Philadelphia, 1847 Free Library of Philadelphia

commercial possibilities of refinement and taste. With this choice came the inevitable tension between truly genteel behavior and business.

That great figure in English landscape gardening and prolific author, John Claudius Loudon (1783-1843), claimed the scarcity of nurseries was the reason for America's lack of a rich landscape tradition. He wrote in a brief history that:

> Four or five public nurseries are all that are recollected of note existing in the states in 1810, and these were by no means profitable establishments. About 1815, a spirit of improvement in horticulture as well as in agriculture began to pervade the country, and the sphere of its influence has been enlarging and the force of example increasing down to the present time. . . . *Nursery establishments in America,* Mr. Buel observes, are increasing in number, respectability and patronage.[13]

Loudon described numerous nurseries in New York, among others that of Messrs. Downing and Co., at Newburgh: "In the city are the extensive seed establishments of Messrs. Thorburn and others. . . . At and near Philadelphia are Bartram's botanic garden, now the nursery of Colonel Carr, and accurately described by his foreman, Mr. Wynne. . . . Messrs. Landreth and Co.'s nursery; and that of Messrs. Hibbert and Buist."[14]

Because gardens were well established as areas of refinement, the nursery garden capitalized on the sense that it was where polite society met to walk, talk, and learn. They attracted the best customers, offering constant reassurance of one's gentility. The desire to exploit that need commercially was a major force in the growth of the profession. As William Wynne noted when speaking of Bartram's Garden: "This garden is the regular resort of the learned and scientific gentlemen of Philadelphia."[15]

In order to enhance the illusion of gentility, the nursery establishment had to be an acceptable location for ladies to visit. Flowers, whose delicacy and beauty reflected the refined viewer's own character, played a huge part in this campaign to attract women as clients.[16]

How did the nurseryman achieve this new character of respectability? I would like to suggest they did it by embracing two

Figure 4. Map of Mr. Andrew Parmentier's Horticultural and Botanic Garden, at Brooklyn, Long Island, Two Miles From the City of New York, c. 1828, Brooklyn Historical Society, Brooklyn, NY

Figure 5. Rustic Prospect-arbor, A. J. Downing, *Theory and Practice of Landscape Gardening* (1849), p. 460, National Gallery of Art

powerful vehicles for the transmission of prestige or refinement: gardens and literature. Elsewhere, I have shown gardens were not only where the socially prominent but also where the intellectual leadership gathered.[17] To name a few Philadelphian sites, certainly John and William Bartram's Garden, Lemon Hill Belmont, and Belfield were prime examples. The great private collections at the estates of William Hamilton, Woodlands, and Nicholas Biddle, Andalusia, were displayed with significant art and book collections. As others have noted, books have long had a place in refined households.[18] Together, these exemplars of refinement, gardens and books, were utilized in the promotion of nurseries. Downing's later achievement was possible because of the early nurserymen in America who wrote essays, treatises, and catalogues in the service of educating a new class of consumers. And in many cases, they also maintained botanic gardens and well-groomed nurseries that exemplified the breadth of their collecting and depth of knowledge as well as taste. Their gardens were not simply storehouses but became, as Richard Bushman has called them, elevated environments.[19] And they taught by example, based on the currently entrenched notion that emulation was the key to improvement.[20]

These demonstration gardens acted as sites to stimulate and instruct those who might wish to acquire the taste and knowledge of rural improvement.[21] As Loudon put it, "Some commercial gardens . . . are the resort of many of the citizens of Philadelphia, more especially the gardens of M. Arran, and M. d'Arras; the first having a very good museum, and the latter a beautiful collection of large orange and lemon trees . . ."[22] (Fig. 2).

The various functions of the nursery in the antebellum period were indicative of the multivalent role of the nurseryman. He served not simply as a purveyor of plant material but also as a conveyor of style and scientific information. We know that students of Benjamin Smith Barton, professor of Botany at the University of Pennsylvania, held classes in Landreth's nursery

(Fig. 3). At M'Mahon's Nursery, seeds and plants brought back by the Lewis and Clark's and Nuttall's expeditions could be studied. Propagation at public nurseries made new plants available for documentation and study, and also, for the first time in many cases, made many native plants available to the public for sale.

The nurseryman for whom Downing has the highest praise and the only one he said practiced landscape garden design with any sophistication was André Parmentier. Downing dedicated several paragraphs in his *Treatise* to this nurseryman whose business in Brooklyn, New York, was known as the Botanic Garden and Nursery (Fig. 4). He wrote, "We consider M. Parmentier's labors and examples as having affected, directly, far more for landscape gardening in America, those of any other individual whatever."[23]

Parmentier, Downing claims, was almost constantly applied to for plans for laying out the grounds of county seats, by persons in various parts of the Union, as well as in the immediate proximity of New York. In many cases he not only surveyed the property to be improved but also furnished the plants and trees necessary to carry out his designs. Laid out by his own hands and stocked from his nursery grounds, he arranged the hardy trees and shrubs that flourished in this latitude in classes, according to their height and growth habits, and published a short treatise on the superior claims of the natural, or modern style, over the geometric, or ancient style, of laying out grounds.

One of Parmentier's most important landscape designs was at Hyde Park, New York, for the botanist Dr. David Hosack (1769-1835), founder of the Elgin Botanical Gardens in New York City (1801). An early description of Parmentier's Brooklyn garden from the *New England Farmer* in 1828 gives some sense of the high standards he promoted:

> "It contains many beautiful plants exhibited with the same tasteful arrangement which characterizes the whole of Mr. Parmentier's establishment: even the method in disposing the pots according to some principle of grouping or contrasting the color and size of the flowers, entertains the eye, and shows the variety of ways in which a skillful gardener may distribute his materials to produce picturesque effect. He will communicate to gentlemen who wish to see him, a collection of his drawings and cottages, rustic bridges, Dutch, Chinese, Turkish Pavilions, temples, hermitages, etc."[24]

In 1828, Parmentier's own article entitled "Landscapes and Picturesque Gardens" appeared in Thomas Fessenden's *New American Gardener,* in which he addressed the link between landscape and refinement. "Where can we find an individual, sensible to the beauties and charms of nature, who would prefer a symmetric garden to one in the modern taste; who would not prefer to walk in a plantation irregular and picturesque rather than in those straight and monotonous alleys, bordered in mournful box, the resort of noxious insects? Where is the person, gifted with any taste, who would not choose those alleys that wind without constraint."[25]

Nurserymen adapted the language of gentility in the promotion of their trade and sites of business, emphasizing such terms as sensibility, gift of taste, of cultivated gentlemen and ladies. The reference to intellectual improvement or cultivation of the mind paralleled deliberately the notions of cultivated and improved grounds.

Parmentier's garden also was featured in J. C. Loudon's *Gardener's Magazine*, in an article that claimed his garden should "be looked upon as an epoch in the history of American horticulture . . . In short, this establishment is well worthy of notice . . . it shows the art of laying out a garden so as to combine the principles of landscape-gardening with the conveniences of the

Figure 6. Lewis Miller, "Botanic garden," Princeton, New Jersey, September 9, 1847. York County Heritage Trust, York, Pennsylvania

Figure 7. unsigned, Elgin Botanic Garden, New York, c. 1815?, New York Botanical Garden, Gift of Rebecca Harvey, 1902.

nursery or orchard."[26]

Downing, in his *Treatise* cited this garden again in his chapter on embellishments and illustrated the rustic prospect arbor, or tower, which was situated on the extremity of Parmentier's place (Fig. 5). "It was one of the first pieces of rustic work of any size and displaying and ingenuity that we remember to have seen here."[27] Downing explained why such ornament was important, "To place before men reasonable objects of ambition, and to dignify and exalt their aims, cannot but be laudable in the sight of all."[28] Parmentier's establishment had it all: taste, style, pleasure, it offered scientific instruction and practical knowledge. Downing later would change the name of his business to Botanic Nurseries for the same reason. The added scientific and educational character enhanced social value in this era of self improvement (Figs. 6 and 7).

Although Parmentier's essay in the *New England Farmer* was one of the first learned discussions on the naturalistic garden to be published in America, an earlier nurseryman and author Bernard M'Mahon (ca. 1775-1816) included a long discussion of laying out gardens in the modern taste and the most approved plans in his book, *American Gardener's Calendar* (The Pleasure of Flower Garden) in 1806, to which is appended a catalogue of seeds, books and tools sold by him.[29] Also in 1816, John Haviland, architect (1792-1852) discussed the basic principles of landscape garden design in his book *The Builder's Assistant*.

One of Landreth's employees, Robert Buist (1805-1880), began his own nursery business with Thomas Hibbert in the 1830s in Philadelphia (Fig. 8). He also was the author of at least three books that served as inspiration if not direct sources for Downing. For example, *The American Flower Garden Directory* had a chapter "On Laying Out a Flower Garden," that was very close to Downing's publications. The Buist firm seems to have impressed Downing in the struggle to establish themselves in their new role as educators, as well as businessmen; he wrote:

Look at the Buists who import hundreds of things that they never get a cent for. . . . A gardener must have zeal, taste, and money to spare, before he can set about leading and improving the popular taste in arboriculture.[31]

In 1839 Buist addressed his new book to a burgeoning clientele: "A work like this present has been a desideratum to aid the very rapid advancement of the culture of flowers among the intelligent of our flourishing republic."[32]

A visit to the Nursery of Thorburn & Co. in Astoria, New York, by "a subscriber" in the *Horticulturist* praised the polished look of that refined establishment (Fig. 9).

> The grounds are by no means like a nursery and would not be taken for one by the stranger . . . entering the front gate you imagine yourself into the grounds of a private gentleman who had been affected with a monomania for Dahlias. We were perfect strangers both to the grounds and proprietor. But this makes no difference here; all are welcome."[33]

Another description of a nursery visit, this one to a nursery in New Jersey refers to the establishment as equal to "any ornamental garden," that is, noncommercial, private pleasure ground:

> A Polished Nursery. The most neatly kept nursery of fruit and ornamental trees that we have ever seen in this country, by all odds, is that of WM. REID, of Elizabethtown N.J. It occupies about thirty acres; and every portion of it appears to be as smoothly combed and brushed, as the most finished parts of other people's grounds. The broad alleys used as cart tracks, and for turning about the horse which cultivates the rows, are smoothly covered with a beautiful turf, kept closely shaven by mowing once a fortnight, and the edges are kept as smoothly trimmed as the walks of any ornamental garden.[34]

It might be argued that the influence of nurserymen in the development of landscape gardening in this country resulted in an emphasis on the plant material rather than on the broad sweep of the park. The smaller scale of gardens in this country as opposed to England also contributed to this trend. Both Humphry Repton and nurseryman John Claudius Loudon had moved in the direction of horticultural diversity, influenced by advances in the availability and increasing taste for rare and exotic

Figure 8. A. Hoffy, View of Robert Buist's City Nursery and Greenhouses, 1846, Library Company of Philadelphia

Figure 9. Title-page in Grant Thorburn, *Catalogue of Kitchen Garden, Field, and Flower Seeds,* 1822, New York Botanical Garden

Figure 10. A. J. Downing's residence, Newburgh, NY, A. J. Downing, *Theory and Practice of Landscape Gardening* (1849), p. 398, National Gallery of Art

plants. Advances in hothouse cultivation and the evolution of glasshouse architecture supported this trend. The term *Gardenesque,* coined by Loudon to describe a new style that emphasized individual specimens and exotics, strongly influenced nurserymen's practice, for obvious reasons: it was their trade to promote plant material.[35] Their writings and their showcase gardens created a demand they met readily with supply. The horticulturally specialized focus continued in Downing's work, not without notice. William Saunders (1822–1900), who worked in Baltimore and then Philadelphia, was critical of Downing, who he said was a good horticulturist but weak in his practical knowledge of site planning and architecture.[36]

A. J. Downing

Downing was born in 1818 in Newburgh, New York, into a family that owned a successful nursery in the Hudson Valley. He joined his brother Charles in running the family business bequeathed to them by their father, bringing it to national attention as the Botanical Gardens and Nurseries of Newburgh (Fig. 10).

Downing's first article appeared when he was still a teenager.[37] Although most of the early essays were botanical in subject, by 1835, he began to discuss broader issues of taste and the branch of horticulture called landscape gardening. In a seminal essay, entitled "Remarks on the Fitness of the different Styles of Architecture for the Construction of Country Residences, and on the Employment of Vases in Garden Scenery," Downing argued that the idea of adapting the style of architecture to the character of the scene in which it is placed, "is a point quite lost sight of in works upon architecture (and which, as there is no practice of landscape gardening in this country, may be very properly inculcated through a medium like the present)."[38] Thus, it is through the medium of horticultural and nursery periodicals, such as Hovey's *American Gardeners Magazine*, that the theory of rural architecture was first regularly promoted. In this article, Downing elucidates principles that underlie his writings for the rest of his career. They include fitness to the end desired and harmony of expression with the landscape. In order to achieve this, he argued, there can be no slavish imitation of European art. This is where he began to enunciate his independence from Europe, a theoretical stance that perhaps earned him the charge by Fisher of being a Yankee. He was clearly speaking on behalf of the landed proprietors in the Northern and Middle States, where he claimed "the march of civilization and the refined stages of society" saw rapid movement.[39]

In addition to Downing's numerous articles in all the horticultural and agricultural periodicals of the day, (in *Horticultural Register* and *Gardener's Magazine, Hovey's American Gardener* Magazine and *Register,* and *American Gardener's* Magazine and Register). Downing, with botanist Asa Gray, brought out the first American edition of John Lindley's *Theory of Horticulture,* and

4 *Advertising Sheet of Mag. of Horticulture.*

ADVERTISEMENT.

PROFESSIONAL LANDSCAPE GARDENING.

The numerous applications and inquiries made of the undersigned respecting his professional terms in giving designs for laying out grounds, in forming new residences, or in embellishing or improving old ones in accordance with correct principles of taste, have induced him to publish his list of prices, as below. It is proper, also, to state, that when extensive improvements are required, in places at a distance, within a certain time, it is desirable to receive notice by letter, a week or two in advance, as, otherwise, previous engagements might give rise to disappointments.

For a design for the grounds of a villa residence, (the survey, or plan of the place, in its existing state, being furnished) . . .	$50 00
For the same, with the working drawings, details of ornamental structures, &c. complete, and the principal outlines staked out on the ground, and lists of trees and shrubs furnished . . .	100 00
A design for the grounds of a cottage residence	30 00
For the same, with details, working drawings, lists of trees, &c., &c.	60 00
*** The travelling expenses are to be added to the above, when the place to be visited is more than fifty miles from New York.	
A design for a flower garden	10 00
A sketch, with working drawings for any ornamental garden building of rustic work, a gate lodge, gardener's house, or green-house	15 00

When a professional visit is required at any place, to give an opinion, to suggest, superintend, or carry out any improvements in the buildings, garden, or grounds, the charge is $15 a day (of six hours,) while at the place, and travelling expenses.

A. J. DOWNING.

Botanic Garden and Nurseries,
Newburgh, N. Y., June 1, 1842.

Garden, Grass, and Vegetable Seeds, Agricultural Implements, &c.

We are agents for Hovey's Patent Hay Cutter, and Vegetable Cutter, two of the most superior machines yet invented for those purposes.

Seeds, tools, plants, &c. packed so as to be safely transported to any part of the country.

Catalogues of Seeds, Bulbs, Garden Tools, Books, &c., will be furnished *gratis*, on application.

HOVEY & CO.

March 1, 1842.

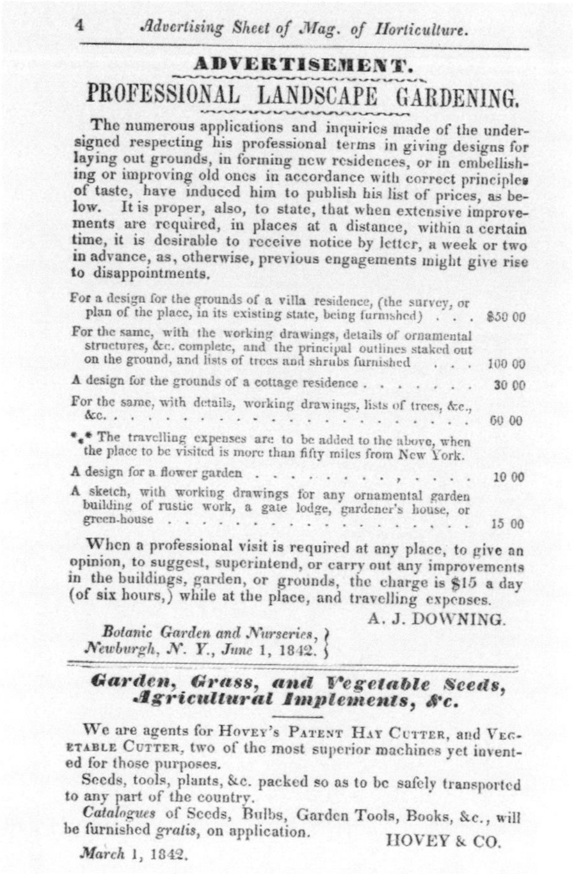

Figure 11. A. J. Downing, Advertisement in Hovey's *Magazine of Horticulture* 8, Library of Congress

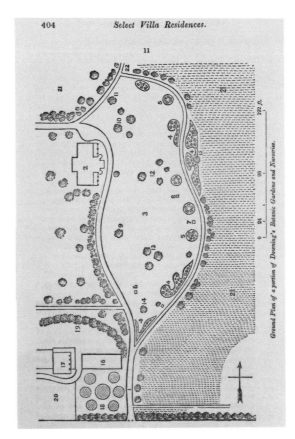

Figure 12. Ground Plan of a portion of Downing's Botanic Gardens and Nurseries in *Magazine of Horticulture*, vol. 7 (November 1841), p. 404, Library of Congress

his *Natural System of Botany*. This was part of an effort to apply science to American agriculture and horticulture and bring American botany abreast of current European theory.[40]

Downing was a frequent contributor to the *Magazine of Horticulture,* a journal that lasted thirty-four years, edited by Charles M. Hovey, of the famous Cambridge, Massachusetts, nurseries. Charles and his brother P. B. Hovey had the most extensive nurseries and plant houses in New England. Before he had his dedicated journal, Charles Hovey, himself, published in the *Silliman's Journal of Arts and Sciences.* Within a few years of his first article appearing in Hovey's *Magazine of Horticulture*, Downing placed an ad for Professional Landscape Gardening in which he published his fee scale (Fig. 11).

What becomes clear in this brief overview is the interconnectedness of members of the emerging landscape profession and the related fields of horticulture, collecting, publishing, natural history, and architecture as well. Whether they were involved with the same projects or publications as collaborators, they seemed to support each other professionally and socially. The paths of these major figures crossed and re-crossed throughout their lives. They formed friendships, influenced each other both through their art and personally, sometimes competed for clients, and sometimes became bitter enemies. I am only touching on very few of their professional encounters and interactions to illustrate how they affected each other's work.

Figure 13. Humphry Repton, des. Trade Card, 1788 From Repton's *Wilton Red Book* (1801) George A. Smathers Libraries, University of Florida

Downing and Hovey's careers are particularly intertwined. Hovey published an eleven-page description of the residence of A. J. Downing, Highlands, which exemplified the principle of emulation as the means of improvement (Fig. 12). He begins the description of Highlands with a historical note about George Washington's headquarters, in lamentable disrepair, nearby. The article concluded with a proof of the powerful effect of this exemplary site:

Since the erection of Mr. Downing's house, and the publication of his work, he has had numerous applications from gentlemen of high standing, for his assistance in laying out gardens and grounds. . . . We are glad to learn from Mr. Downing, that his *Treatise* on the subject has succeeded beyond his expectations, and we hope it will be the means of awakening the public taste to the importance of art.[41]

Hovey developed a pattern of features that Downing would repeat in his *Horticulturist,* which included visits to various gardens throughout the states. These descriptions serve the garden historian as an invaluable source of information on American gardens, how they were perceived, who visited them and details of planting schemes, creation of views and vistas, and architecture. They also present ideas on taste, fashion, and polite manners that may not be explicit elsewhere. A typical title is: "Select Villa Residences, with Descriptive Notices of each; accompanied with Remarks and Observations on the principles and practice of landscape Gardening: intended with a view to illustrate the Art of Laying out, Arranging, and Forming Gardens and Ornamental Grounds."

Downing's 1845 treatise *Fruits and Fruit Trees of America* was praised during his short life time and became a standard reference.[42] He was at the top of the pomological game so to speak when, in 1847, Downing announced his decision to leave the nursery business:

I have sold out my nursery interest . . . I shall now devote my time to literary pursuits altogether, and my home grounds, as the nursery stock is gradually withdrawn, to experimental purposes.[43]

Thus the literary pursuits and the scientific gardening, once added to enhance his nurseryman profession, now became Downing's primary occupations.

The Emergent Field

Scholars have written about professional landscape gardeners such as Humphry Repton and J. C. Loudon and their struggles as they tried to forge a career in England in a highly competitive field (Fig. 13).[44] In America, throughout the first of the century, landscape gardeners and designers struggled to create even a viable field, one in which they could make a living. Just two examples will serve to illustrate the situation: William Birch, well known as a painter and engraver of scenes around 1800 in Philadelphia, was also a trained landscape designer from England. His garden at Springland, a country house that he built to be a showcase of his landscape design, failed to bring him the commissions needed to establish a practice (Fig. 14).[45]

In New England, Henry A. S. Dearborn (1783-1851) provides a second example of efforts to promote training in the field of landscape design. A cemetery designer and horticulturist, he wanted to address the lack of nurseries and develop profitable ornamental plants in the region. He oversaw the initial landscape design of the first rural cemetery,

Figure 14. William Russell Birch, The Grove in Springland, before 1805, The Barra Foundation, McNeil Collection, Wyndmoor, DE

Mount Auburn in Cambridge; one of the most important monuments in American landscape history (Fig. 15). Dearborn failed to make this burial ground a site of experiment and education. He had proposed "an Institution of Education of Scientific and practical Gardeners," a school of horticulture and design, which would teach a science and art (combining "Natural History and Physics, Botany, Mineralogy, Hydraulics, Mechanics, Architecture, Chemistry, and Entomology").[46] These are just two instances of failed attempts to launch a profession.

There were personal conflicts as well. Downing called Gervase Wheeler (ca. 1815-1870), author of a villa treatise, a "pseudo architect from abroad," and his book, "un-American."[47] Downing also wrote in the 1844 edition of his *Treatise* called "Notes on Professional Quackery;" railing indirectly against Hans Jacob Ehlers (1804-1858), a competitor who had emigrated from Denmark. He has an interesting choice of words, quackery, borrowed from another current problem caused by those practicing medicine without training. He wrote:

> Landscape Gardening, alike all other arts, is not free from ignorant pretenders to knowledge . . . We have seen one or two examples lately where a foreign *soi-disant* (self-proclaimed) landscape gardener has completely spoiled the simple grand beauty of a fine river residence . . . In this case he only followed a mode sufficiently common and appropriate in a level inland country, like that of Germany, from whence he introduced it, but entirely out of keeping with the bold and life-like features of the landscape which he thus made discordant.[48]

Ridiculing newcomers or foreigners was a common tactic taken by these newly established professionals, one that suited well those men promoting native trees and shrubs over exotics. But it was an unpleasant theme that appeared routinely in their

Figure 15. Thomas Chambers, Mount Auburn Cemetery, Mid-19th century, National Gallery of Art

writings. Downing wrote in his 1852 essay on "American versus British Horticulture" in the *Horticulturist*:

> Anybody who knows the effect of habit and education on character, knows that it is as difficult for an Irishman to make due allowance for American sunshine and heat, as for a German to forget sour-krout, or a Yankee to feel an instinctive reverence for royalty . . . Rapid as the progress of horticulture is at the present time in the United States, there can be no doubt that it is immensely retarded by this disadvantage, that all our gardeners have been educated in the school of British horticulture.[49]

He called for a "native school of horticulture to which even foreign gardeners are obliged to yield."[50] Downing and colleagues presented landscape gardening as a patriotic pursuit, using bold rhetoric to obscure their wholehearted imitation of European models and the contradiction this wrought in their republican world.

In the transition from tradesmen to professionals, these practitioners yearned for genteel customers because it confirmed the propriety of their products and services. Gentrification of their practice, through the application of theory and refining associations, such as literature and art, stabilized their identity amid the social confusion that existed at this time.

In the nineteenth century, the architectural landscape was dominated by master builders and carpenters and not architects.[51] Similarly, the designed landscape was dominated by homeowners, gardeners, and nurserymen, and not by professional landscape gardeners. As scholars have noted, the American landscape was the complex product of owner's aspirations and tastes shaped by reading, travel and experience. It resulted from gardeners and nurserymen's capabilities, pattern books, periodical literature, and not by professional designers. Even Downing acknowledged this:

> Almost all the improvements of the grounds of our finest country residences, have been carried on under the directions of the proprietors themselves, suggested by their own good tastes, in many instances improved by the study of European authors, or by a personal inspection of the finest places abroad.[52]

Ironically, the proliferation of books and periodicals promoting gardening and landscape affairs enabled homeowners to take the improvement of their property into their own hands. Even so, it did create an ideal of rural taste and consequently some demand for services and products. Entrepreneurial nurserymen were able to meet the demand with plants, and the required embellishments.[53]

The following letter to the *Horticulturist* attests to a certain level of success for Downing and his predecessors in 1847:

> While perusing the pages of the beautiful work, I no longer feel myself an isolated being far out upon the borders of the cultivated portions of our land, but in the midst of highly gifted and refined minds sensibly aligned to the best interests of our common country.[54]

Finally and to conclude, let us jump to the end of the Civil War, when Frederick Law Olmsted was deeply engaged in the design of parks and gardens across the United States. The appropriate title of this newly emerged profession still had not been determined. In 1865, Olmsted argued over it with his partner Calvert Vaux (who had been Downing's first partner). Olmsted wrote:

> I am all the time bothered with the miserable nomenclature of landscape architecture. Landscape is not a good word. Architecture is not; the combination is not. Gardening is worse . . . the art is not gardening, nor is it architecture. What I am doing here . . . especially is neither. It is the "sylvan art," fine art in distinction from agriculture, or "sylvan useful art." We want distinction between a Nurseryman and a market gardener and an orchardist and an artist . . . If you are bound to establish this new art-you don't want an old name for it.[55]

Ultimately and reluctantly Olmsted replaced the title "landscape gardener," coined in 1764 by poet William Shenstone, with the phrase promoted by Calvert Vaux, "landscape architect."

One of Olmsted's earliest appearances in print was a letter to the editor of *Horticulturist,* Downing. It appeared in August 1847, entitled "Queries on Sea-Coast Culture." In the piece, he refers to Downing's *Fruits of America* and a recent issue of the *Horticulturist* suggesting he was a regular reader of Downing's works.[56] Years later, in a review of Olmsted's *Walks and Talks of an American Farmer in England* (1852), Downing introduced Olmsted as, "one of our original Yankee farmers."[57] Little did Sidney George Fisher know Downing would use that same epithet, Yankee, which in Fisher's case was derogatory but for Downing was the highest compliment.[58]

[1] (June 8, 1849) Nicholas B. Wainwright, ed. *A Philadelphia Perspective: The Diary of Sidney George Fisher Covering the Years 1834-1871.* (Philadelphia: The Historical Society of Pennsylvania, 1967), 224. Fisher's assessment of Downing has been discussed by George Tatum, David Schuyler, and Judith K. Major, on whose important work on Downing I have depended. They are each cited specifically below.

[2] *The Magazine of Horticulture* (November 1841): 411.

[3] Downing to John Jay Smith, Nov. 1841, cited in Judith K. Major, *To Live in the New World. A. J. Downing and American Landscape Gardening* (Cambridge, Mass.: MIT Press, 1997), 32.

[4] (May 20, 1846) Wainwright, *A Philadelphia Perspective,* 189. The book Fisher refers to is Downing's *A Treatise on the Theory and Practice of Landscape Gardening,* first published in 1841.

[5] (June 21, 1846) Ibid., 201.

[6] (November 1, 1847) Ibid., 202. The architect of the new house was Gervase Wheeler, who would later receive harsh criticism by A.J. Downing for his treatise on villa architecture.

[7] (June 8, 1849) Ibid., 227-28.

[8] (October 28, 1849) Ibid., 228. One of Fisher's last mentions of Downing was very informative: "On Thursday drove out to dine at Fisher's to meet Downing, the landscape gardener, who was there for a few days to help him lay out the grounds of his place . . . Engaged Downing to give Henry a few hours on Saturday to determine the site for his house . . . Downing goes in this way to give advice, professionally. His charge is $20 per diem. It shows improvement in the taste of the country that he can find such employment."

[9] H. L. Mencken, *The American Language: An Inquiry into the Development of English in the United States.* 2nd ed. (New York: A. A. Knopf, 1921).

[10] November 1849, quoted in David Schuyler, *Apostle of Taste. Andrew Jackson Downing,* 1815-1852 (Baltimore: Johns Hopkins University Press, 1996), 90.

[11] Dell Upton, *Architecture in the United States* (Oxford, 1998), 249-59. Upton argues that the drive towards professionalization in the related discipline of architecture revolved around autonomy or self-definition and the packaging or public recognition of the architect's distinctive claim.

[12] Luther Tucker, "Mr. Downing and the Horticulturist." *Horticulturist,* 7 (Sept. 1852): 393.

[13] J. C. Loudon, *An Encyclopaedia of Gardening* (London: Longman, Brown, Green, and Longmans, 1850), 335-339.

[14] Ibid., 339.

[15] William Wynne, *Gardener's Magazine,* 8 (1832): 272-273.

[16] David Schuyler has argued that "Downing recognized women as natural allies in his crusade to improve American taste: 'in all countries, it is the taste of the mother, the wife, the daughter, which educates and approves, and fixes, the tastes and habits of the people'" (Schuyler, 100-101).

[17] Therese O'Malley, "Cultivated Lives, Cultivated Spaces" in *The Culture of Nature: Art and Science in Philadelphia, 1740-1840.* Amy Meyers, ed. forthcoming (Yale, 2006).

[18] Richard Bushman, *The Refinement of America: Persons, Houses, Cities* (New York: Knopf, 1992), 280.

[19] Ibid., 132.

[20] Ibid., 403-6.

[21] Neil Harris, *The Artist in American Society* (New York: Clarion, 1966), 171.

[22] Loudon, *Encyclopaedia,* 339.

[23] Downing, *A Treatise on the Theory and Practice of Landscape Gardening,* 41.

[24] "Parmentier's Horticultural Garden." *New England Farmer,* 7 (3 October 1839): 85.

[25] Parmentier, "The Art of Landscape Gardening" in Fessenden, *New American Gardener* (Boston: J. B. Russell, 1828), 184-185.

[26] J. W. S. *Gardener's Magazine* (1829): 70; 72.

[27] Downing, *Treatise,* 459-60.

[28] See "Remarks on the Fitness of the Different Styles of Architecture for the Construction of Country Residences and on the Employment of Vases in Garden Scenery," *The Magazine of Horticulture* (August 1836).

[29] U. P. Hedrick, *A History of Horticulture in America to 1860; with an Addendum of Books Published from 1861-1920 by Elisabeth Woodburn.* (Reprint, Portland, Ore.: Timber Press, 1950), 474-75.

[30] Hovey published a description of Buist's City Nursery and Greenhouse in 1842: "On it is a green house forty feet long; a camellia house facing the north, forty feet; a hot-house; and a geranium-house; about forty feet, the whole being connected range. In addition to this there is a rose house, lately erected, about forty feet long. The whole we found well filled, for the season of the year, with a choice collection of healthy and well-grown plants. The camellia were in excellent health; they are kept in the house the year round" (*Magazine of Horticulture* [April 1842]: 124).

[31] Downing to Smith, 15 Nov. 1841 cited in Major, *To Live in the New World,* 59-61.

[32] Robert Buist, *The American Flower Garden Directory.* 2nd ed. (Philadelphia: E. L. Carey & A. Hart, 1839), introduction.

[33] *Horticulturist,* 5 (October 1850): 180-81

[34] *Horticulturist,* 7 (Oct. 1852): 483

[35] On Loudon's Gardenesque style see Melanie Simo, *Loudon and the Landscape: From Country Seat to Metropolis,* 1783-1843 (New Haven, Conn.: Yale University Press, 1988).

[36] William Saunders and Thomas Meehan established a distinguished partnership including the designs of Fairmount Park. Saunders also invented the "fixed roof" for greenhouses making them more efficient. During the Civil War he came to Washington, D.C., as superintendent of the experimental gardens for the Department of Agriculture (*Pioneers of American Landscape Design.* Charles A. Birnbaum and Robin Karson, eds. [New York: McGraw Hill, 2000], 329).

[37] Signed "X.Y.Z. *Newburgh,* Sept. 12, 1832," Downing's first article, "Rural Embellishments" appeared in the *New-York Farmer and Horticultural Repository,*

5 (Sept. 1832): 329-330. It is important to point out that Downing did not use his full name in print but chose to avoid reference to President Andrew Jackson, that president of the people. In her book on Downing, Judith Major explains this as a tactic to remove the association at a time when the landed elite of the young country was attempting to establish themselves apart from southern rural population Jackson championed. She links it specifically to his marriage to the grandniece of John Quincy Adams, our third president, who was ousted by Andrew Jackson (Major, 2-3). The political and social climate at the time of his birth in the year of General Jackson's military victory was significantly changed when he first published as a young man of seventeen. This reduction of his full name to initials, I suggest was also a specific strategy, especially important in his move from the laboring class of the nurserymen to the intellectual realm of the author on aesthetic style and taste in which he was embarking.

[38] Downing, *American Gardener's Magazine,* 2 (Dec. 1836): 283.

[39] Ibid., 284.

[40] For an important study of A. J. Downing's publication history and horticultural literature in general, see Charles B. Wood II, "The New 'Pattern Book' and the Role of the Agricultural Press," in *Prophet with Honor: The Career of Andrew Jackson Downing,* 1815-1852. George B. Tatum and Elisabeth Blair MacDougall, eds. (Washington, D.C.: Dumbarton Oaks Research Library and Collection, 1989), 165-90.

[41] *Magazine of Horticulture* (Nov. 1841): 401-11.

[42] Gordon P. D. E. Wolf Jr., "Andrew Jackson Downing and Pomology" in *Prophet with Honor,* 125-64.

[43] Quoted in George B. Tatum, "Introduction: The Downing Decades" in *Prophet with Honor,* 34. David Schuyler suggests that Downing was forced to sell the nursery business because of financial distress (Schuyler, 89-90).

[44] For studies of Repton's and Loudon's professional careers, see Stephen Daniels, *Humphry Repton: Landscape Gardening and the Geography of Georgian England* (New Haven: Yale University Press, 1999) and Simo, *Loudon and the Landscape.*

[45] Emily Cooperman, "William Russell Birch (1755-1834) and the Beginnings of the American Picturesque." (Philadelphia: University of Pennsylvania, unpublished Ph.D. thesis, 1999).

[46] *Pioneers,* 82-85.

[47] W. Barksdale Maynard, *Architecture in the United States,* 1800-1850 (New Haven, Conn.: Yale University Press), 30-31.

[48] Downing, *Treatise,* 527; *Pioneers,* 105

[49] Downing, "American versus British Horticulture," *Horticulturist* (June 1852): 250.

[50] Ibid., 251.

[51] Maynard, 24.

[52] Downing, *Treatise,* 40.

[53] The genteel world created by this explosion of popular literature had at its center women. Women, as the most avid consumers of the literature of rural affairs, and deeply engaged in the domestic realm where garden theory and practice predominated, were early practitioners of the discipline. It is no surprise that one of the founding members of the American Society of Landscape Architects was Beatrix Farrand.

[54] *Horticulturist,* 2 (August 1847): 96.

[55] Calvert Vaux Papers, August 1, 1865. Noted in Pregill and Volkman, *Landscapes in History,* 449.

[56] Olmsted, *Horticulturist,* 2 (August 1847): 100.

[57] *Horticulturist* 7 (March 1852): 135.

[58] George B. Tatum points out that on Downing's death Fisher wrote in his diary entry for 9 December 1852 a more conciliatory note: "Downing was one of the numerous victims of a dreadful accident last summer. . . . He has reformed the habits of the country, cultivated and extended a fondness for rural life and rural embellishment, diffused sound knowledge on the subject and produced practical results in countless cottages and villas, lawns, and gardens, built and planned in accordance with his instructions and scattered throughout the land" (Tatum, 40-41).

Botanical Progress, Horticultural Innovations and Cultural Changes

Botanical Progress, Horticultural Innovations and Cultural Changes

Botanical Progress, Horticultural Innovations and Cultural Changes

Botanical Progress, Horticultural Innovations and Cultural Changes

Turning Over a New Leaf:
The Impact of Qât (Catha edulis) in Yemeni Horticulture

Daniel Martin Varisco

Satan cropped up from the ground the night they were out eating wild plants. Thus a great calamity took hold of the Yemeni people. The half-consumed wild plants stirred in their bellies as the sewer-like veins of the Devil now carried their blood. Imagine that a thief enters your storehouse and in the morning chases you to the mountaintops, where you will not sleep peacefully that night. Dreams of ghosts unsettle your mind and body, suspending you between joy and sadness, boldness and retreat, defeat and victory, wealth and poverty, a sound mind and insanity. Yemenis are under a diabolical spell half their life, living in a full-color picture something like an adventure film or an exciting plot as in a child's fairy tale. This is what comes from the power of that abominable plant *qât,* the highest authority in Yemen, the tyrant that rules absolutely the life of Yemenis, elite and commoner. Such is the lyrical view of the famous Yemeni revolutionary, Muhammad al-Zubayrî, for whom the plant known as *qât* was as much an impediment to his country's liberation as the medieval religious autocrat in power at the time, which was 1958.[1]

Compare this sober assessment of a political philosopher to the poetic praise of an earlier Yemeni religious scholar:

The twigs of *qât* on the slopes of Hadanân

Are better than precious coral that is renowned.

When the righteous see them, to and fro they sway,

Comfortably, getting a taste in the best way.

And with their chewing energy is increased

With the loveliest variety, not for a bladder's plague

Upon you at any time. For this *qât* must

Be a safeguard against affliction, that is safety

And protection from the Devil's temptation to lust,

Such dangers to the soul as pretty, flirtatious girls.[2]

An instrument of the Devil or protection from the wiles of the Devil? What is this plant that evokes such starkly different judgments? (Fig. 1) In terms of scientific nomenclature, this is *Catha edulis*, a dicotyledon of the celastraceae family. It has been

Fig. 1. *Qât* trees irrigated by tubewells in Bani Hubaysh, central Yemen, 2005. Photo credit: D. Varisco

cultivated in Yemen, at the southwestern corner of the Arabian peninsula, since around the the latter part of the fifteenth century. The sole reason for its importance is the tender young leaves and new shoots, which are chewed to produce a stimulant effect. Thus we have a relatively recent exotic, easily grown, with widespread cultural significance and no obvious nutritional benefit.[3]

The history of *qât* in Yemen is intertwined with another stimulant, far better known. This is coffee (*Coffea arabica*), imported to Yemen from East Africa at about the same time and probably for the same reason. By the end of the sixteenth century, coffee had conquered the seemingly invincible Ottoman empire and was about to spread to Europe. The Red Sea Yemeni port of Mocha not only became the main entrepot for the bean but gave the beverage a lasting name. No such international interest was generated by Yemeni *qât*, which has been limited to local significance. Yet the cultural history of both plants within Yemen cannot be separated. There is a popular Yemeni legend that both *qât* and coffee were first grown in a southern mountain town called al-'Udayn (literally, the two twigs). Although the historical accuracy is suspect, the symbolic association is well worth exploring.

Catha edulis: The Plant and its Care

Starting with the botanical details, the *Celastracae* family is widely distributed with some eighty-five genera and 850 species. The designation *Catha edulis* stems from the work of the Swedish botanist Pehr Forsskål, who first identified it in 1763 during an ill-fated Danish expedition to Yemen.[4] The genus name was Latinized by Forsskål from the common Yemeni plant name, *qât*. The epithet of *edulis* indicates that the leaves were consumed, although not as food. The most recent botanical research indicates that this plant belongs to a monospecific genus. Forsskål originally suggested the term *Catha spinosa* for a similar looking shrub, but this is now recognized as *Maytenus parviflora* (Vahl) Sebsebe.[5]

Catha edulis is an evergreen, cultivated either as a shrub, not unlike the tea plant in overall appearance, or a tree reaching up to ten meters in height.[6] The trunk is slim and straight with a smooth, white bark. The leaves, ranging from pale green to red in color, are serrated and ovate-lanceolate to elliptical in shape. The flowers are small and white. The root system can extend up to three to five meters. In Yemen considerable variation in appearance has been noted, depending on elevation, soil, climate, type of irrigation, and cultivation methods.

Most *qât* in Yemen is cultivated in the temperate southern and central highlands between one thousand and twenty-five hundred meters elevation. (Fig. 2) The tree has remarkable endurance in dry spells and can endure limited frost on sheltered

Fig. 2. Extensive *qât* production in al-Ahjur, central Yemen, 2005. (Photo credit: D. Varisco)

terraces. As the plant is produced exclusively for its new growth of leaves, there are few damaging pests or diseases, in contrast to coffee production. Sheep and goat may occasionally eat leaves from lower branches, but this is not a major problem. One of the few wild animals known to attack *qât* plants is the porcupine (*qumayra*). *Qât* is resistant to termites, hence its wood is used in construction in Ethiopia. Locusts, once a major plague in Yemen, ignore *qât*. Traditionally, farmers dusted the leaves with loose soil to counteract occasional blight on the leaves. In recent years, there has been heavy use of imported pesticides, resulting in numerous cases of minor poisoning if the leaves are not properly cleaned before chewing.

Yemen's well-drained and fertile highland soils, with limited clay content, are ideal for growth of the plant. Almost all production in Yemen takes place from cuttings or suckers, which are usually taken from mature trees of at least ten years of age. Although seed production is theoretically possible, there is a poor germination rate. The sucker is planted about twenty centimeters deep, generally at a meter's distance from the next plant. The distance may be greater if a plough and draft animal are used to turn the soil.[7] In the central highland Yemeni valley of al-Ahjur, the favorite time for planting *qât* is during the spring. In the past, intercropping with grain crops would sometimes be practiced, but there appears to be a shift toward mono-cropping with the increase in tubewell plantations on flat, level land.[8] *Qât* has also been intercropped with coffee trees in highland terraces.[9] Farmers do not plant *qât* near apricot trees or grapevines, but many plant on terraces with the few standing acacia and wild fig trees.[10] Because of the continual harvesting of the young leaves, it is useful to apply manure or fertilizer to replenish soil nutrients. On spring-irrigated land in al-Ahjur, weeding and turning soil in terrace plots are usually done about

Fig. 3. Drying of *qât* leaves due to lack of rainfall, al-Ahjur, central Yemen, 2005. Photo credit: D. Varisco

four times per year. One of the major problem weeds is *Cynodon dactylon* Pers., locally called *wabl* in Yemen.

Catha edulis will thrive on rainfed land with at least five hundred millimeters of rainfall per year. In some areas the traditional Yemeni practice of rainwater harvesting increases the amount of moisture available after a rain. The two main rain periods in the highlands are in early spring and late summer. Applying irrigation water can stimulate new growth in less than two weeks, especially if the weather is warm. Under irrigation, the young seedlings need more water right after planting. Practice varies, but in general water should be applied at least once every five days for the first month and then about every ten days in the second month. By the third month there may be only one or two waterings. In the past two decades there has been heavy use of irrigation water from tubewells, causing rapid drawdown rates in much of the country. Some estimates indicate that as much as fifty-five percent of all groundwater currently extracted for irrigation goes to growing *qât*.[11] The tree should reach about a meter in height within four years and be ready for production. The geographer Horst Kopp thinks that *qât* trees can yield for up to fifty years, but little research has been done on how long marketable *qât* leaves can be produced from the same tree.[12]

The harvesting process is selective and can be done at almost any season, depending on the last time of harvesting and rainfall or the application of irrigation water.[13] A branch will be cut off about forty to fifty centimeters from the tip. In most areas, a single tree can be harvested twice a year, but under irrigation this may be increased to five times a year. In Hadanân on Jabal Sabr, for example, *qât* is picked three times, first after the spring rains. The preferred time to harvest the leaves is in the morning. If left on the branch, the *qât* is usually bundled in the field to minimize damage to the leaves and preserve moisture (Fig. 3).

Study of the pharmacological aspects of *qât* leaves has been underway for more than a century. One of the more useful studies of the stimulant influence of *qât* in Yemen is that of John Kennedy, an anthropologist who carried out field research in the Yemen Arab Republic in the 1970s. Kennedy's *The Flower of Paradise* surveys the previous literature and concludes that ethnocentric reporting by Westerners has greatly exaggerated the dangers of *qât* use in Yemen, although there is probably cause for a range of minor health problems.[14] *Catha edulis,* at least for the leaves in a fresh state, is a stimulant rather than a narcotic. One of the main active ingredients is cathinone, which is similar to d-amphetamine in stimulating the body's central nervous system, blood pressure and heartbeat. In a complex discussion of whether or not Yemeni users are addicted, Kennedy agrees with a number of earlier researchers that there is little evidence for physiological withdrawal symptoms, although he suggests that "a mild form of physiological dependence does result from extremely heavy use."[15] Suffice it to say that most Yemenis can stop chewing, as they often must do when working abroad, more easily than they can stop smoking.

Early History in Yemen

Although the biochemistry of the leaves of *Catha edulis* has been studied in some detail, there has been relatively little field research on the derivation and spread of this cultivar. Today, or at least within the last century, this species is found both in wild and cultivated varieties throughout eastern, central and southern Africa. Of special relevance for the Yemeni varieties is its presence in Ethiopia, where it is called *c'at* in Amharic and *jimaa* among the Oromo. I am inclined to think that reports, which I have not confirmed, of *Catha edulis* in Afghanistan and India reflect a recent historical introduction from Yemen rather than an earlier spread from East Africa.

The relative lack of field study on Yemeni varieties is exacerbated by uncritical acceptance of a theory proposed by Raman Revri in 1983. Based on faulty cytological research, Revri argued that the progenitor of the Yemeni species was a widely distributed plant identified by Forsskål as *Catha spinosa*.[16] Revri's basic argument is quite simple. The alleged species *Catha spinosa* is diploid and propagated by seed, whereas the seeds of *edulis* are sterile, and it requires cultivation by cuttings. Thus, he thinks the primary origin of the cultivated plant in Yemen is from the wild Yemeni species, which he postulates was taken to Ethiopia in the sixth century, domesticated there and only later returned across the Red Sea to Yemen. In his important survey of *qât* use in Yemen, John Kennedy was inclined to support this proposition, at the very least suggesting that the Ethiopian origin had been put into question.[17]

Revri's speculation is totally without merit. First and foremost, Forsskål's designation *Catha spinosa* is invalid. The type of specimen he collected is equivalent to *Celastrus parviflora* Vahl (1790), now superceded as *Maytenus parviflora* (Vahl) Sebsebe.[18] Revri's point for a direct ancestral link between the two plants is thus genetically impossible. Second, his argument ignores compelling indirect evidence available at the time. *Catha edulis* was known to be widely distributed in Africa, including wild varieties. Indeed, the distribution of tropical flora and fauna in Yemen suggests that the flow was usually from the African continent into tropical pockets of the drier Arabian Peninsula.[19] The third major problem, at least for an Occamist, is that it requires a bizarre historical migration. There is no documentary evidence in Yemen of the term *qât*, cultivation of the plant, or use of its leaves as a stimulant or medicant, before the fourteenth century at the very earliest. It is hard to believe that "Ethiopian conquerors" more than half a millennium earlier would have recognized a medicinal quality in a local ancestor of *Catha edulis*. There is no reason to doubt the diffusion of *Catha edulis*, like that of *Coffea arabica*, into Yemen directly from East Africa in relatively recent history.[20]

The historical and linguistic evidence for the appearance of *qât* in Yemen is meager and contradictory. The earliest reference in a historical text is by the non-Yemeni historian Ibn Fadl Allâh al-'Umarî (d. 1349), who said the choicest leaves are "eaten" in order to increase mental alertness (*dhakâ'*), improve memory, relax and lessen the need for food, sleep and sex. Because this plant allows one to endure sleeplessness, it was used in traveling.[21] The historian further describes the plant as having both large and small varieties with its leaves resembling those of the orange tree. There is no doubt he is describing *Catha edulis*.

Al-'Umarî remarked that he was amazed at a story related to him about Ethiopian Muslims who went to the court of the Rasulid sultan al-Malik al-Mu'ayyad (r. 1296–1332). While in Yemen, they had a craving for *qât* and planted some. As they were picking the choice leaves, the sultan inquired about the benefit. After it was explained that the leaves lessened desire for food, sleep, and sex, al-Malik al-Mu'ayyad responded, "What delight in the world is the equal of these? By God, I will not eat it. I do not spend for anything but these three things. So how can I use what comes between me and my delight for those things?" If this encounter really happened, it is strange that there is no mention of *qât* in any of the Yemeni chronicles; nor is there any reference to *qât* in the rather extensive Rasulid corpus on Yemeni plants and agriculture.[22] Nor does the Egyptian historian, al-Maqrîzî, mention this story in his copying of al-'Umarî's account a century later.[23]

It is worth noting that al-'Umarî's anecdote about Yemen occurs in a discussion of the cultivated plants in the Ethiopian region of Ifat. There is a documented reference in Amharic for the year 1329 to an Ethiopian Muslim king who boasted that he would plant *c'at* in the capital town of the Christian king Amda Seyon.[24] Although more historiographic research on the recognition of *qât* in Ethiopia is needed, it is most probable that the stimulant properties were known by Muslims in Ethiopia by the first quarter of the fourteenth century before the plant was cultivated in Yemen.

The plant name *qât* is unknown in the Arabic lexicons. Despite attempts by a few modern Yemenis to derive the term from an Arabic root concerned with strength, there is reason to suspect that the plant name is not originally Arabic. In Yemeni sources the term *qât* is not recorded with a direct link to *Catha edulis* until the mid-sixteenth century legal text of Ibn Hajar al-Haytamî (d. 1565 or 1587), who quotes earlier sources including a reference to its use in Ta'izz during the reign of the Tahirid sultan 'Amîr ibn 'Abd al-Wahhâb (r. 1489–1517).[25] The earliest mention in Yemeni poetry appears to come from Muhammad ibn Sa'îd ibn Kabbân (d. 1438).[26] Several scholars have pointed to use of the term in the pharmaceutical treatise compiled by Abû al-Rayhân al-Bîrûnî (d. 1051), which would make it the earliest reference anywhere. However, the description provided by al-Bîrûnî and the fact that it is said to be exported from Turkestan make it a very unlikely candidate for Yemen's *Catha edulis*.[27] The major late medieval Arabic herbals, including the widely consulted text of Dâwûd al-Antâkî (d. 1599), do not mention *qât*.[28]

It would make sense that a plant imported from Ethiopia into Yemen would have the name given it in Ethiopia. This is clearly what happened with Ethiopian teff, a grain that became *tahaf* in Arabic. But here arises a linguistic problem. The Amharic term is *c'at,* with cognates in all the Ethio-Semitic languages, Oromo, and the Highland East Cushitic languages.[29] Unfortunately, as the linguist Maxime Rodinson noted several decades ago, a direct transference from Amharic *c'* to Arabic *q* is not linguistically viable.[30] Several scholars have been seduced into looking for scenarios to explain an Arabic origin despite the absence of textual evidence. Thus, Chaim Rabin assumed ancient Yemenis took the name to Ethiopia because in the local Yemeni Azd dialect the *q* can become *j* or "tsh."[31]

I believe I have now solved the linguistic puzzle. Everyone, Arab and Western scholars alike, has been looking in the wrong place, written lexicons, rather than considering the ways in which plant names are communicated orally.[32] The clues have been there all along. In the Egyptian texts by Ibn Fadl Allâh al-'Umarî and al-Maqrîzî, the reference is not to *qât* but to *jât*. Indeed, the earlier author specifically says that the first letter in his transcription is pronounced between the Arabic *j* and the Arabic *sh*.[33] So how did *jât* become *qât*? In Egyptian dialect the letter *j* is often pronounced as a hard *g,* a linguistic shift also found in contemporary southern dialects of Yemeni Arabic. I propose that the Arabized Amharic plant name arrived in Yemen in oral form as *gât*. We simply do not have any textual evidence about its arrival, nor is this unusual in the history of plant name diffusion. The earliest written record is from Yemenis in the Zaydi north, where it is the letter *q* that is pronounced as a hard *g*, not the letter *j*. Legal scholars in the north wrote about a newly introduced plant called *gât*, which they would have rendered as *qât* in formal written Arabic. Hence the etymological trajectory of the term is Amharic *c'at* to Arabic *jât*, pronounced as *gât* and later written down as *qât*.[34]

Yemeni legend and folkore suggest that the *qât* plant originally came from Ethiopia. On the folk level, the story is told throughout Yemen that a goatherd discovered the stimulant properties of *qât* leaves after watching their effect on goats who had browsed on the plant.[35] A variant of this legend, told to me in 1979 by a poet from Husn al-'Arûs, suggests that an Ethiopian came along to explain to the goatherd what was happening. Another tall tale speaks of its introduction to Yemen from China by the enigmatic al-Khadîr, a companion of Dhû al-Qarnayn, a mythical namesake of Alexander the Great.[36] A legend seemingly more based in history is recounted by the Yemeni historian Yahyâ ibn al-Husayn, who says that in the year 1543 the Zaydi imam Sharaf al-Dîn banned the use of *qât* in Yemen.[37] But clearly the plant had to be cultivated already in order to be banned.

The most likely vector for introducing *Catha edulis* into Yemen is the same as that for coffee. The Yemeni poet 'Abd Allâh al-Baraddûnî speaks for most Yemeni scholars in attributing the origin to Sufi mystics who used the plant for its stimulant qualities.[38] A number of specific candidates have been suggested, including Shaykh 'Alî ibn 'Umar al-Shâdhilî (d. 1418), who is said to have preferred drinking coffee to a concoction made from the *qât* leaves.[39] Other candidates for introducing coffee include Yemeni Sufis such as Ahmad ibn 'Alwân (d. 1267),[40] the thirteenth century al-'Arasî[41] and Ibrâhîm Abû Zarbîta, one of forty-four thirteenth-century missionaries sent from the Yemeni Hadramawt to Ethiopia.[42] Archaeological evidence from Zabid in the Yemeni coastal system indicates that individual drinking cups for coffee were being manufactured by the start of the sixteenth century, suggesting that coffee drinking was then becoming a social habit.[43] In a recent assessment, Michel Tuchscherer argues that coffee production did not take hold in Yemen until the middle of the sixteenth century.[44] Unfortunately, there is no comparable archaeological evidence for the earliest *qât* chewing.

The spread of *qât* into Yemen would have started in the southern highlands, as the shrub would not survive in the coastal heat of Aden. The Yemeni poet al-Baraddûnî suggests that Jabal Sabr, located just above the important southern town of Ta'izz, would have been one of the earliest locations for major production, along with nearby al-'Udayn and 'Utma. Jabal Sabr is recognized today as producing some of the best *qât* in Yemen and has been frequently noted as an important *qât*-growing region by poets and later travelers.[45]

How would the earliest *Catha edulis* have been planted in Yemen? This raises an important issue in distinguishing horticulture as cultivation in "gardens" from the more generic sense of agriculture. There is an extensive literature on the range of agricultural cultivation in Rasulid Yemen, the period (thirteenth to mid-fifteenth century) in which *qât* may have been first introduced. The sultans of Yemen not only encouraged local production of basic food crops, but were themselves avid gardeners and personally introduced exotic fruits (such as the coconut) and aromatic plants into their gardens. The most important royal garden was at the spring of Tha'bât, located near the Rasulid highland capital of Ta'izz.[46] Begun as early as the middle of the twelfth century as a royal retreat, it became a kind of retirement home for the long-reigning Rasulid monarch al-Muzaffar.[47] Al-Muzaffar's son and heir, al-Malik al-Ashraf 'Umar, who no doubt knew this garden residence well, wrote the first major Yemeni treatise of agriculture at the end of the thirteenth century.[48] A later Rasulid sultan, al-Malik al-Afdal al-'Abbâs wrote a more extensive agricultural handbook, borrowing from the writings of earlier Yemeni authors but also incorporating advice about arboricultural techniques from the influential ninth-century Iraqi *Kitâb al-Filâha al-Nabatiya* and from the Andalusian agronomist Ibn Bassâl.

None of these Rasulid texts mentions *qât*, nor is *qât* necessarily a plant limited to small, irrigated gardens. However, the introduction of both *qât* and coffee occurred in Yemen at a time when detailed knowledge of planting, pruning, and caring for non-Yemeni tree crops was already available. I assume that those who brought the first seedlings to Yemen would have had prior knowledge of how the plant was grown in Ethiopia at the time. There was no need to propagate a domestic variety, as might have been the case with fruit trees or even coffee, as it was only the natural growth of leaves that was desired. I suggest that the earliest cultivation would have been in small irrigated gardens near to dwellings. Before the evolution of a large market beyond the needs of the original Sufi users, there would only have been limited plantation for local use. Thus, *qât* shrubs could have been planted during the latter part of the Rasulid period and may not have been noticed by those writing texts, either agricultural or medical.

When the chewing of *qât* leaves expanded and demand for the plant created a market, it would have been relatively easy to grow *Catha edulis* on Yemen's network of agricultural terraces. In the southern highlands, these need not have been irrigated because rainfall is more than sufficient for successful propagation of this rather hardy shrub. Although the formal Arabic knowledge of arboriculture would have been available to the educated elite, it would not have been necessary. Yemeni farmers would have been able to grow *qât* far easier than coffee, which we know soon developed a major market both within and beyond Yemen's borders. Thus, the earliest history of *qât* in Yemen was probably as a garden variety exotic, which served a limited clientele with a personal devotional interest rather than a commercial one. The fact that both *qât* and coffee eventually became major cash crops in Yemen is clearly a function of their respective stimulant properties mapped onto the developing institution of afternoon gatherings and the rise of the "coffeehouse" for travelers.

Economic Impact

In the past three decades, since the end of the civil war in the former North Yemen, *qât* has become the most important cash crop in the local economy. This also has been the case in Ethiopia, which has endured even longer civil strife and despite periodic government attempts to discourage its production.[49] Statistical information on the spread of *qât* cultivation in Yemen is hard to come by and official government reports usually contain gross estimates. It is generally assumed that in 1972 about forty-three thousand hectares, three percent of cultivated land at the time, were planted with *qât* in the Yemen Arab Republic and only a very little in suitable areas of the People's Democratic Republic of Yemen, the former Aden colony of Britain (Table 1).

Table 1. Estimated *Qât* Production in the Yemen Arab Republic (1972)[50]

Province	Area (ha)
Ibb	15,000
Sanaa	10,000
Ta'izz	10,000
Hajja	3,000
Others	5,000
Total	43,000

The first official countrywide agricultural census in the north, conducted primarily in 1978–1979, suggested a total hectarage of about forty-seven thousand hectares (Table 2), but a recent set of official government statistics notes that the area of *qât* expanded from 45,000 hectares in 1980 to 110,873 hectares by 2003 (Table 3). For comparison, in Ethiopia 79,000 hectares were under *qât* production in 1999/2000.[51]

Table 2. Estimated *Qât* Production in the Yemen Arab Republic (1978–1983)[52]

Province	Area (ha)	Trees	Holdings
Dhamâr	1,329	2,226,000	15,856
Al-Hudayda	114	197,000	1,338
Hajja	3,301	4,196,000	16,702
Al-Mahwît	4,800	6,501,000	7,750
Ta'izz	3,997	4,585,000	21,464
Ibb	4,920	13,221,000	28,388
Sa'da	9,249	6,722,000	7,771
Al-Bayda'	1,791	20,606,000	6,711
Sanaa	16,248	45,996,000	63,667
Al-Jawf	68	766,000	1,143
Ma'rib	1,396	4,608,000	4,909
Total	47,213	109,624,000	175,705

Table 3. Estimated Crop Production (ha) in Yemen (1980–2003)[53]

Crop	1980	1985	1990	1994	2003
Qât	45,000	56,000	80,000		110,873
Coffee	8,000	17,000	25,000	25,000	33,662
Sorghum	644,000	604,000	507,000	448,000	378,036
Maize	33,000	42,000	52,000	44,000	29,982
Wheat	73,000	71,000	98,000	100,000	86,520
Fruits	44,000	46,000	57,000	70,000	98,317
Vegetables	13,000	35,000	53,000	51,000	71,210
Legumes	75,000	32,000	50,000	53,000	114,862

It has long been assumed that the expansion of *Catha edulis* has come at the direct expense of *Coffea arabica*. This was the surmise of a 1955 FAO mission to Yemen, while the country was still under control of the Zaydi imam.[54] World Bank assessments and the major USAID Agricultural Sector Assessment in 1982 continued to echo this view.[55] Some analysts have claimed that Yemenis started tearing down coffee trees and replacing them with *qât* as early as the 1950s.[56] In a thorough study conducted in the late 1970s, Horst Kopp concluded that coffee was being replaced by *qât* in some regions, but there was no simple shift to a new cash crop.[57] First, the *qât* plant can be cultivated at a greater elevation and much of the land being put under *qât* cultivation had not previously been planted with coffee. Second, the issue is less the production area in Yemen than the fact that almost two centuries ago Yemen lost its share of the world coffee market. Although Yemen exported coffee

almost exclusively in the seventeenth and early eighteenth centuries, the growth of colonial coffee plantations in Indonesia, Africa and South America resulted in Yemen producing only one percent of the world's coffee by 1850.[58] With the decline in coffee export profits, *qât* cultivation was probably encouraged by the Zaydi imams for taxation revenue.[59] Coffee production decreased dramatically during the civil war and through the 1970s. From 1980 to 1990, the total coffee area in the north increased from 8,000 to 25,000 hectares at the same time as *qât* increased from 45,000 to 80,000 hectares (Table 3). Most of this has been at the expense of traditional grains, especially sorghum.[60]

The switch to *qât* as a major crop is entirely for its local market value. There is no export value, apart from minor illicit trade to Saudi Arabia and in fact there is some minor import from Ethiopia. High demand allows for highly inflated prices for a commodity consumed solely

Fig. 4. Branch of *qât* in al-Ahjur, central Yemen, 1986. Photo credit: D. Varisco

for its stimulant qualities and not as a food or drink. *Catha edulis* is an ideal cash crop for the small farmer since it can be grown on rainfed land, where appropriate, has low water and fertilizer requirements, few pests or diseases, and limited labor needs. Even after frost damage *qât* leaves can be sold as an inferior quality called *zabîb*.[61] Under irrigation the timing of an application can stimulate production of new leaves when the market has low supply. As numerous analysts have noted, *qât* is not an easy crop to monopolize.[62] Thus, the decision to grow *qât* is quite rational for Yemeni farmers from an economic perspective.[63]

The *qât* leaves are sold by bundle, often called a *rubta,* either on the branch or with the leaves packed together in a banana leaf or plastic (Fig. 4). A bundle is generally the amount needed for a day's average chew. Production yield is generally measured by the bundle, although this unit is not standardized in the literature. In his mid-1970s survey, Schopen noted that better quality *qât*, like Bukhârî, might have between fifteen and twenty branches in a bundle, while a bundle of a lesser variety like Sawtî would have forty to fifty branches.[64] Overall production for consumption in Yemen is said to have reached 592 million bundles in 1996.[65]

One of the more frequently cited studies is an estimate made by 'Abbâs al-Sa'dî for *qât* Yemeni production in 1980 (Table 4).

Table 4. Estimated *Qât* production in the YAR in 1980[66]

Item	Unit	Number
Total Area	ha	90,000
Qât bundles	per ha	2,200
Market value	riyals/bundle	54
Total Production	bundles	198,000,000
Total market value	riyals	10,692,000,000
Production costs	riyals/ha	4,000
Total production costs	riyals	360,000,000
Total tax (10 percent)	riyals	1,069,200,000
Total net return	riyals	9,262,800,000
		[$2,030,000,000]

Al-Sa'dî, for reasons that are unclear, doubles the government estimate of area under *qât* production. With this caveat, he assumes that a hectare would yield about 2,200 bundles per year, resulting in total marketable bundles of 198,000,000. Other available government statistics suggest that annual production in 1980 was closer to 400,000,000 bundles.[67] Compounding the difficulty in working with al-Sa'dî's analysis is a misunderstanding of the nature of production. The number of harvests per year varies from one to five, not all trees are spaced at the same distance and plots with young plants are generally not harvested for the market until the third or fourth year. There is, in effect, no average production unit because the commodity is simply natural growth of the leaves rather than a specific number of fruits or berries.

The farm profit on *qât* is equally hard to measure, given wildly variable data in the literature, regional differences and evidence of a reluctance by farmers to provide detailed and accurate production data to researchers.[68] In a study in 'Utma of the southern highlands in the mid-1970s Kennedy found per hectare net profits to range from $24,038 to $12,019 on irrigated land and $6,010 to $1,202 on rainfed land.[69] One of the more reliable field studies was conducted in 1983 by Hywel Rees-Jones in the central highland plain of Qâ' Bakîl, located about seventy kilometers south of the capital, Sanaa. At this time Rees-Jones estimated that about ninety percent of dryland and irrigating farmers grew some *qât*. The dryland *qât* was harvested once or twice a year and sold directly to a trader or middleman. Assuming two harvests per year, average gross riyal return on dryland per ha was 265,522 (ca. $59,000) and on irrigated land was 376,843 (ca. $83,743), although production costs of water would be high.[70] For the Jabal Râzi' area near the Saudi border, Shelagh Weir estimated that per hectare gross profit around 1980 was $90,000 for two harvests and $130,000 for three harvests.[71] These figures are almost double the widely quoted estimate of Raman Revri that gross profit was around 213,900 riyals (ca. $47,530) per hectare in 1983 on irrigated land but only 35,650 ($8,000) on rainfed land.[72] Production costs vary, with al-Sa'dî assuming about 4,000 riyals (ca. $890) per hectare. Al-Sa'dî arrives at a per hectare net profit of 103,000 riyal (ca. $22,890), but this is not disaggregated according to type of production.

Average yields are problematic not only because the estimates vary widely but also because there can be major fluctuation from one year to the next, especially on rainfed land. Kennedy describes a farmer in Radâ' who earned about $4,400 per hectare of *qât* in 1975, but in a previous year he earned double this amount.[73] This particular farmer made more money selling irrigation water from his well than for the *qât* he harvested and sold. It should not be assumed that the profit for the farmer is equivalent to the actual consumer cost in the market. The cost differential between farm and market can result in as much as a 900 percent markup in Ethiopia.[74]

As a marketable product, even though it is not a foodstuff, *qât* is subject in theory to both the *zakât* production tax and market taxes. The general production tax would be ten percent, but this appears to have been rarely applied in full in Yemen. A ten percent sales tax has been applied in local markets. During the early 1980s, the annual tax revenue from the market of Kusma in Jabal Rayma was said to be about 900,000 riyals ($200,000), of which about 18,000 riyals ($4,000) was allocated directly to the local development association.[75]

Overall production data for Yemen is again highly variable in the literature. Al-Munibari indicates current net profit of 400,000–1,800,000 riyals ($3,400–$15,000) per hectare; this is almost three times the return on coffee.[76] On the market side, Al-Munibari estimates annual *qât* sales at $765,000,000 per year or about twenty percent of GDP and half as much as all other agricultural production.[77] A recent FAO study found that from 1990 to 2000, *qât* contributed almost half as much to Yemen's GDP as oil revenues.[78]

Social Importance and Cultural Significance

No study of the horticultural significance of *Catha edulis* could be concluded without considering the social importance of chewing the leaves and the cultural significance of chewing as a national marker, for good or ill, of Yemeni identity.[79] It is generally assumed that before the middle of the nineteenth century most Yemenis were too poor to chew on a daily basis and that the major consumers were members of the religious and tribal elite.[80] It should be pointed out, however, that travelers and administrators in British-controlled Aden often complained that the poor in the port wasted much of their meager income on the plant. By 1970 the vast majority of Yemeni men, and probably a simple majority of women, chewed *qât* in social gatherings at least once a week if not more. Rising demand in the 1970s through the present thus represents less a situation in which people who never chewed have chosen to do so than a result of increased chewing frequency brought on by the influx of cash remittances from Yemeni workers abroad and an improving transportation network for marketing.[81]

By the mid-1970s, John Kennedy estimated that overall about 80–85 percent of men and 50–60 percent of women chewed more than once a week.[82] In a recent sociological survey by al-Munibari, it is reported that 74 percent men chew in urban areas and 86 percent of men do so in rural areas; the percentages for women are 44 urban and 75 rural.[83] The impact on household budget is hard to gauge in Yemen. Farmers can produce surplus for their own consumption and often have access to *qât* from relatives or neighbors for less than market price. It is commonly estimated that family expenditure on *qât* is around a quarter of the total household budget[84] (Fig. 5.). This figure, which would be alarming if accurate, is mitigated by a number of factors usually left out of the economic analysis. In many cases, there are alternative forms of income or chewers are provided *qât* as clients of rich patrons. In some rural areas, daily laborers may be given a day's supply as part of their wage. It is doubtful that many Yemeni families are on the brink of starvation because the men have spent a large amount on *qât*.

The market price of *qât* depends to a large extent on variety, season, and overall demand. It is impossible to document the full range of *qât* varieties in Yemen. Armin Schopen estimated that there were at least two thousand in North Yemen during the 1970s.[85] Most of the varietal names are geographical in origin, but some are known as consistently producing the best effect, not unlike vintages of wine. Historical evidence for the price of *qât* is limited. A market document for Sanaa from about 1750 C.E. indicates that the price of *qât* was 1/8 *qirsh* (5 *buqsha*) for a *rubta* of 10 ounces (350 grams) of leaves; this was equivalent at the time to the daily wage of an unskilled laborer in construction.[86] In 1974–1975, a bundle of good quality *qât* sold for 10–25 riyals and the poor quality for only 4–6 riyals in Sanaa.[87] For the years 1979–1980, Shelagh Weir estimates that the price range for high quality *qât* was up to 80 riyals ($17.70), but low grade *qât* could be purchased for as

Fig. 5. Political cartoon showing the economic impact on the family of *qât* use. Source: *Yaman bila qat* #20-21, p. 7, 2004.

little as 20 riyals ($4.40)[88] She estimates that the average bundle of *qât* for a daily chew would have been 45 riyals or $10, which was about three quarters of the average daily wage of an unskilled worker. In 1983 the average price for a bundle of *qât* in Sanaa rose to $15–20. At this time, average quality *qât* sold for 30 riyals ($6.67) in the remote region of Jabal Rayma and could be found for as little as 10 riyals ($2.22).[89]

The context of chewing is preeminently a social institution. Yemenis usually chew *qât* in social gatherings after the main meal and mid-afternoon prayer. In the fasting month of Ramadan, chewing commences in the evening after breaking the fast. Most sessions are segregated by gender and relatively small in size with up to a dozen men present. Much larger chews occur in the houses of important political figures or during celebrations of weddings and other major social events. In most settings an individual brings his or her own supply, but sharing some of this is common.

The amount chewed depends on the variety and the desired effect, not unlike social drinking in America. Kennedy's data from the mid-1970s shows that the average amount of leaves actually chewed was about one hundred grams, with only a few people chewing over two hundred grams. The leaves and tender shoot ends are carefully cleaned by the fingers and placed into a wad in the side of one cheek. The leaves are "stored," referred to by the verb *khazana* in Yemen, rather than chewed. Most Yemenis spit out the remaining pulp when the session is over. Three stages of influence are generally recognized. After about a quarter of an hour from the start of the juice descending to the stomach a sense of heightened alertness can usually be felt. As the central nervous system is stimulated by the alkaloids, with an amphetamine-like effect, the heartbeat increases and the pupils dilate. This energizing stage lasts for up to two hours. This is followed by a mellowing out and sense of calmness, called *kayf*, but is not always reached. Conversation in the session generally ceases and often the only sound that can be heard is the gurgling sound of the waterpipe, which many Yemenis smoke as they chew. The final effect is a listless feeling and a mild depression. *Qât* tends to reduce appetite, so little or no food is eaten in an evening meal, and it often produces insomnia.

Fig. 6. Poster advocating that *qât* not be chewed. Source: National Society to Combat Qât Addiction, Yemen.

Students claim that chewing allows them to stay awake and study, thus completing the circle that began with Sufi mystics using *qât* to say their prayers.

For the most part Yemeni scholars have approved of *qât* chewing as a practice not prohibited in Islam.[90] A legal debate in the sixteenth century essentially decided that *qât* does not have the same intoxicating effect of opium, hashish, or wine and that its use should be determined by scientific, that is, medical, evidence. Chewing is done both as a daily social gathering and to celebrate important social events. There is a wealth of folklore and poetry in Yemen about the plant. Much of this reflects the traditional religious view that *qât* is allowed for Muslims. It is reported that some Yemenis claim to have found the first part of the *shahada* (*lâ ilaha ilâ Allâh,* There is no god but God) written on *qât* leaves.[91]

Why do Yemenis chew *qât*? There is no single answer to this intriguing cultural question. Although the practice is said to have begun with religious adepts intent on staying awake to recite, the vast majority of Yemenis today chew for more mundane social reasons. In some cases the recent increase in chewing may be a result of what Shelagh Weir terms conspicuous consumption, in which *qât* serves as a medium of prestige.[92] My own view, advanced in 1986, is that in the 1970s *qât* served as a profound identity marker for Yemenis at a time when their sense of pride in the Arab world was being questioned. Other Arab countries, especially Saudi Arabia, banned the use of *qât* and considered it backward. But Yemenis invigorated this custom, associated with urban elites in the past, as a form of national identity along with markers of tribal dress. Of course, the use of a stimulant like *qât* is done for the most obvious of reasons: it is seen as pleasurable and the medical and social drawbacks are minimal. Indeed, *qât* was perceived as a beneficial medicant in Yemeni practice.

Concluding Remarks

In this article I have attempted to trace the history of two exotic stimulants in Yemeni context. The history of *qât* cannot be told apart from the history of its more famous twin, coffee. Coffee expanded out of its original Red Sea niche and soon served the colonial plantation agriculture of the emerging European nation states. Yemen's brief monopoly of the coffee trade came to an end in the seventeenth century as Europe began to plant it on appropriated foreign soil in Indonesia, Africa, and Latin America. The social institution of the coffeehouse, both in the Middle East and Europe, reshaped the social and political landscape, ironically about the same time that New World tobacco was introduced into a not-yet-nicotine-addicted Old World. So this morning you might have walked into a Starbucks and ordered the brew from berries that Juan Valdez picked. But you will not be chewing *qât* leaves as you read this, unless you are Yemeni or East African.

Why did *qât* not catch on outside Yemen, where it has been more important as a local stimulant than coffee? Indeed, why is *qât* banned in a conservative country such as Saudi Arabia? Both stimulants came to Yemen at the same time and probably for the same reason. Both caused concern among Muslim scholars, who debated whether each was a forbidden intoxicant (like wine or hashish) or an important new medicant. Like coffee beans, the leaves of *qât* can be crushed and made into a less stimulating tea. Both seedlings were available to the Dutch and other Europeans who transplanted them to their colonies. Theoretically we could be drinking Nesqat instead of, or as well as, Nescafe. There but for the grace of historical accident in the horticultural history of stimulants is a puzzle not yet solved.

Finally, to return to the question posed at the start of the essay, is the *qât* plant comparable to the devil's poison or an antidote? If the devil is in the details, such a judgment depends on balancing the benefits and detriments. On the positive side, *Catha edulis* is a hardy plant with few pests or diseases to limit production. Harvesting the natural growth of leaves is a lot less problematic than growing fruit. Although leaves must be marketed fresh, they can be stored on the tree until the market is ready. Apart from supplemental irrigation water, farm inputs are minimal so that even poor farmers can make profit on small plots. In some areas, *qât* trees play a role in stabilizing terraces and thus retarding terrace erosion.[93]

The negative consequences are mostly a matter of presumed excess (Fig. 6). There is a danger that cultivation of *qât* prevents increased production of cash crops with domestic or international market potential. Chewing is often assumed to have negative health consequences and drain household budgets. Yet these criticisms involve how the plant is used rather than any specific quality of the plant agronomically, pharmacologically or economically. On a cultural level, the *qât* chewing session provides a forum for increased social solidarity. The seating arrangement of the session itself usually breaks down overt hierarchical barriers. There is also evidence for social mobility out of the traditional structure of aristocrats, tribesmen, and inferior client classes. Indeed, *qât* farming can be considered a social equalizer and thus a contribution to Yemen's nascent political democracy.

The story of *qât* is an intriguing example where botany engages with anthropology in the study of a cultural as well as horticultural and agricultural icon. People chew for various reasons, including the physiological effect, but it is important to keep in mind that *qât* chewing is by and large a social phenomenon. The stimulant properties of *qât* leaves, not unlike that in drinking coffee, provide more than a personal feeling. The elaboration of the plant as a symbol of Yemeni Muslims makes use of the leaves a legal and moral issue. The struggle of a young nation state to develop from near-medieval isolation only four decades ago focuses attention on *qât* as a relic of underdevelopment. Yet *qât* is not just a drug to be dismissed, a plant diffusion to be relegated to historical footnotes, nor an exotic that civilization can assume to be a sign of the primitive or less-developed other. Whatever the future holds for Yemen, it is sure to turn over a new leaf as this dynamic people continue to come to terms with this relatively old and exotic leaf.

NOTES

[1] M. al-Zubayrî, "Al-Hâkim al-awwal fi al-Yaman," in *Al-Qât fi Hâyat al-Yaman wa-al-Yamâniyín* (Sanaa, Yemen center for research and studies, 1981–82), 33. This was originally published in 1958. I provide a close paraphrase here rather than an exact translation. Al-Zubayrî's anecdote is widely quoted in Yemeni texts, for example by 'Abd al-Malik al-Maqramî, *Al-Qât bayn al-siyâsa wa-'ilm al-ijtimâ'* (Beirut, 1987), 39.

[2] Al-Nimârî, quoted in *Al-Qât fi Hâyat al-Yaman wa-al-Yamâniyín* (Sanaa, 1981–82), 86. Hadanân is a famous *qât*-growing area on Jabal Sabr near Ta'izz.

[3] There is a wide variety of information available about *Catha edulis* and its use, but the vast majority is on its pharmacological aspects. Literature on the history and use of the plant is widely variable and should be used with caution. For example, the brief, updated article of Hess in *The Encyclopaedia of Islam* (4:741, 1978) is fraught with error. The two main sources in English are now out of date, although still useful in parts: John Kennedy, *The Flower of Paradise* (Dordrecht, 1987), Shelagh Weir, *Qât in Yemen* (London, 1985). There is a wealth of Arabic source material for Yemen; some of this has been analyzed by Armin Schopen, *Das Qât* (Wiesbaden, 1978). For a delightful description of a Yemeni *qât* chew, see Tim Mackintosh-Smith, *Yemen: Travels in a Dictionary Land* (London, 1970), 16–28.

[4] Information on Forsskål's research and specimens is provided by F. N. Hepper and I. Friis, *The Plants of Pehr Forsskål's 'Flora Aegyptiaco-Arabica.'* Kew, 1994), 100–101.

[5] See Demissew Sebsebe and N. K. B. Robson, "*Celastraceae.*" In I. Hedberg and S. Edwards, editors, *Flora of Ethiopia* (Uppsala, 1989), 3:331–47.

[6] In Ethiopia there are reports of trees as high as twenty-five meters, as reported by Dechassa Lemessa, "Khat [*Catha edulis*]: Botany, Distribution, Cultivation, Usage and Economics in Ethiopia" (Addis Adaba, 2001), 3. Hess, "KÂT," 741 is wrong in saying the height reaches only 12 feet.

[7] Shelagh Weir, *Qât in Yemen*, 33, provides an illustration of this for Jabal Râzi'.

[8] For an example of an intercropped field system in the Harar region of Ethiopia, see Clarke Brooke ('Khat [*Catha edulis*]: It's Production and Trade in the Middle East' *Geographical Journal* 126[1960]:53), who points out that intercropping generally ceases after the tree is five to six years old.

[9] I saw this in the central highland valley of al-Ahjur in 1978. The same kind of interplanting is reported by Carsten Niebuhr, *Travels through Arabia* (Edinburgh, 1792), 2:351.

[10] Reported by Schopen, *Das Qât,* 71.

[11] Walid Al-Saqqaf (1999) "The Fight against *Qât* is the Fight for the Future." *Yemen Times* IX(42), October 18–24, 1999. Electronic document: www.yementimes.com/99/iss42/view.htm. Accessed January, 2004. An FAO (2002) survey, conducted in 2001, of 378 Yemeni *qât* farmers found that only 56 percent of the total *qât* hectarage in the sample was irrigated.

[12] Horst Kopp, *Agrargeographie der Arabischen Republik Jemen* (Erlangen, 1981), 236. Informants in Ethiopia told Brooke, "Khat (*Catha edulis*)," 53 that trees can last up to seventy-five years.

[13] For a good summary of the process, see Weir, *Qât in Yemen,* 93–97.

[14] John Kennedy, *The Flower of Paradise,* 233–39. On aspects of *qât* not associated with its stimulant effect and cultural use, Kennedy often uses sources uncritically, especially those in Arabic.

[15] Kennedy, *The Flower of Paradise,* 193.

[16] Raman Revri, Catha Edulis *Forsk. Geographical Dispersal, Botanical, Ecological and Agronomical Aspects with Special Reference to Yemen Arab Republik.* (Göttingen, 1983), 4, 30–31

[17] Kennedy, *The Flower of Paradise,* 64.

[18] I. Friis, Botanical Museum and Library, University of Copenhagen, personal communication, January, 2004. The reference to this plant in Hepper and Friis, *The Plants of Pehr Forsskål's,* 100–101, does not reflect the dropping of *Catha spinosa* as a valid classification.

[19] Some of the Yemeni plant species found in East Africa are documented in A. Al-Hubaishi and K. Müller-Hohenstein, *An Introduction to the Vegetation of Yemen* (Eschborn, 1984), 81.

[20] Several Yemeni authors had speculated that Ethiopians may have brought *qât* to Yemen before Islam, but there is no historical evidence for this on either side. For a useful discussion on this and other dates before the fourteenth century, see al-Sa'dî, *Al-Qât fi al-Yaman,* 11–13.

[21] Ibn Fadl Allâh al-'Umarî, *Masâlik al-absâr fi mamâlik al-amsâr min al-bâb al-thâmin ilâ al-bâb al-râbi' 'ashr* (Cairo, 1988), 40, 48. The later Egyptian historian, al-Maqrîzî (d. 1442), also discusses *qât* use in Ethiopia, but this is taken directly from the earlier text of al-'Umarî.

[22] This point was made by the noted Yemeni bibliographer 'Abd Allâh al-Hibshî, "Al-Qât fi al-adab al-Yamanî" in his *Dirâsât fi al-turâth al-Yamanî* (Beirut, 1977), 141. Nor is there any mention in the fourteenth-century travel book of Ibn Battûta.

[23] Al-Maqrîzî, *al-Imâm bi-akhbâr man bi-ard al-Habash min mulûk al-Islâm* (Leiden, 1790), 11. The editor, Rinck, renders the term as *jât* in Arabic transcription and "Jaat" in Latin.

[24] G. W. B. Huntington, *The Glorious Victories of 'Amda Seyon, King of Ethiopia.* Oxford, 1965), 55–56. 'Amda Seyon reigned from 1314–1344 C.E.

[25] There are several editions of Ibn Hajar's *fatâwa* collection; I have used that edited by 'Abd Allâh al-Hibshî, *Thalâth rasâ'il fi al-Qât* (Beirut, 1986), 19–49. For an English translation of part of Ibn Hajar's comments, see R. B. Serjeant, 'Qât.' In R. B. Serjeant and Ronald Lewcock, editors, *San'â': An Arabian Islamic City* (London, 1983), 173–74.

[26] 'Abd Allâh al-Hibshî, Mushkilat zuhûr al-Qât . . . wa-fuqahâ' al-qarn al-'ashr. In *Al-Qât fi Hâyat al-Yaman wa-al-Yamâniyín* (Sanaa, 1981–82), 78; see also Schopen, *Das Qât,* 176, note 4.

[27] Al-Bîrûnî, *Al-Biruni's Book on Pharmacy and Materia Medica* (Karachi, 1973) 1:264. The editor does not attempt to identify the plant, which is clearly not *Catha edulis*. For discussion of the probable botanical identification in al-Bîrûnî's account, see Powels, "Zur Herkunft des Wortes Qât (Catha Edulis Forsk.)," *Zeitschrift für arabische Linguistik* 24 (1992):13–19. Unlike Powels, I do not think there is any linguistic link between the term used by al-Bîrûniî

and the Yemeni term derived from Amharic. Schopen, *Das Qât,* p. 45, claims that the earliest reference to *qât* is in Najîb al-Dîn al-Samarqandî's (died 1222) *Kitâb al-Qarâbâdîn,* but he is quoting from a poorly informed source. This is no doubt in reference to the same plant as described by al-Bîrûnî, unless it is a misreading or emendation as suggested by Maxime Rodinson, "Esquisse d'une monographie du *qât*," *Journal Asiatique* 265(1977):76.

[28] Al-Antâkî, *Tadhkirat ûlâ al-bâb wa-al-jâmi' li-al-'ajab al-'ujâb* (Beirut, 1951), I:86 does mention Yemeni coffee.

[29] I thank Gene Gragg, Oriental Institute, University of Chicago, for providing me with an update on the usage of the term in East Africa. The various forms of the term in Ethiopian dialects are listed by Sylvia Powels, "Zur Herkunft des Wortes *Qât* (Catha Edulis Forsk.)," 8–9.

[30] Maxime Rodinson, "Esquisse d'une monographie du *qât*," 75.

[31] Chaim Rabin, *Ancient West-Arabian* (London, 1951), 55.

[32] The most bizarre hypothesis was proposed by Rémy Cottevieille-Giraudet, Le *Catha Edulis* fut-il connu des Égyptiens? *Bull. de l'Institute Français d'Archaeologie Orientale* 35:110, 1935, who derives the Arabic *qât* from Hieratic Egyptian *qt(qd)* in reference to a sacred plant of the cult of Isis at Philae.

[33] Al-'Umarî, *Masâlik al-absâr,* 40. Note that the editor, Mustafâ Abû Dayf Ahmad, misreads the term as *jân*.

[34] The key to my interpretation lies with continuity of the pronunciation in both Egyptian and Yemeni dialects for at least six hundred years of usage. There are indications that the *j>g* transition was practiced in Egypt, but I do not yet have conclusive evidence for the historical depth of the transition in Adeni dialect. However, given the long historical trajectory of other pronunciation shifts in Yemen, I think this is the most likely scenario.

[35] Goats are one of the few animals that can be seen occasionally eating *qât* leaves, but the story is surely apocryphal. A variant with a camel is recorded by Cesare Ansaldi, *Il Yemen. Nella storia e nella leggenda* (Rome, 1933), 202.

[36] Ahmad ibn 'Alî al-Madhhajî, "Al-*Qât* fi al-shi'r al-Yamanî," *al-Yaman al-Jadîd* 3(2), 1972, 25. This is referenced by Maxime Rodinson, "Esquisse d'une monographie du *qât*," 81.

[37] Yahyâ ibn al-Husayn, *Ghâyat al-amânî,* II:69; see also Muhammad al-Zabâra, *A'immat al-Yaman* (Ta'izz, 1952), 422.

[38] 'Abd Allâh al-Baraddûnî, Al-*Qât* . . . min zuhûrih ilâ ista'mâlih. In *Al-Qât fi Hâyat al-Yaman wa-al-Yamâniyîn* (Sanaa, 1981–82), 44.

[39] Ralph S. Hattox, *Coffee and Coffeehouses: The Origins of a Social Beverage in the Medieval Near East* (Seattle, 1985), 18,24. This is reported in a text on coffee written about 1556 by al-Jazîrî. In contemporary Yemen the leaves only appeared to be chewed rather than dried and made into a tea.

[40] Al-Baraddûnî, "Al-*Qât*," 44.

[41] Mentioned in Kennedy, *The Flower of Paradise,* 63, as a local story from Ta'izz. I am not sure of the spelling and have not identified the individual.

[42] Heard in Harar by Sir Richard Burton, *First Footsteps in East Africa* (1856), 75.

[43] Edward J. Keall, "The Evolution of the First Coffee Cups in Yemen," in Michel Tuchscherer, editor, *Le commerce du café avant l'ère des plantations coloniales* (Cairo, 2001), 44.

[44] Michel Tuchscherer, "Commerce et production du café en mer Rouge au xvie siècle," in Michel Tuchscherer, editor, *Le commerce du café avant l'ère des plantations coloniales* (Cairo, 2001), 69.

[45] Al-Maqramî, *Al-Qât bayn al-siyâsa wa-'ilm al-ijtimâ',* 58. In 1837 Paul Botta found it to be the dominant crop of Jabal Sabr; see Botta, *Relation d'un voyage dans l'Yemen* (Paris, 1841), 125. The same is reported for 1938 by Hugh Scott, "A Journey to the Yemen," *Geographical Journal,* 1939, 93(2), 107.

[46] See G. Rex Smith, "The Yemenite Settlement of Tha'bât: Historical, Numismatic and Epigraphic Notes." *Arabian Studies* 1:119–134, 1974.

[47] For a historical study of this sultan, see Varisco, "Texts and Pretexts: The Unity of the Rasulid State in the Reign of al-Malik al-Muzaffar." *Revue du Monde Musulman et de la Méditerranée* 67(1)13–21. 1993.

[48] For an analysis of the agricultural knowledge of al-Ashraf, see Varisco, *Medieval Agriculture and Islamic Science. The Almanac of a Yemeni Sultan.* (Seattle, 1993).

[49] Dechassa Lemessa, "Khat (*Catha edulis*): Botany, Distribution, Cultivation, Usage and Economics in Ethiopia" (Addis Adaba, 2001), 2. By the late 1950s *qât* had surpassed coffee production in Harar region, according to Brooke. "Khat (*Catha edulis*): It's Production and Trade in the Middle East," 52.

[50] Al-Sa'dî, *Al-Qât fi al-Yaman,* 59, quoting statistics of the YAR's Central Planning Organization.

[51] Lemessa, "Khat (*Catha edulis*): Botany, Distribution, Cultivation, Usage and Economics in Ethiopia," 4.

[52] Ministry of Agriculture and Fisheries, YAR, *Summary of the Final Results of the Agricultural Census in Eleven Provinces.* (Sanaa, 1983), table 16. The census was conducted primarily in 1978–1979, but extended to 1983 and is an estimate based on limited field sampling.

[53] Muharram, *Zâhirat al-Qât fi al-Yaman* (Dhamâr, 2000),18. It is unclear if the figures for 1994 reflect the unification of North and South Yemen in 1990.

[54] A. W. R. Joachim et al. *Report of the FAO Mission to Yemen* (Rome, 1960), 23.

[55] Edward B. Hogan et al. *Agricultural Sector Assessment, Yemen Arab Republic* (Sanaa, 1982), 122.

[56] Ahmad Fakhrî, *Al-Yaman: Mâdiha wa-hâdirhâ* (Cairo, 1957), 21. Given the conservative nature of Yemeni farmers, I find this hard to believe as a general trend. There is evidence, however, for cutting down coffee trees in the 1960s civil war. It is reported that fifty thousand trees of Hâshid shaykh Husayn al-Ahmar were destroyed as a penalty, according to al-Sa'dî, *Al-Qât fi al-Yaman,* 19.

[57] Kopp, *Agrargeographie der Arabischen Republik Jemen,* 237. See also Kennedy, *The Flower of Paradise,* 159–163.

[58] Kopp, *Agrargeographie der Arabischen Republik Jemen,* 370

[59] This point is made by al-Sa'dî, *Al-Qât fi al-Yaman,* 180.

[60] Milich and Al-Sabbry, "The 'Rational Peasant' vs. Sustainable Livelihoods." See also Weir, *Qât in Yemen,* 36.

[61] This term literally means "raisin," and stems from the sense that raisins are shriveled grapes similar to the damaged *qât* leaves drying out and turning brown on the edges.

[62] Although the leaves must be marketed fresh, they can last up to a week if properly bundled, according to Everard B. Britton, "Appendix I: The Use of Qât." *Geographical Journal* 93(1939):121.

[63] I argued this point in Varisco, "The *Qât* Factor in North Yemen's Agricultural Development" *Culture & Agriculture* 34 (1988):11–14.

[64] Schopen, *Das Qât,* 76.

[65] Muharram, *Zâhirat al-Qât fî al-Yaman,* 18. I think the 1980 figure is underestimated.

[66] al-Sa'dî, *Al-Qât fî al-Yaman,* 72. His breakdown by province can be found on p. 74.

[67] Lenart Milich and Muhammand Al-Sabbry, "The 'Rational Peasant' vs. Sustainable Livelihoods." Development #3 Electronic document, http://ag.arizona,edu/~/milich/yemen.html. Accessed January, 2004.

[68] I noted this problem when trying to collect accurate data on land ownership in al-Ahjur during ethnographic fieldwork in 1978. A similar problem was mentioned by Hywel Rees-Jones, *Farm Systems Survey Qa Bakil.* (Dhamar, 1983), 26.

[69] Kennedy, *The Flower of Paradise,* 157. Kennedy estimates production costs at 30 percent, which might be on the high side for dryland farming.

[70] Rees-Jones, *Farm Systems Survey Qa Bakil,* 26. In his field survey the average holding with *qât* was circa .13 hectares. For estimates of Yemeni riyals from 1978 to 1983 I use the official government rate circa 4.5 riyals per U.S. dollar.

[71] Shelagh Weir, "Economic Aspects of the *Qât* Industry in Northwest Yemen." This was a paper presented at the Conference on Contemporary Yemen, held in Exeter in July. Weir notes that profits were high this year.

[72] Revri, Catha Edulis *Forsk.,* 84. For the Radâ' region, a profit of $37,000 per ha has been estimated; see Peter de Lange, "Rada Integrated Rural Development Project Study into Water Resources in Al Bayda Province, Progress Report for Period April to August 1983" (Rada, 1983), appendix 5.

[73] Kennedy, *The Flower of Paradise,* 154.

[74] Lemessa, "Khat (*Catha edulis*): Botany, Distribution, Cultivation, Usage and Economics in Ethiopia," 10.

[75] Swagman, *Development and Change in Highland Yemen* (Salt Lake City, 1988), 90.

[76] Al-Munibari, "The Socioeconomic, Agricultural and Environmental Implication of Qât Production/Consumption in Yemen." (Sanaa, 2002).

[77] For roughly the same period, Walid Al-Saqqaf, "The Fight against Qât is the Fight for the Future" estimates that 36 billion riyals ($300,000,000) are spent on *qât* in Yemen annually.

[78] FAO (2002). This same report indicates that the contribution of *qât* to the GDP was closer to 10 percent.

[79] For more information on the cultural aspects, see Varisco, "On the Meaning of Chewing . . ."

[80] For example, Robert Finlay on an 1823 trip described *qât* as an expensive luxury in Sanaa; P. J. L. Frankl, "Robert Finlay's Description of San'â' in 1238/1239/1823." *British Society for Middle Eastern Studies Bulletin* 17(1)1990, 26.

[81] For a discussion of this relative affluence in the rural sector, see Varisco and Adra, "Affluence and the Concept of the Tribe in the Central Highlands of the Yemen Arab Republic," In Richard F. Salisbury and Elisabeth Tooker, editors, *Affluence and Cultural Survival* (Washington, D.C., 1984).

[82] Kennedy, *The Flower of Paradise,* 78. About 80 percent of adults chew *qât,* according to Kopp, *Agrargeographie der Arabischen Republik Jemen,* 236.

[83] Al-Munibari ("The Socioeconomic, Agricultural and Environmental Implication of Qât Production/Consumption in Yemen") administered a questionnaire to 3,402 urban families and 1,455 rural families with interviews of 200 families in five regions of Yemen.

[84] Kopp, Agrargeographie der Arabischen Republik Jemen, 238.

[85] Schopen, *Das Qât,* 66. For listing of the major varieties see al-Sa'dî, *Al-Qât fî al-Yaman,* 35–38,41–53 and al-Maqramî, *Al-Qât bayn al-siyâsa wa-'ilm al-ijtimâ',* 48–55.

[86] Serjeant and al-Akwa', "The Statute of San'â' (*Qânûn San'â'*)," 189. The trader's profit would be 1 1/2 *buqsha* per bundle.

[87] Schopen, *Das Qât,* 82. A 1973 study in Sanaa showed daily expenditure for *qât* and coffee at about 3.5 riyals (70 cents), as quoted in Weir, *Qât in Yemen,* 101.

[88] Weir, *Qât in Yemen,* 101. Weir compiles a list of other documented prices on p. 178, note 10.

[89] Swagman, *Development and Change in Highland Yemen,* 50.

[90] For a discussion of the legality issue in Yemen, see Varisco, "The Elixer of Life or the Devil's Cud: The Debate over *Qât* (*Catha edulis*) in Yemeni Culture." In Ross Coomber and Nigel South, editors, *Drug Use and Cultural Context: Tradition, Change and Intoxicants beyond 'The West'* (London: 2004).

[91] Al-Maqramî, *Al-Qât bayn al-siyâsa wa-'ilm al-ijtimâ,'*89

[92] Weir, *Qât in Yemen,* 154–65.

[93] This also has been observed in the Hararghe region of Ethiopia, according to Lemessa, "Khat (*Catha edulis*): Botany, Distribution, Cultivation, Usage and Economics in Ethiopia," 4.

Botanical Progress, Horticultural Innovations and Cultural Changes

Botanical Progress, Horticultural Innovations and Cultural Changes

Botanical Progress, Horticultural Innovations and Cultural Changes

Botanical Progress, Horticultural Innovations and Cultural Changes

The Role of Horticulture in a Changing World

Peter Del Tredici

It's easy to enumerate hot-button issues in contemporary American culture: gun control, abortion, globalization, terrorism, and immigration reform are just a few. One thing that the debates about these issues have in common is that they are highly polarized, with neither side paying much attention to what the other is saying. Another is that they often have an overarching moralistic tone that pits good against evil, with little regard for facts. How our society will manage to move forward on such contentious issues remains to be seen.[1]

Within my own narrow field of expertise, plant ecology, the use of exotic versus native species in designed landscapes is an issue that seems to bring out the worst in people, not unlike the debates over gun control and abortion. As a representative of the Arnold Arboretum of Harvard University, I have served as a member of the Massachusetts Invasive Plant Advisory Group, a voluntary collaboration of nursery professionals, conservationists, botanists, land managers, and representatives from various government agencies that reports to the state's Office of Environmental Affairs. Over the course of the past three years, the group has produced a list of species that are invasive in "minimally managed" habitats and developed a strategic plan with recommendations for how to cope with the problem.

On the national scale, researchers have determined that invasive species, including plants, animals and microbes, are an ongoing threat to native ecosystems as well as to rare and endangered species and are the cause of economic losses totaling approximately \$137 billion annually.[2] In 1999, President Clinton issued an executive order establishing a National Invasive Species Council to investigate the problem and develop a comprehensive plan for dealing with it.[3] The Council's report, which was issued in January 2001, strongly recommended a strategy that focused on the early detection of invasions followed by rapid response eradication on the ground.[4]

Implicit in the proposals that call for the control or eradication of invasive species is the assumption that the native vegetation will return to dominance once the invasive is removed, thereby restoring the "balance of nature." That's the theory; reality can be quite different. Land managers and others who have to deal with the invasive problem on the ground know that often as not the old invasive species comes back following removal—reproducing from root suckers, stump sprouts or seeds—or else a new invader moves in to replace the old one. Oftentimes the only thing that seems to turn this dynamic around is cutting down the invasives, treating them with herbicides, and planting native species in the gaps where the invasives once were. Following this, the site requires weeding invasives for an indefinite number of years, until the natives are big enough to hold their ground without human assistance.[5]

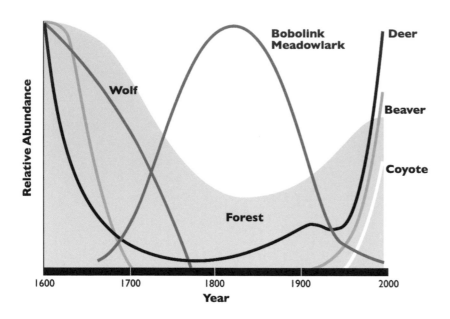

1. Changes in the relative abundance of selected animal species in Massachusetts over the past 400 years as a function of forest cover. Note that the wolf has been eliminated from the landscape, the beaver has been successfully reintroduced, and the coyote represents a new addition to the fauna. The rarest species, bobolink and meadow lark, are associated with open fields, which have been steadily diminishing since the mid-1800s (modified from Foster et al., 2002).

What's striking about this so-called restoration process, with its ongoing need for planting and weeding, is that it looks an awful lot like gardening. Call it what you will, but anyone who has ever worked in the garden knows that planting and weeding are endless tasks.[6] So the question becomes: Is "landscape restoration" really just gardening dressed up with jargon to simulate ecology, or is it based on scientific theories with testable hypotheses? To put it another way: Can we put the invasive species genie back in the bottle, or are we looking at a future in which nature as we know it becomes a cultivated entity?[7]

The answer to this question lies in an understanding of the concept of ecological succession, the term used to describe the change in the composition of plant and animal assemblages over time. In the good old days, before World War II, ecologists generally tended to view succession as an orderly process leading to the establishment of a "climax" or steady-state community that, in the absence of disturbance, was capable of maintaining itself indefinitely. I like to refer to this as the Disney version of ecology, stable and predictable, with all organisms living in perfect harmony. Following the war, a younger generation of ecologists began challenging this orthodox view, eventually formulating what is known as the theory of patch dynamics, which views natural disturbances as an integral part of a variable and unpredictable succession process.[8] The key concept here is that the nature, timing, and intensity of the disturbances are critical factors—together with climate and soil—in determining the composition of successive generations of vegetation. From the modern ecological perspective, the apparent stability of current plant associations is an illusion; the only certainty is that things will be substantially different within fifty years.[9]

When one broadens the traditional definition of disturbance to include the effects of acid rain on the earth's surface and carbon dioxide enrichment on its atmosphere, it becomes clear that there is no place on the planet that has not experienced some level of disruption as a result of human activities.[10] The absurd position that global warming has not yet been "proven" is based on the assumption that people have the capacity to understand—at a detailed level—how the world's climate system actually works. From my perspective, the scariest thing about climate change is the uncertainty about how it will actually play out on the ground. When and if scientists get around to accurately predicting the effects of pumping massive amounts of carbon dioxide and nitrous oxide into the atmosphere, it will be far too late to do anything about it.

A cursory glance at the land-use history of New England, as documented by researchers at the Harvard Forest in

2. A dead American chestnut *(Castanea dentata)* in the town of Harvard, Massachusetts, photographed in 2002. The stump, which is thirty-one inches in diameter, was killed by the introduced chestnut blight between 1920 and 1930 (left). On the right is a "sprout-clump" of the same species as it now appears in the forests of eastern North America (right). Essentially the chestnut blight has diminished the role of the American chestnut from a canopy tree to an understory shrub. Photographs by P. Del Tredici

Petersham, Massachusetts, makes it clear that the scope of landscape transformation over the past three hundred years has been vast and its outcomes unanticipated and irreversible (Figure 1).[11] As such, these studies provide a picture not only of how the past has influenced the present, but also of how the landscapes of the Northeast are likely to change in the future. Unfortunately, this approach, which has been referred to as "stepping back to look forward,"[12] does not account for the fact that the scale and pace of today's environmental change far outstrips anything seen in the past.

One particularly problematic aspect of the restoration concept is its denial of the inevitability of ecological change. Implicit in much of the popular writing on the subject is the assumption that the plant and animal communities that existed in North America prior to European settlement can be returned to some semblance of original composition. The fact that the environmental conditions that led to the development of these pre-Columbian habitats no longer exist—and can never be recreated—does not seem to matter to strict restorationists, whose "faith-based" notions of restoration clearly have more to do with philosophical and ethical values than with ecological reality (Figure 2).

3. The common reed, *Phragmites australis,* growing in the Back Bay Fens of Boston, Massachusetts, was probably planted in the 1880s, during F. L. Olmsted's Muddy River restoration. Photo by P. Del Tredici

Experience from eastern North America clearly shows that even when the individual components of former ecosystems make successful comebacks, they tend to function differently than they did in the past because of irreversible changes in other parts of the ecosystem. This is best exemplified by the extensive herds of white-tailed deer (*Odocoileus virginianus*) that were formerly controlled by the hunting activities of Woodland Indians,[13] but today they roam the countryside selectively browsing plants they find palatable—both cultivated and native—while ignoring unpalatable invasive species. In the process, they manage not only to annoy homeowners but also to alter long-established patterns of forest succession.[14] The dynamic nature of interactions among people, plants, and animals in today's world are producing novel ecological associations with unpredictable consequences for all parties concerned.[15]

The concept of ecological restoration in an urban or suburban context is particularly problematic, given the abundance of pavement, road salt, heat buildup, air pollution, and soil compaction that characterize metropolitan centers. Indeed, the critical question facing landscape professionals is not what plants grew there in the past, but which ones will grow there in the future? Starting in the early 1800s and continuing through the present, plants from around the world have been brought together in our cities and suburbs in order to enhance their livability.[16] Like it or not, a relatively small percentage of these ornamental introductions have adapted well to the urban habitat and begun reproducing on their own. Regardless of the disparaging labels we apply to such "naturalized" species, they are actually performing significant ecological functions including water filtration, mineral cycling, and carbon fixation and storage.[17]

A good example of this functionality is seen in the common reed, *Phragmites australis*, which is native to Europe and central Asia, as well as to North America where it grows in brackish wetlands up and down the east coast, most dramatically in the meadowlands bordering the New Jersey Turnpike west of Manhattan (Figure 3). Although *Phragmites* is often portrayed as the ultimate invasive species because of its tendency to crowd out other vegetation, it is actually mitigating pollution by absorbing a great deal of the nitrogen and phosphorous that typically accumulates in disturbed wetlands.[18] From the functional perspective, the presence of *Phragmites* in this landscape can be viewed as a symptom of environmental degradation rather than as its cause. It turns out that many invasive plants have a similar kind of "Jekyll and Hyde" impact on the local ecology, pushing out some native species, while simultaneously providing food and shelter for a variety of native animals, especially pollinating insects and migrating birds.[19]

Regardless of how one feels about the unique assemblages of plants that populate our sprawling cities, they have become its de facto native vegetation. In a very real sense, the diversity and spontaneity of these new "immigrant" communities mirrors

4. The tree of Heaven, *Ailanthus altissima,* growing in its native habitat near the Great Wall of China (left); and the same species making itself at home along a roadway wall in Boston (right). Photos by P. Del Tredici

that of our own society (Figure 4). Indeed, the very same processes that have led to the globalization of the world economy—unfettered trade and travel among nations—have also resulted in the globalization of our environment.[20]

So what can landscape professionals, working on the ground, do to cope with the widespread environmental devastation and ecological uncertainty that have become such an integral part of the modern world? I have developed a practical, three-step approach that covers the construction, design, and maintenance of urban landscapes. The first step is to make sure that the soil can support the growth of plants selected for the site. Typically, this means coming to grips with the fact that the heavy equipment used during the construction process inevitably compacts the surrounding "soil" to a density approaching concrete. Such compacted soils, with their low oxygen tensions, high bulk-density and impeded drainage,[21] are lethal to many landscape plants, especially native species that come from upland habitats. Without adequate remediation of the compaction and drainage problems that abound in urban landscapes, invasive species are pretty much the only plants that will survive.

My second suggestion is not to limit planting designs to a palette of native species that once grew on the site. Imposing such a limitation not only reduces the aesthetic possibilities for the landscape but also its overall adaptability to future environmental change. Instead, I propose that sustainability be the standard for deciding what to plant. According to my definition of this overused word, sustainable landscape plants are tolerant of the conditions that prevail on a given site; require minimal applications of pesticides, herbicides, or fertilizers to look good; have greater drought resistance and winter hardiness than other plants; and, finally, they do not spread aggressively into surrounding natural areas. Landscapes that are designed with plants that fit this criteria—including both native and introduced species—will not only be less costly to maintain over time but also will be better able to tolerate the unpredictable weather patterns that loom in the future.[22] In this regard, American designers have much to learn from their European counterparts, who have developed a tradition of using cosmopolitan plant associations to create naturalistic landscapes.[23]

My third and final suggestion is to recognize, early on in the design process, the ongoing need for maintenance on all constructed landscapes. All too often the concept of sustainability is misinterpreted to mean self-sustaining, a fantasy that is as false in horticulture as it is in ecology. Landscape maintenance, including irrigation, weeding, mulching and replanting, is necessary in order to promote the successful establishment of new plantings, as well as to counteract the effects of disturbance—both natural and human—which continually threaten the integrity of mature plantings. From the horticultural perspective, a truly sustainable landscape design is one that is in balance with the financial resources available to maintain it.[24]

My recommendations for creating sustainable landscapes are really little more than a plea for the practice of sound horticulture. It's not a glamorous message, but to ignore it means that beautiful designs will fall apart and romantic restorations disappear. By grounding their work in in the realities of the modern world, landscape professionals of every stripe can make a small, but significant, contribution to cleaning up the mess that people have made of the planet.

NOTES

1. A substantially different version of this article was published under the title "Neocreationism and the Illusion of Ecological Restoration," in *Harvard Design Magazine* 20 (2004): 87–89.

2. David Pimentel, Lori Lach, Rudolfo Zuniga, and Doug Morrison, "Environmental and Economic Costs of Nonindiginous Species in the United States," *BioScience* 50 (2000): 53–65.

3. William J. Clinton, "Invasive Species," Executive Order 13112, February 3, 1999, *Weekly Compilation of Presidential Documents* 35 (5) (February 8, 1999): 157–210.

4. National Invasive Species Council Management Plan, "Meeting the Invasive Species Challenge." January 18, 2001 (http://www.invasivespecies.gov/council/nmp.shtml).

5. In 2003, the Limahuli Garden of the National Tropical Botanical Garden on the island of Kaua'I in Hawaii was spending approximately $30,000 per acre on removing invasive species and replanting native vegetation. How successful this treatment will be over time remains to be seen.

6. Peter Del Tredici, "Nature Abhors a Garden," *Pacific Horticulture* 62, no. 3 (2001): 5–6.

7. Daniel Janzen, "Gardenification of Wildland Nature and the Human Footprint," *Science,* 279 (1998): 1312–13; Mark A. Davis and Lawrence B Slobodkin, "The Science and Values of Restoration Ecology," *Restoration Ecology* 12 (2004): 1–3.

8. Michael G. Barbour, "Ecological Fragmentation in the Fifties," in *Uncommon Ground,* ed. William Cronin (New York: W. W. Norton, 1995), 233–55.

9. Jean Fisk and William A. Niering, "Four Decades of Old Field Vegetation Development and the Role of *Celastrus orbiculatus* in the Northeastern United States." *Journal of Vegetation Science* 10 (1999): 483–92.

10. Peter Vitousek, "Beyond Global Warming: Ecology and Global Change," *Ecology* 75 (1994): 1861–76.

11. David R. Foster and John D. Aber, *Forests in Time: the Environmental Consequences of 1,000 Years of Change in New England* (New Haven, Conn.: Yale University Press, 2004).

12. Charles H. W. Foster, ed., *Stepping Back to Look Forward: A History of the Massachusetts Forest* (Petersham, Mass.: Harvard Forest, 1998).

13. Shepard Krech III, *The Ecological Indian: Myth and History* (New York: W.W. Norton, 1999).

14. David Foster, Glen Motzkin, D. Bernardos and J. Cardoza, "Wildlife Dynamics in the Changing New England Landscape," *Journal of Biogeography* 29 (2002): 1337–1357. James P. Sterba, "Landscape Architects: Deer Are Designing Future Look of Forests." *Wall Street Journal,* December 1, 2004.

15. Jillian W. Gregg, Clive G. Jones and Todd E. Dawson, "Urbanization effects on tree growth in the vicinity of New York City." *Nature* 424 (2003): 183–87.

16. Sarah H. Reichard and Peter S. White, "Horticulture as a pathway of invasive plant introductions in the United States." *BioScience* 51 (2001): 103–13.

17. Andrew P. De Wet, Jonathan Richardson and Catherine Olympia, "Interactions of land-use history and current ecology in a recovering 'urban wildland.'" *Urban Ecosystems* 2 (1998): 237–62.

18. Arthur F. M. Meuleman, Hans Ph. Beekman, and Jos T. A. Verhoeven, "Nutrient retention and nutrient-use efficiency in *Phragmites australis* stands after wastewater application." *Wetlands* 22 (2002): 712–21.

19. Paul D. Thacker, "California Butterflies: at Home with Aliens," *BioScience* 54 (2004): 182–87; Carla D'Antonio and Laura A. Meyerson, "Exotic Plant Species as Problems and Solutions in Ecological Restoration: A Synthesis," *Restoration Ecology* 10 (2002): 703–13.

20. Dennis Normile, "Expanding trade with China creates ecological backlash." *Science* 306 (2004): 968–69.

21. Philip J. Craul, *Urban Soils: Applications and Practices* (New York: John Wiley and Sons, 1999).

22. Peter Del Tredici, "Survival of the most adaptable." *Arnoldia* 60, no. 4 (2001): 10–18.

23. Nigel Dunnett and James Hitchmough, eds., *The Dynamic Landscape* (London: Spon Press, 2004).

24. Hein Koningen, "Creative management," in *The Dynamic Landscape,* Nigel Dunnett and James Hitchmough, eds., (London: Spon Press, 2004), pp. 256-292.

Botanical Progress, Horticultural Innovations and Cultural Changes

Botanical Progress, Horticultural Innovations and Cultural Changes

Botanical Progress, Horticultural Innovations and Cultural Changes

Botanical Progress, Horticultural Innovations and Cultural Changes

Contributors

Mauro Ambrosoli has always looked at the study of European history as a sound base for the solution of contemporary problems. Between 1972/73 and 1978/79 he focused his research interests on the agricultural revolution in Europe and what was then called the "green revolution" and decided to follow forgotten connections. These regarded mainly early agricultural books, seed exchanges, and the connection between botany and agriculture. In 1980, a sabbatical year allowed extensive research on these topics in libraries and archives in Italy, France, and England. In 1987/88, Magdalen College, Oxford offered him a Visiting Fellowship and the opportunity of delivering the Waynflete Lectures to present to the public the whole story of the introduction of fodder grasses in European agriculture from 1350 to 1850. This book came out in 1992 as *Scienziati, proprietary e contadini* and its English version, *The Wild and the Sown,* in 1997. Although Dr. Ambrosoli developed other fields of research in textile and institutional history, he has always been active in the study of plant innovations in early modern centuries. The role of the garden as a seed bed of development is his most recent research topic.

Nurhan Atasoy is a renowned scholar of Turkish art and culture. She received her B.A., M.A., and Ph.D. from the Department of Fine Arts and Art History at Istanbul University. She was a professor at Istanbul University until her retirement in 1999. During her academic life she attended and lectured at many international congresses and symposia, and participated in research and international meetings on Turkish and Islamic art throughout the world. She has mounted several important exhibitions and in 2000 her book *Otag-i Hamayun: The Ottoman Tent Complex* won the Textile Society of America's R. L. Shep Book of the Year Award. Among other honors, she has also received the Council of Europe's prestigious Pro Merito Medal and an award from the Economic Cooperation Organization (ECO) in recognition of her outstanding performance in the field of fine arts. Atasoy has published over seventy articles and seventeen books.

Michel Conan, Director of Garden and Landscape Studies and Curator of the Contemporary Design Collections at Dumbarton Oaks, is a sociologist interested in the cultural history of garden design. He contributed to a renewal of garden history in France in the mid-1970s with the publication of several reprints. He recently published the *Dictionnaire Historique de L'Art des Jardins* (1997) and *L'Invention des Lieux* (1997); and edited three Dumbarton Oaks symposia, *Perspectives on Garden Histories* (1999), *Environmentalism and Landscape Architecture* (December 2000), and *Aristocrats and Bourgeois Cultural Encounters in Garden Art* (February 2002), and

Landscape Design and the Experience of Motion (May 2003). His most recent publications are: *The Quarries of Crazannes by Bernard Lassus, An Essay Analyzing the Creation of a Landscape*, Washington DC: Spacemaker Press, 2004; and *Essais de Poétique des Jardins*, Firenze: Daniele Olschki, 2004, which received the Grinzani Hanbury prize for the best garden book in Italy in 2005.

Peter Del Tredici holds a B.A. in Zoology from the University of California, Berkeley (1968), an M.A. in Biology from the University of Oregon (1969), and a Ph.D. in Biology from Boston University (1991). Peter has worked at the Arnold Arboretum of Harvard University since 1979 as a plant propagator, Editor of *Arnoldia*, Director of Living Collections, and, most recently, Senior Research Scientist. Since 1984, he has been the Curator of the famous Larz Anderson collection of bonsai plants, housed at the Arboretum. Dr. Del Tredici has been a Lecturer in the Department of Landscape Architecture since 1992, with a strong interest in urban ecology. He is the winner of the Arthur Hoyt Scott Medal and Award for 1999, presented annually by the Scott Arboretum of Swarthmore College "in recognition of outstanding national contributions to the science and art of gardening." Dr. Del Tredici has worked on various aspects of both botany and horticulture over the last twenty-five years. His interests are wide ranging and include such subjects as plant introduction from China, the root systems of woody plants, stress tolerance in urban trees, and the cultural and natural history of the Ginkgo tree.

Mohammed El Faïz, economist and historian of Agronomy and Arab gardens, is currently researching and teaching at the Université Cadi Ayyad in Marrakech (Morocco). He is known worldwide for his contribution to the defense of the historic landscape of the city of Marrakech, and for the protection of its gardens and oasis landscapes. His publications cover a variety of topics concerning water, agronomy and gardens in the Arab-Muslim civilization. Most importantly: *Agronomie de la Mésopotamie antique: Analyse de l'Agriculture Nabatéenne de Qûtâmä*, édition J. Brill, Leiden-Köln-New-York, 1995; *Les jardins historiques de Marrakech: mémoire écologique d'une ville impériale*, édition Edifir, Florence, 1996; *Les jardins de Marrakech*, édition Actes Sud, Arles, 2000; Ibn al-Awwâm, *Livre de l'Agriculture (Kitâb al-Filâha)*, édition revue et corrigée avec introduction de Mohammed El Faïz, édition Sindbad/Actes-Sud, 2000; *Marrakech: patrimoine en péril*, édition Actes-Sud/ Eddif, 2002; *Jardins du Maroc, d'Espagne et du Portugal: un art de vivre partagé (ouvrage collectif)* édition Malika/Actes Sud, Madrid, 2003; and *Histoire de l'hydraulique arabe: conquêtes d'une école oubliée*, édition Actes Sud, Paris, 2005.

Susan Toby Evans is an archaeologist specializing in the cultures of ancient Mexico and Central America. She began her research career studying the agricultural land use patterns of Aztec commoners and excavated the only systematically studied Aztec village in the Basin of Mexico, the heartland of the Aztec empire. But when those excavations uncovered a palace, home of the village headman, she turned her research interest toward elite Aztec culture, spending a year in the library at Dumbarton Oaks to uncover other Aztec palaces. Research on Aztec palaces led to her discovery of the importance of monumental gardens in Aztec culture. Dumbarton Oaks recently published *Palaces of the Ancient New World*, a volume that Evans edited with Joanne Pillsbury, her co-organizer of the 1998 Dumbarton Oaks Pre-Columbian symposium on that topic. Evans is also the author of *Ancient Mexico and Central America: Archaeology and Culture History* (Thames & Hudson, 2004), an overview of Mesoamerican cultural evolution that received the Society of American Archaeology's 2005 Book Award.

Yizhar Hirschfeld is currently an Associate Professor at the Institute of Archaeology of the Hebrew University of Jerusalem. He was formerly a Senior Archaeologist with the Israel Antiquities Authority. He has directed excavations at numerous archaeological sites, including Ramat Hanadiv, Ein-Gedi, and, currently, Tiberias. His other research interests include the monasteries and churches of the Holy Land and the archaeology of the Dead Sea region.

W. John Kress was born in Illinois and received his education at Harvard University (B.A., 1975) and Duke University (Ph. D. 1981) where he studied tropical biology, ethnobotany, and plant systematics. Since then he has traveled to tropical areas around the world studying and collecting heliconias, gingers, and bananas. Before coming to the Smithsonian, Dr. Kress was the Director of Research at the Marie Selby Botanical Gardens in Sarasota, Florida, from 1984 to 1988. He currently is Curator and Research Scientist as well as Chairman of the Department of Botany at the National Museum of Natural History, Smithsonian Institution. He is a Fellow of the American Association for the Advancement of Science and currently Executive Director of the Association for Tropical Biology and Conservation.

Wybe Kuitert graduated with a degree in landscape architecture from Wageningen University in the Netherlands. After studying for four years at the graduate school of the Institute of Landscape Architecture of Kyoto University, he returned to Wageningen to earn his Ph.D. with a thesis on the cultural history of Japanese gardens. His thesis was later published as *Themes in the History of Japanese Garden Art* (University of Hawaii Press, 2002). He held a post-doc research fellowship at the International Research Center for Japanese Studies in Kyoto and was later a full-time associate professor in landscape architecture at the Kyoto University of Art and Design, where he is still a visiting professor. Dr. Kuitert currently lives in the Netherlands, serves as adviser to several botanical gardens, and works on the research, design and construction of gardens. Japanese flowering cherries are among his passions; his book *Japanese Flowering Cherries* appeared in 1999 (Timber Press, Portland, Oregon).

Georges Métailié is a research worker (Directeur de recherche) at the CNRS (National Center for Scientific Research), and member of the Centre Alexandre Koyré, Paris, France. His research focuses on the history of the knowledge about plants and animals in China and Japan from an ethnobiological point of view. This work includes the history of botanical knowledge in ancient China, the formation of modern botany in China and Japan, the history of horticulture and its techniques, and the history of flowers in Chinese gardens. He is Honorary Associate of the Needham Research Institute (Cambridge, UK). His membership on editorial board includes *Anthropozoologica* and *East Asian Science, Technology, and Medicine (EASTM)*. He is a former president of the French association for Chinese Studies and Founding Officer and former Treasurer of the International Society for the History of East Asian Science, Technology and Medicine.

Therese O'Malley is the associate dean at the Center for Advanced Study in the Visual Arts at the National Gallery of Art in Washington, D.C. She received her B.A., M.A., and Ph.D. from the University of Pennsylvania in the history of art. Her dissertation, which she wrote as a fellow at Dumbarton Oaks, was on the history of the National Mall from 1791 to 1852. Since 1984 she has been with the National Gallery. Her publications have focused on the history of landscape and garden design primarily in the eighteenth and nineteenth centuries, concentrating on the transatlantic exchange of plants, ideas, and people. Her current projects include a reference work entitled *Keywords in American Landscape Design*, in press with Yale University; and a book on the history of botanic gardens. Among her professional activities, she is currently president of the Society of Architectural Historians, on the editorial board of UPenn's Press Landscape Studies series; and adviser to the Bard Graduate Center program of Garden History and Landscape Studies. She was chair of the Association of Research Institutes in the History of Art from 1994–2000 and Senior Fellow in Landscape Studies at Dumbarton Oaks from 1989-1995. O'Malley lectures internationally and has taught at Harvard, Princeton, UPenn and Temple Universities.

Saúl Alcántara Onofre is Professor Researcher at Universidad Autónoma Metropolitana, Azcapotzalco, Mexico. He is the author of a syllabus of postgraduate courses. His specialties include Conservation and Restoration of Monuments; and Planning, Design and Conservation of Landscapes and Gardens.

Maria Subtelny is Associate Professor of Persian and Islamic Studies in the Department of Near and Middle Eastern Civilizations at the University of Toronto. She received her doctorate from Harvard University in the Department of Near Eastern Languages and Civilizations. She has published numerous studies on the cultural history of medieval Iran, and on the period of Tamerlane's descendants in particular. Subtelny has a long-standing interest in Persian gardens, and has made important contributions to the study of medieval Persian garden design, and Persian agricultural manuals. In her contribution to this volume, she combines her interest in Persian garden culture with Perso-Islamic mysticism.

Alain Touwaide is a Visiting Scientist at the Department of Botany, National Museum of Natural History, Smithsonian Institution. He does research on the history of botany in the Mediterranean world from Antiquity to the Renaissance with a special focus on taxonomy and plant representations. At the University of Louvain (Belgium), he earned a B.A. in Classics in 1975, an M.A. in Oriental Philology and History in 1978 and a Ph.D. in Classics in 1981. He also obtained a "Qualification to direct scientific research programs" in Ancient and Medieval History at the University of Toulouse in 1999. He mainly brings to light, publishes, translates and studies previously unknown or overlooked botanical texts in ancient Greek, Latin, and Arabic, with a particular interest in texts on medicinal plants. Touwaide is preparing an encyclopedic flora of antiquity and building a computerized database on the therapeutic uses of medicinal plants in ancient Mediterranean cultures. To conduct this research, Touwaide visits many universities, laboratories, and research centers all over the world, in such different fields as botany, pharmacology, medicine, history of medicine, history and classics. He has published and lectured extensively, and has organized several exhibitions on the history of botany, and many panels in international conferences on history, the history of science, ethnobotany, and ethnopharmacology. His work has been awarded national and international prizes by the Belgian Academy of Letters and Fine Arts, the Belgian Academy of Medicine, and the Washington Academy of Sciences with the 2003 Award for Scientific Achievement in the Behavioral and Social Sciences. He serves as Chair of the History Committee of the Washington Academy, and in different editorial and scientific committees in the field of history of ancient and medieval sciences.

Daniel Martin Varisco is a cultural anthropologist and historian who has conducted research in Egypt, Qatar, the United Arab Emirates, and Yemen. From 1991 to 2001 he edited "Yemen Update," the bulletin of the American Institute for Yemeni Studies and currently maintains "Yemen Webdate," an online archive. He is currently president of the Middle East Section of the American Anthropological Association. He has published three books on agriculture and folk astronomy in the Arab world. Varisco is chair of the Department of Anthropology at Hofstra University.

Elliot R. Wolfson is the Abraham Lieberman Professor of Hebrew and Judaic Studies at New York University. He is the author of *Through a Speculum That Shines: Vision and Imagination in Medieval Jewish Mysticism* (Princeton University Press, 1994), which won the American Academy of Religion's Award for Excellence in the Study of Religion in the Category of Historical Studies, 1995, and the National Jewish Book Award for Excellence in Scholarship, 1995. His publications include *Along the Path: Studies in Kabbalistic Hermeneutics, Myth, and Symbolism* and *Circle in the Square: Studies in the Use of Gender in Kabbalistic Symbolism,* both published in 1995 (State University of New York Press); *Abraham Abulafia—Kabbalist and Prophet: Hermeneutics, Theosophy, and Theurgy* (Cherub Press, 2000); *Language, Eros, and Being: Kabbalistic Hermeneutics and the Poetic Imagination* (Fordham University Press, 2005); *Alef, Mem, Tau: Kabbalistic Musings on Time, Truth, Death* (University of California Press, 2005); and *Venturing Beyond: Law and Ethics in Kabbalistic Mysticism* (Oxford University Press, 2005).

Index